Evolution of Dam Policies

Waltina Scheumann · Oliver Hensengerth
Editors

Evolution of Dam Policies

Evidence from the Big Hydropower States

d·i·e

Deutsches Institut für
Entwicklungspolitik

German Development
Institute

Springer

Editors
Waltina Scheumann
Department of Environmental Policy and Natural Resources Management
Deutsches Institut für Entwicklungspolitik
Bonn
Germany

Oliver Hensengerth
Department of Social Sciences and Languages
Northumbria University
Newcastle
UK

ISBN 978-3-642-23402-6 ISBN 978-3-642-23403-3 (eBook)
DOI 10.1007/978-3-642-23403-3
Springer Heidelberg New York Dordrecht London

Library of Congress Control Number: 2014932825

Printed on acid-free paper

Springer is part of Springer Science+Business Media (www.springer.com)

Preface

Following a rise in civil society protests and a reorientation of development paradigms towards environmental sustainability, large dams fell out of favour in the developed North in the 1990s. At the same time, international organisations started to codify a range of international norms pertaining to good environmental and social practices for the construction and operation of dams, including the World Bank's Operational Policies, the Equator Principles, the OECD Common Approaches and the International Finance Corporation's Performance Standards. Their aim is to mitigate the negative impacts of large dams on the environment and local communities.

However, the critical debates in the North have not prevented emerging and developing countries from pursuing large-scale hydropower programmes. Moreover, many of the emerging economies have also questioned the relevance of the above-mentioned international norms. China in particular has rejected the universal applicability of norms that contradict its system of governance. Brazil, China, India and Turkey have all argued that these norms might impede their economic and social development.

This raises the question about the status that international environmental and social norms have in emerging economies that have large dam-building programmes and that support dam construction in third countries. The studies on Brazil, China, India and Turkey in this volume seek to understand the extent to which these norms are relevant in emerging countries, how local norms in these countries interact with international norms and what the results of this interaction are. Furthermore, studies on China's involvement in Ghana and Cambodia investigate the interaction between Chinese actors and the respective host governments.

The cases show that each country has relevant laws in place, and that the development of these laws is a result of both domestic policy-learning and international influences. However, the studies also find that the political system in each country determines the way in which norms are interpreted and applied. As such, this book offers important clues as to the future of international environmental and social norms for large dams.

In the meantime, after decades of abstinence, multilateral development banks and donors are re-engaging in hydropower development as a step towards promoting low-carbon strategies. The studies collected in this book show that the environmental and social downsides of hydropower are yet to be adequately addressed and managed. And they show that there is some evidence of policy-learning in dam-related policies that might offer important hints for future dam projects.

We are grateful to the German Federal Ministry for Economic Cooperation and Development for financing the research project and sharing their expertise with us. We also thank the staff of Sektorvorhaben Sustainable Hydropower, with whom it has been a pleasure to cooperate.

Bonn Ines Dombrowsky

Acknowledgments

Narmada, Belo Monte, Atatürk, Three Gorges, to mention but a few names, have brought us—the editors and authors of this book—together. They are the names of large hydroelectric and multipurpose dams—constructions that, like no others, have led to heated debates. In times of climate change, hydropower and reservoir storage have become a dominant discourse in many developing and emerging countries. However, for their opponents, large dams symbolise the wrongs of modernisation. The World Commission on Dams was unable to bridge these differences. It was perhaps naive to believe that this might be possible. However, the WCD has had a lasting effect on the international debate about hydropower and has set high standards for future projects. Yet, again, it would be naive to believe that these standards would simply diffuse and be unequivocally accepted by decision-makers around the world.

Trying to understand the political complexities of—and complex processes in—our countries, and the factors and actors that drive change towards environmentally and socially friendly dams, was a profound and lively learning process for all of us. It allowed us to draw from the many years of experience and knowledge we had gained through field studies and scholarships in the countries concerned and elsewhere.

However, in developing the overall research framework and in realising the studies, we were not only supported financially and logistically by our own institutions—the German Development Institute in Bonn and the University of Northumbria, UK—but by the German Federal Ministry for Economic Cooperation and Development, namely Christoph Merdes, whom we thank for his enormous commitment and active involvement. The staff of the Sektorvorhaben "Policy Advice for Sustainable Hydropower" always provided kind and efficient assistance (Kirsten Nyman, Ana Bucharova, Cathleen Seeger, Michael Fink, Alejandro Garcia). We are grateful to our academic colleagues and appreciate very much their contributions to finding solutions for methodological challenges: Volkmar Hartje (Technische Universität Berlin), Imme Scholz (German Development Institute), Klaus Dingwerth (University of Bremen), Christoph Humrich (Peace Research Institute Frankfurt/Main), Peter P. Mollinga (School of Oriental and African Studies, London, and Centre for Development Research, Bonn), Esther Doehrendahl (Wuppertal Institute for Climate, Environment and Energy), Aysegül Kibaroglu (Okan University Istanbul), Vakur Sümer (Selcuk University

Konya/Turkey), Albert Cordeiro Geber de Melo (Electric Energy Research Centre, CEPEL, Brazil), Ramaswamy R. Iyer (formerly Secretary, Water Resources, in the Government of India), Guoqing Shi (National Research Centre for Resettlement, Hanoi University, Nanjing, China), Tianbao Qin (School of Law, Research Institute of Environmental Law, Wuhan University, China) and Kaan Tuncok (Regional Business Development Manager of Danish Hydraulic Institute, Turkey).

Last, but not least, we would like to thank all those—not listed here but in the respective chapters—who helped in realising our studies and in editing the book.

Contents

Abbreviations

AAI	Avaliação Ambiental Integrada (Integrated Environmental Evaluation, Brazil)
AASU	All Assam Students Union, India
ADB	Asian Development Bank
AIMSU	All Idu Mishmi Student Union, India
AKP	Adalet ve Kalkinma Partisi (Justice and Development Party, Turkey)
ANA	Agência Nacional das Águas (Regulatory Water Agency, Brazil)
ANEEL	Agência Nacional de Energia Elétrica (Electricity Regulatory Agency, Brazil)
BCM	Billion cubic metre
BMU	Bundesministerium für Umwelt, Naturschutz und Reaktorsicherheit (Ministry for Environment, Nature Conservation and Nuclear Safety, Germany)
BMZ	Bundesministerium für wirtschaftliche Zusammenarbeit und Entwicklung (Federal Ministry for Economic Cooperation and Development, Germany)
BNDES	Banco Nacional de Desenvolvivimento Econômico e Social (National Bank for Economic and Social Development, Brazil)
BOO	Build-Own-Operate
BOT	Build-Operate-Transfer
CAO	Compliance Advisor Ombudsman
CCPY	Comissão Pró-Yanomami (Pro-Yanomami Commission, Brazil)
CDM	Clean Development Mechanism
CDU	Christlich Demokratische Union Deutschland (Christian Democratic Union, Germany)
CEA	Central Electricity Authority
CEIA	Cumulative Environmental Impact Assessment
CEO	Chief Executive Officer
CETIC	China Electric Power Technology Import and Export Corporation
CF/88	Constituição Federal de 1988 (Federal Constitution of 1988, Brazil)
CIDA	Canadian International Development Agency

CIMI	Conselho Indigenista Missionário (Indigenous Missionary Council, Brazil)
CIRR	Commercial Interest Reference Rate
CNPE	Conselho Nacional de Política Energética (National Council for Energy Policy, Brazil)
COIAB	Coordenação das Organizações Indígenas da Amazônia Brasileira (Coordination of Indigenous Organisations of the Brazilian Amazon, Brazil)
CONAMA	Conselho Nacional do Meio Ambiente (National Council on the Environment, Brazil)
CPFL	Companhia Paulista de Força e Luz (name of a company, Brazil)
CPPCC	Chinese People's Political Consultative Conference
CPPT	Centro de Pesquisas de Populações Tradicionais Cuniã (Research Center for the Cuniã Traditional Populations, Brazil)
CTF	Clean Technology Fund
DDP	Dams and Development Project
DEG	Deutsche Investitions- und Entwicklungsgesellschaft (German Investment and Development Company)
DIE	Deutsches Institut für Entwicklungspolitik (German Development Institute)
DSI	Devlet Su Isleri (Turkish State Hydraulic Works)
DUP	Declaração de Utilidade Publica (Declaration of Public Utility, Brazil)
ECA	Export Credit Agency
ECHR	European Convention on Human Rights
EFT	Environment Foundation of Turkey
EIA	Environmental Impact Assessment
EIE	Elektrik Isleri Etut Idaresi (Electrical Power Resources Survey and Development Administration, Turkey)
EMRA	Energy Market Regulatory Authority
Eneram	Comitê Coordenador dos Estudos Energéticos da Amazônia (Committee of Power Studies of the Amazon Region, Brazil)
EPC	Engineering, Procurement and Construction
EPE	Empresa de Pesquisa Energética (Energy Research Company, Brazil)
EPFI	Equator Principles Financial Institutions
ERM	Environmental Resources Management
ESI	Electricity Supply Industry
ESIA	Environmental and Social Impact Assessment
ESMAP	Energy Sector Management Assistance Program
EU WFD	European Union Water Framework Directive
EU	European Union
FoE	Friends of the Earth
FUNAI	Fundação Nacional do Índio (National Indian Foundation, Brazil)

GAP	Guneydogu Anadolu Projesi (South-Eastern Anatolia Project, Turkey)
GAP RDA	GAP Regional Development Administration
GDP	Gross Domestic Product
GDRS	General Directorate for Rural Services
GfbV	Gesellschaft für bedrohte Völker (Society for Threatened Peoples, Germany)
GH¢	Ghana Cedis
GIZ	Gesellschaft für Internationale Zusammenarbeit (German Society for International Cooperation)
GoI	Government of India
GTI	Grupo de Trabalho Interministerial (interministerial working group, Brazil)
GW	Gigawatt
GWh	Gigawatt hour
HPP	Hydroelectric Power Plant
HSAF	Hydropower Sustainability Assessment Forum
HSAP	Hydropower Sustainability Assessment Protocol
IADB	Inter-American Development Bank
IBAMA	Instituto Brasileiro do Meio Ambiente e dos Recursos Naturais Renováveis (Brazilian Institute for the Environment and Natural Renewable Resources)
IBRD	International Bank for Reconstruction and Development
ICLS	Idu Cultural Literary Society, India
ICOLD	International Commission of Large Dams
IFC	International Finance Corporation
IHA	International Hydropower Association
IHA-HSAF	IHA-Hydropower Sustainability Assessment Forum
IIRSA	Iniciativa para la Integración de la Infraestructura Regional Suramericana (South American Regional Infrastructure Integration Initiative)
INPA	Instituto Nacional de Pesquisas da Amazônia (Amazonian National Research Institute, Brazil)
IPHAN	Instituto do Patrimônio Histórico e Artístico Nacional (National Institute of Historical and Artistic Heritage, Brazil)
IPK	Information Point Kurdistan
IPPNW	International Physicians for the Prevention of Nuclear War
IR	International Rivers
ISO	International Organization for Standardization
IUCN	International Union for Conservation of Nature
IWMI	International Water Management Institute
KfW	Kreditanstalt für Wiederaufbau (German Development Bank)
KHRP	Kurdish Human Rights Project
km^2	Square kilometre
kV	Kilovolt

kW	Kilowatt
kWh	Kilowatt hour
LSP	Lower Subansiri Project, India
m	Metre
MAB	Movimento dos Atingidos por Barragens (Movement of Dam Affected People, Brazil)
MCM	Million cubic metre
MEF	Ministry of Environment and Forestry, Turkey
MMA	Ministério do Meio Ambiente (Ministry of Environment, Brazil)
MME	Ministério de Minas e Energia (Ministry of Mines and Energy, Brazil)
MoEF	Ministry of Environment and Forestry, India
MoU	Memorandum of Understanding
MPWS	Ministry of Public Works and Settlement
MRC-GTZ	Office of the German Development Cooperation in the Mekong River Commission (GTZ, since January 2011 GIZ)
MST	Movimento dos Trabalhadores Rurais Sem Terra (Landless Workers Movement, Brazil)
MW	Megawatt
NABU	Naturschutzbund Deutschland (the German section of Birdlife International)
NCIWRD	National Commission for Integrated Water Resource Development, India
NDRC	National Development and Reform Commission, China
NGO	Non-governmental organisation
NHPC	National Hydroelectric Power Corporation
ODA	Official Development Assistance
OECD	Organisation for Economic Cooperation and Development
OMS	Operational Manual Statement
PBA	Plano Básico Ambiental (Basic Environmental Plan, Brazil)
PDEE	Plano Decenal de Expansão de Energia (10-Year Energy Expansion Plan, Brazil)
PMSBV	People's Movement for Subansiri Brahmaputra Valley, India
PND	Programa Nacional de Desestatização (National Privatisation Programme, Brazil)
PNE	Plano Nacional de Energia (National Energy Plan, Brazil)
PNMA	Política Nacional de Meio Ambiente (Brazilian Environmental Policy)
PPA	Power Purchase Agreements
R$	Brazilian Real
RAP	Resettlement Action Plan
RBC	Review Board of Consultants
RESEB	Reestruturação do Setor Elétrico Brasileiro (Restructuring the Brazilian Electricity Supply Industry)
RETA	Regional Technical Assistance

RIMA	Relatório de Impacto Ambiental (Environmental Impact Report, Brazil)
RPF	Resettlement Planning Framework
RVC	Rural Volunteers Centre, India
SEA	Strategic Environmental Assessment
SISNAMA	Sistema Nacional de Meio Ambiente (National Environmental System, Brazil)
SPD	Sozialdemokratische Partei Deutschland (Social Democratic Party, Germany)
SPO	State Planning Organisation, Turkey
TA	Technical Assistance
TAC	Termo de Ajuste de Conduta (Protocol for Adjustment of Public Conduct)
TAN	Transnational Advocacy Network
TEMA	Turkiye Erozyonla Mucadele, Agaclandirma ve Dogal Varliklari Koruma Vakfi (Turkish Foundation for Erosion Control, Afforestation and Nature Protection)
TICO	Takoradi International Company
TMMOB—CMO	Turk Muhendis ve Mimar Odalari Birligi—Cevre Muhendisligi Odasi (Union of Chambers of Turkish Engineers and Architects—Chamber of Environment Engineers)
TNC	The Nature Conservancy
TOOR	Transfer of Operating Rights
ToR	Terms of Reference
TSKB	Turkiye Sinai Kalkinma Bankasi (Industrial Development Bank of Turkey)
TUSEV	Turkiye Ucuncu Sektor Vakfi (Third Sector Foundation of Turkey)
UHE	Usina Hidrelétrica (Hydropower plant)
UN	United Nations
UNECE	United Nations Economic Commission for Europe
UNEP	United Nations Environment Programme
UNESCO	United Nations Educational, Scientific and Cultural Organization
US$	United States Dollar
VBDF	Volta Basin Development Foundation
VMT	Verba de Manutenção Temporária (Temporary Maintenance Allowance, Brazil)
WCD	World Commission on Dams
WEED	World Economy, Ecology and Development
WFD	Water Framework Directive
WWC	World Water Council
WWF	World Wide Fund for Nature
YXK	Yekîtiya Xwendekarên Kurdistan (Kurdish Students Association)
ZKB	Züricher Kantonalbank (Cantonal Bank of Zurich)

About the Editors

Waltina Scheumann holds her master's in Political Science and a Ph.D. in engineering. She has been a faculty member of the Environmental and Land Economics chair at Technical University of Berlin, and later worked as senior researcher at the Helmholtz Centre for Environmental Research (UFZ), Leipzig. She is presently employed with the German Development Institute (DIE) in Bonn. Her work on water-related topics includes cooperation on transboundary waters and cross-sectoral coordination of the water and energy sectors; governance issues in irrigated agriculture, including drainage; and the water rights implications of large-scale land acquisitions.

Oliver Hensengerth is a Lecturer in Politics at Northumbria University in Newcastle in the United Kingdom. Before joining Northumbria, he worked for the Universities of Southampton, Warwick and Essex and has conducted research for think tanks and academic institutions in the United States, Germany and the United Kingdom for a UK-based NGO and for the German development cooperation. He works on Chinese foreign and environmental policies; Chinese hydropower investment in continental Southeast Asia; transboundary water cooperation in the Mekong River Basin; and Cambodian domestic politics. He has extensive fieldwork experience in the region, particularly in China and Cambodia.

About the Contributors

Andreas Atzl who holds a master's degree in Sociology, Geography and applied Hydro-Chemistry from the University of Mainz, is a research fellow and Ph.D. candidate at the Karlsruhe Institute of Technology, researching the implications of uncertainty in climate change projections on water infrastructure planning and management in southern Africa. His diploma thesis focused on "The role transnational NGO-networks play in the implementation of international standards: The anti-Ilisu dam campaign in Turkey". He worked at the German Development Institute (DIE) on a paper that traced the spread of the World Commission on Dams' norms in the international arena.

Vera Baumann is head of the Economic and Development Cooperation Division at the German Embassy in Tirana. From 2009 to 2012, she was responsible for public relations in the Federal Ministry for Economic Cooperation and Development (BMZ). She holds a master's degree in Applied Geography and Business Administration. As a young professional at the German Development Institute (DIE), she has cultural sciences from the Leuphana University of Luneburg and specialised in development work on resettlement and expropriation in dam projects in Turkey.

Nirmalya Choudhury completed his postgraduate studies in Economics at Calcutta University and received his postgraduate diploma in Forest Management from the Indian Institute of Forest Management, Bhopal. From 2005 to 2007, he worked at the International Water Management Institute in the IWMI-Tata Water Policy Programme. Since 2007, he has worked with the Sir Ratan Tata Trust and Collectives for Integrated Livelihood Initiatives as a research consultant on water-centric tribal livelihood enhancement strategies in central India. In September 2008, he joined the German Development Institute (DIE) research project with a Ph.D. scholarship from the Federal Ministry of Education and Research (BMBF).

Agnes da Costa is Senior Economist at the Brazilian Ministry of Mines and Energy and specialises in the economics of the energy and mining sectors. She graduated with a degree in Economics at the Federal University of Rio de Janeiro and completed her MS studies in Energy Economics at the University of São Paulo. Her professional experience also includes working at a Brazilian bank with project finance in the energy sector. Her main research interests comprise electricity market reform and liberalisation; environmental and energy policy; sustainable development; financing of energy projects in developing countries; and private sector development. In September 2008, she joined the German Development Institute (DIE) research project with a Ph.D. scholarship from the Federal Ministry of Education and Research (BMBF).

Sara Eichert received a master's degree in Latin American studies from Complutense University of Madrid, where she has recently become a research assistant in the Department of Political Science and Public Administration II. She also holds a bachelor's degree in Political Science from the Free University of Berlin. Her research primarily focuses on new forms of governance, environmental and water policy, and the influence and participation of collective actors in the context of natural resource governance.

Klaus Horstmann is a member of the Environment and Sustainability Department of KfW Development Bank. He works on the environmental and social assessment of projects, internal guideline preparations and training. He is also responsible for the follow-up on international environmental and social standard-setting developments, including those related to dams. At the start of his professional life, he worked at a university research centre for 10 years and was a lecturer on environmental impact assessment in an international postgraduate course. Before he joined KfW in 2005, he had worked as an independent consultant for 15 years with clients that included German government authorities and their subordinate bodies. He holds an engineering diploma in Urban and Regional Planning and a Ph.D. in Economic and Social Sciences.

Sylvia Ismail works as a desk officer at the Education Division of the German Federal Ministry for Economic Cooperation and Development (BMZ) since July 2011. Before that, she was assistant to the Director at the Centre for International Migration and Development, a consortium of the GIZ GmbH and the German Federal Employment Agency. She holds two master's degrees in Development Management and Administration (Ruhr University Bochum, Germany, and the University of the Western Cape, South Africa). As a young professional at the German Development Institute (DIE) in Bonn, Sylvia Ismail researched environmental planning procedures and instruments in Turkey.

Anna-Lena Mueller works as a project manager at KfW Development Bank, where she is focusing on governance issues in sub-Saharan Africa. Before that, she carried out research on expropriation and resettlement in Turkish dam projects within a programme for young professionals at the German Development Institute (DIE). She holds a master's in International Economics from the Johann Wolfgang Goethe University in Frankfurt as well as a master's from the University Dauphine in Paris.

Dennis Mutschler holds a master's in Political Science and Social Anthropology. As a young professional at the German Development Institute (DIE) in Bonn, Mutschler researched environmental planning procedures and instruments in

Turkey. Since 2009 he has been a project manager at the Deutsche Gesellschaft für Internationale Zusammenarbeit GmbH within the corporate unit AgenZ (Agency for market-oriented concepts). His focus areas are global governance, political communication and strategic marketing.

Kirsten Nyman is trained as an environmental engineer. She started her career within the thematic preparations team for the International Freshwater Conference in Bonn (2001). She worked as a water policy advisor for the Deutsche Gesellschaft fur Technische Zusammenarbeit (GIZ) to the German Federal Ministry for Economic Cooperation and Development (BMZ) on international water policy issues at the global level, until joining the Global Water Partnership in 2005 and working as a programme manager. Building on the experience of integrating different sectors and stakeholders in water resources management (IWRM) at the national level, she has been heading the sector project titled "Policy Advice for Sustainable Hydropower" since 2008. In that capacity, she also represented BMZ in the Hydropower Sustainability Assessment Protocol process.

Christiane Rudolph is senior environmental and social specialist in the Sustainable Development and Environment Department of the Deutsche Investitions- und Entwicklungsgesellschaft (DEG). As a member of the KfW Bankengruppe, DEG finances investments of private companies in developing and transition countries. The main focus of her work is applying international environmental and social standards with a sector focus on hydropower and agriculture and a regional focus on Asia and Africa. In doing so, she follows closely trends in international good environmental and social practices. She studied geography and has a master's in International Politics, History and Economics, whereby she focused on the developing world.

Thomas Walenta works as a project manager at KfW Development Bank, where he is responsible for financing infrastructure projects in developing countries and emerging markets throughout Asia. He holds two master's degrees in International Business Administration, awarded by European University Viadrina, Germany, and the Catholic University of Cordoba, Argentina. As a young professional at the German Development Institute (DIE) in Bonn, Thomas Walenta was a member of an interdisciplinary research team that worked on the implementation of international environmental and social standards for large hydropower projects in Turkey.

Chapter 1
Dams and Norms: Current Practices and the State of the Debate

Waltina Scheumann and Oliver Hensengerth

1.1 Background

No other infrastructure has attracted more criticism than large reservoir-based hydropower and multi-purpose dams. Their detrimental social and ecological effects make them especially contentious, with opponents and proponents often unable to reach compromise solutions. In December 2000, the World Commission on Dams (WCD) proposed a new framework for decision-making in its report of the same name (WCD 2000), and this has become a landmark for sustainable practices.

The WCD recommendations were the result of a unique process. Set up by the participants in the Gland Initiative in 1998, the Commission had twelve members selected for their expertise and regional background, not as representatives of specific stakeholders. The Forum advising the Commission, on the other hand, comprised a large number of stakeholders representing, for example, industry, government agencies and the people affected. However, the consensus reached within the WCD on the ambitious guidelines did not translate into a broader normative consensus outside the Commission (Dubash et al. 2001; Dingwerth 2005; Dubash 2009). One reason for this was that governments were not represented in the WCD's norm-and standard-setting process. The Commission's mandate ended with the release of the *voluntary* guidelines, and since the WCD had no implementation mechanism, it remained to be seen whether governments not represented in the Commission would eventually accept its recommendations and guidelines them.

W. Scheumann (✉)
Department of Environmental Policy and Natural Resources Management, German Development Institute—Deutsches Institut fuer Entwicklungspolitik (DIE), , Bonn, Germany
e-mail: Waltina.Scheumann@die-gdi.de

O. Hensengerth
Department of Social Sciences and Languages, Northumbria University, Newcastle, UK
e-mail: Oliver.Hensengerth@northumbria.ac.uk

W. Scheumann and O. Hensengerth (eds.), *Evolution of Dam Policies*,
DOI: 10.1007/978-3-642-23403-3_1, © Springer-Verlag Berlin Heidelberg 2014

It was hoped by many that the WCD report would bring an end to the protracted debate on the pros and cons of dams. Gone were the days, it seemed, when dams were praised by some as the new temples (Jawaharlal Nehru) and the modern pyramids (Gamal Abdel Nasser), while their opponents saw them as everything that could go wrong with modernization, whether capitalist or communist. Media reports of struggles and campaigns against dams regularly capture the headlines and inflame passions. Recently, it has been the Belo Monte storage hydropower dam in Brazil: although the technical design has been altered several times and now entails less resettlement, and although the environmental authority has laid down strict environmental requirements for its construction, environmentalists, small farmers and Indian tribes living in the region have announced that they will continue campaigning to stop the project (Guardian 2011; Phillips 2011).[1] Another case is the Ilisu Dam on the River Tigris in South-Eastern Turkey, from which international financial consortia withdrew because of the Turkish government's failure to satisfy all the environmental and resettlement requirements to which it had contractually committed itself (Scheumann 2008; Warner 2012).

The list of names of disputed dam projects is endless, and it must be realized that the release of the WCD report was not the end of the dam discourse, but just one more contribution to the ongoing debate about what sustainable development and decision-making mean in relation to dams (Conca 2002).

Despite the persistent controversy, it must also be realized that the WCD's norms and standards matter. To give but a few examples, some construction firms have agreed to adhere to the recommendations (Park 2009), the African and Asian Development Banks have responded positively, and the Asian Development Bank (ADB) even refers to one of the most contested strategic priorities (free prior and informed consent[2]). Even the World Bank, which was at first extremely cautious because of its client governments' reservations, has promised to consider how the WCD guidelines might inform its safeguard policies (Park 2009, p. 120). The International Hydropower Association (IHA) convened the Hydropower Sustainability Assessment Forum, which updated the IHA's Hydropower Sustainability Assessment Protocol (Locher et al. 2010; Bosshard 2010; IHA 2010).[3] The draft Protocol also detailed how it corresponded to the WCD's recommendations. The same cannot, however, be said of the final Protocol. The European Union's Linking Directive[4] concerning dam projects in the context of the Clean Development Mechanism (CDM) explicitly refers to the WCD (EU 2004, Recital 14 and

[1] For updates and NGOs' views see the Belo Monte blog at International Rivers, http://www.internationalrivers.org/blogs/244/letting-the-xingu-run-freely. Accessed 29 July 2013.

[2] For more information on the consent and its form as recognized by the ADB in its 2009 Safeguard Policy Statement, http://www.adb.org/sites/default/files/pub/2009/Safeguard-Policy-Statement-June2009.pdf. Accessed 29 July 2013.

[3] See also the website for the Hydropower Sustainability Assessment Protocol at http://www.hydrosustainability.org. Accessed 29 July 2013.

[4] Directive 2004/101/EC of the European Parliament and of the Council of 27 October 2004 amending Directive 2003/87/EC establishing a scheme for greenhouse gas emission allowance

Article 11b(6)). The German Federal Ministry for Economic Cooperation and Development—Bundesministerium für wirtschaftliche Zusammenarbeit und Entwicklung (BMZ)—and its implementing agencies, namely the Gesellschaft für internationale Zusammenarbeit (GIZ) and Germany's development bank—Kreditanstalt für Wiederaufbau (KfW)—have promoted the dissemination of the recommendations and made financial support for dam projects conditional on compliance with the strategic priorities. Furthermore, five countries (Nepal, South Africa, Thailand, Uganda and Vietnam) started more extensive dialogues with a view to understanding precisely what the WCD report means in their national context (Park 2009, p. 121).

However, the impact of the WCD is not straightforward. Its strategic priorities and, in particular, its guidelines have been criticized for representing ideal-type norms from the perspective of affected people (Briscoe 2010). The governments of Brazil, China and India have been sceptical about the applicability of the seven strategic priorities, and some of them have rejected the 26 guidelines outright. But some small countries, too, such as Nepal and Ethiopia, have accused the WCD of vetoing dam construction, thus hindering their social and economic development.

Against this background, research for this book began when the German Development Institute—Deutsches Institut für Entwicklungspolitik (DIE)—won a research grant from the BMZ. The Federal Ministry had an immediate interest in in-depth research on the effects of the WCD: it had supported the WCD process from the very beginning, it had been involved in disseminating its recommendations, and it had facilitated multi-stakeholder dam dialogues in a number of countries. **Kirsten Nyman, Klaus Horstmann** and **Christiane Rudolph** (in this book) consider how the WCD recommendations have been integrated into German development cooperation and what mechanisms have been used to promote their dissemination and implementation. Their article takes a practical view and discusses some of the challenges posed by the implementation of the WCD's strategic priorities and guidelines from experience gained in German development cooperation.

The Ministry's major concerns have prompted academic interest in:

(i) whether the WCD has triggered changes to norms and policies related to dams in such emerging countries/economies as Brazil, China, India and Turkey;
(ii) the role played by non-governmental organisations (NGOs) and transnational advocacy networks (TANs) and their capacity to influence changes at domestic level;
(iii) the interaction between Chinese actors and host developing countries, and whether domestic changes to environmental regulations are reflected in China's overseas activities.

(Footnote 4 continued)
trading within the Community, in respect of the Kyoto Protocol's project mechanisms (Text with EEA relevance).

While the project originally examined norm diffusion and policy change in Brazil, China, India and Turkey, it developed beyond the initial concerns to examine the role of NGOs and TANs and the interaction between Chinese and host government institutions. The three sections of this book are devoted to this development.

1.2 Conceptualizing Norm Change: Studying Brazil, China, India and Turkey

The brief exposition above indicates that normative frameworks like those developed by the WCD, a global platform, do not travel well. The salient questions are whether and how norms—defined as "shared expectations held by a community of actors about appropriate behaviour for actors with a given identity" (Khagram et al. 2002, pp. 13–14; Finnemore and Sikkink 1998, p. 891)—"actually reach a domestic area", "get on the domestic agenda" (Checkel 1997, pp. 476, 478–479) and become part of domestic normative frameworks.

National appropriation of international norms often occurs in a process in which new norms are tested and eventually re-defined and re-shaped to conform to the social and political context of the countries concerned. The resulting *norm empowerment*, a term coined by Checkel (1997), can be the consequence either of instrumental adaptation to external pressure (the rationalist approach) or of elite learning in the form of internal institutional learning (constructivist approach). According to Checkel (1997), both processes relate to and are shaped by the domestic political structure of the country in question, and neither they nor the outcomes are an either-or process. Cortell and Davis (2000, pp. 71–72) rightly emphasized that merely observing that state behaviour is consistent with an international norm does not reveal whether that behaviour is the result of domestic normative change or a rational response to a certain international or domestic situation (Checkel 2000, pp. 69–70). Indeed, Goldstein and Keohane (1993) showed that state elites may change their strategies if it is in their national interest to do so. In a twist to this argument, Katsumata (2011) points out that states may resort to the "mimetic adoption" of international norms to gain legitimacy in the international community.

Risse-Kappen (1995) focused on transnational activities and their impact at domestic level, which would depend on the nature of the domestic structure and the extent to which an issue area was the subject of international rules. This, he claimed, determined the extent to which transnational actors could "penetrate the social and political system" of the "target" country (1995, p. 6). The more an issue area was governed by international norms, the more it could be permeated by transnational activities. Similarly, when faced with a state-controlled/state-dominated system—characterized by a combination of centralized political institutions, a weak societal structure and consensual policy networks—transnational actors might find initial access and penetration difficult, but once access was obtained, it was easier for them to bring about what they considered to be a desirable change, if they could form winning policy coalitions with the powerful centralized state actors. In general, opportunities for gaining access changed with and depended on particularities of domestic structures.

While works by Risse-Kappen (1995), Checkel (1997), Khagram (2004) and Keck (1998) reflected the importance of domestic structures and institutions in the process of norm diffusion, transnational actors (TANs and NGOs, international organisations such as the World Bank and western governments) drove and maintained control of the process. An influential contribution to the early norm diffusion debate is Keck and Sikkink's (1998) *boomerang model.* This model entails a non-linear move of an international norm from the international to the domestic level via non-state actors who successfully lobby a national (western) government to exert pressure on another government. Yet the model also emphasizes that domestic norm change is not possible when dominant elites refuse to adopt an international norm. Risse and Sikkink (1999) then developed a five-phase spiral model to explain differences in the domestic internalization of an international norm. Their key argument is that norm internalization must go hand in hand with domestic policy change. The spiral model therefore acknowledges that there is never—or rarely—a process of pure, linear norm diffusion. Similarly, the norm life cycle model developed by Finnemore and Sikkink (1998) argues that, while a norm diffusion process consists of norm emergence, norm cascade and internalization, the norm life cycle may remain incomplete and not extend beyond norm emergence.[5]

What this early debate did not take into account was the real-life practice of normative change in which an outcome somewhere between wholesale adoption and rejection might be possible.

Meanwhile, the norm diffusion debate has moved on and assumed the form of a norm contestation debate, as a number of authors have introduced culture into it: Reus-Smit (1997, p. 563) pointed out that "primary social institutions shape identity, in turn influencing basic institutional practices". Cortell and Davis (2000) argued that the cultural match between international and domestic norms is one of a set of conditions for successful norm diffusion. Similarly, Wiener (2007) pointed out that the transnationalization of political processes shifts constitutional norms from one socio-cultural environment to another, where it leads to a process of de-linking indigenous cultural and organisational practices. As culture underpins domestic institutions, it enhances the capacity of domestic institutions to filter international norms, a process better characterized as norm localization than norm diffusion.

Transnationalization of norm diffusion thus gives way to norm localization, an international norm interacting with domestic norms (the "cognitive prior") in order to be adapted to local political, economic, social and cultural circumstances

[5] For a recent application of this model see Kelley (2008).

(Acharya 2004). In the process of norm localization it is the local actors who play the lead role in reconstructing international norms to enhance their legitimacy, authority and prestige. During the process of norm localization norms are subject to "pruning" to make them a "better fit with existing local beliefs and practices" (Acharya 2004, p. 246). This process results in an "institutional change", with existing institutions taking on new functions, including new tasks and processes, and so increasing their scale of operations and constituents (Acharya 2004, p. 253).

In norm localization, which takes a more bottom-up view of norm diffusion, domestic actors, structures and institutions are equally important, as are access points for transnationally operating actors who advance global normative frameworks. Risse and Sikkink (1999) and Khagram (2004) hypothesized that democratic regimes and a high degree of domestic social mobilization enable transnationally allied advocacy groups (or, in the absence of such mobilization, limit their ability) to have deeper access to domestic processes. Where these enabling conditions exist, international actors are able to promote the institutionalization of international norms within a country by demanding access to information, respect for civil rights and greater public involvement in decision-making (Khagram 2004, p. 206). The difference between these two approaches is important: while Khagram (2004) and Risse and Sikkink (1999) conceptualize transnational actors as drivers of norm diffusion, with domestic actors being no more than constraining-enabling agents, Acharya (2004) sees the local actors as the key to norm diffusion. In the former case domestic norms are replaced by global norms, while in the latter case international norms are reconstructed to fit the domestic norms and contexts.

1.3 Basic Assumptions About Triggers and Drivers

On the basis of the above literature, country studies were undertaken to examine processes of change in a number of emerging economies that are significant regional and/or global players: Brazil, China, India and Turkey. We considered the following three hypotheses:

Hypothesis_1 International Embeddedness
A country's embeddedness in global politics and its efforts to be or become a global player has repercussions for its decision whether or not its government refers to (dam-related) international norms and standards not only rhetorically but in practice. Embeddedness as understood here concerns a country's participation in international regimes, its ratification of international treaties, the socialisation of its elites in international epistemic communities and its membership of international governmental organisations (see e.g. Johnston 2007).

The assumption was that, with increasing international embeddedness in an issue area, governments would refer to the international norms and principles valid in that issue area and could therefore be held accountable for complying with international norms. In this respect, international embeddedness would be one of the possible pathways through which international norms would "actually reach a domestic area" and "get on the domestic agenda", as Checkel (1997, pp. 476, 478–479) put it.

Hypothesis_2 Social Mobilization
In the case of social mobilization the hypothesis was that countries where there is a thriving civil society and the action they take is tolerated, it will be able to diffuse international social and environmental standards in policy and in practice through various combinations of information, symbolic politics, leverage and accountability politics (see, for example, Tarrow 2011). The access that NGOs/civil society have to the political and juridical systems and to decision-making processes in various policy fields may differ with the nature of the political regime.

Closely associated with social mobilisation and normatively recognised as an important component for gaining public acceptance (WCD 2000) for hydropower dam projects is the role of the public in the decision-making process. Public involvement is motivated by three rationales: democratic, instrumental and management (Dore and Lebel 2010; Petts 1999; Kunreuther and Slovic 1996; Fiorino 1990; King et.al. 1998).[6] At process level public involvement can be decomposed into public information, public consultation and public participation[7] (Rowe and Frewer 2005). At outcome level public involvement could be seen either as a spectrum of increasing public impact (IAP2 2007) or as a ladder of public empowerment where the affected public would have an increasing impact on the final decision (Arnstein 1969).

Hypothesis_3 Reliance on Foreign Financial Services and Technology
If the implementation of national dam programmes, or projects, relies on international financial services and expertise, both international financial institutions and business corporations can act as norm carriers and standard setters. On the premise that these international actors attach conditions to their financial services, and benefit from applying international standards the hypothesis was that they would be able to act as drivers in the diffusion of more ambitious social and

---

[6] Democratic rationale flows from the argument that people whose life and livelihood are (negatively) affected by a decision have a right to take part in the decision. The instrumental rationale, most applicable to contentious infrastructure projects like dams, is based on the argument that if different stakeholders are involved during the decision-making a decision is more likely to be acceptable for different stakeholders and projects are less likely to suffer delays from blockages. The management rationale is based on the argument that with more stakeholders in the decision-making much more information will be processed which would improve the quality of the decision.

[7] Rowe and Frewer's (2005) information-flow model of public involvement defines public information as one way flow of information from the organizers of public involvement exercise to the participants; public consultation as one-way flow of information from the public to the organizers of public involvement exercise; public participation as a simultaneous two-way flow of information. The ideal case under this information flow-model is when maximum bidirectional flow of information takes place between the participants and the organizers.

environmental norms in various target countries by applying technical and financial leverage.

1.4 Research Design of Country Studies in Brazil, China, India and Turkey

The country studies on Brazil (**Agnes da Costa**), China (**Oliver Hensengerth**), India (**Nirmalya Choudhury**) and Turkey (**Waltina Scheumann et al.**) followed a qualitative research design using a case study approach. A two-level—national and project—research design was adopted for each of the country studies. This decision was made in the knowledge that developments relating to hydropower and multi-purpose dams at national and project level were often interrelated. It was assumed that the two-level analysis would adequately cover the complexity of decision-making in a multi-actor and multi-level setting. At national level, the analysis would focus on planning and decision-making laws and procedures and their changes over time in the policy fields of environment and resettlement. We wanted to understand the role of public, private and civil society actors, their interactions and the causes of change. However, to determine whether or not normative change had taken place over and above the formal incorporation of new norms into legislation, we examined whether these changes affected the planning and implementation of specific dam projects.

This research comprised both within-case and cross-case analyses (Mahoney 2007; Ragin 1987). The within-case studies took a historical perspective spanning a period of more than three decades and including process-tracing and causal-process observation (Mahoney 2007, p. 17). The research was expected to provide not only insights into causal factors and mechanisms but also contextualized information that would identify both temporal changes and the role of the various drivers (Geddes 2003, p. 119). Thus, in the context of this research, the within-case analysis would result in both questions being answered: (i) the nature of changes (towards greater sustainability) and (ii) the causality of changes (the drivers).

The social and environmental norms under observation were operationalized through two issue areas. Historically, the two most contentious and challenging issues confronting the sustainable development of dams have been the resettlement and rehabilitation of the displaced population and the mitigation of adverse environmental impacts. It was postulated that a resettlement and rehabilitation policy and an effective environmental governance system incorporating Environmental Impact Assessment (EIA) procedures were two fundamental requirements for sustainable dam development. The within-case analyses forming part of the country studies were to focus on the evolution of resettlement and rehabilitation policy and environmental governance over the years and on the roles played by governmental and non-governmental actors in this process.

Given that dam planning is related to broader economic and administrative processes, the analyses had to consider changes in policy fields outside the narrow sphere of dam planning, with particular reference to the liberalization of the energy/electricity sector and changes in the water sector which, together with fundamental political shifts, would affect the procedural and substantive aspects of planning and decision-making in the countries concerned. The country studies therefore included a third topic, the decision-making processes that govern the construction of dams, the aim being to analyse the complexity of multilevel dam-related decision-making processes so that it might be determined if, why and how planning and decision-making have been opening up to diverse societal interests. All three topics—resettlement and rehabilitation, environmental governance, and decision-making on dams with particular reference to public involvement—were considered at both national level and project level.

In order to attribute norm changes (or their absence) to the WCD principles/norms, we had to choose a period for analysis which would "reflect the full range of outcomes in the relevant universe" (Geddes 2003, p. 117), thus avoiding incorrect causal attributions. Any change identified as a case of diffusion of WCD norms might have been a mere continuation of a pattern of behaviour that existed before those norms came into being. The period covered by the analysis therefore preceded the advent of the WCD recommendations.

A cross-case analysis was also undertaken as part of each country case study. In Brazil's case, the hydropower projects selected had been built under different political regimes. In China's case, purposive sampling was undertaken to enable a detailed case analysis to be made of two dam projects in terms of their respective importance for resettlement and environmental impact assessment. In India's case, the cross-case analysis was based on purposive sampling, cases being chosen on the basis of the presence/absence of international actors and the level of social mobilisation. In Turkey's case, dam projects were selected before and after the enactment of environmental legislation and on the basis of the presence/absence of international actors. In the final Chap. 11, Waltina Scheumann, Oliver Hensengerth and Nirmalya Choudhury draw conclusions from the country studies which they hope will prove helpful to decision-makers.

1.5 Non-Governmental Organisations as Vehicles of Norm Diffusion

Both Andreas Atzl and Sara Eichert paid tribute to the role played by NGOs in the dam discourse and as vehicles of norm diffusion. **Andreas Atzl** analyses the campaigning of a transnational NGO network against the Ilisu Dam (South-Eastern Turkey) on the River Tigris and the influence it is having on the Turkish government's decision-making. He takes a close look at the composition of the network, the interrelationships and alliances between local and non-domestic NGOs, the campaigning levels (local, national, European) and the politics applied.

He also considers how the transnational NGO network has changed and adapted its strategies to gain influence on Turkish decision-makers, placing greater emphasis on environmental aspects and forging closer alliances with Turkish dam opponents. **Sara Eichert** explores the diverging strategies pursued by two international NGOs to encourage private-sector actors to promote sustainable dam development and discusses why some NGOs choose to collaborate with the dam industry to a greater or lesser degree. To this end, she compares the strategies of the World Wide Fund for Nature International and International Rivers in a recent industry-driven multi-stakeholder dialogue—the Hydropower Sustainability Assessment Forum (HSAF)—and their positions on its outcome, the International Hydropower Association's 2010 Hydropower Sustainability Assessment Protocol.

1.6 Interaction Between Chinese Actors and the Governments of Host Developing Countries

Two case studies then investigate the interaction between Chinese actors and Ghanaian and Cambodian agencies (both **Oliver Hensengerth**) with a view to assessing Chinese engagement from evidence obtained in the field. While not strictly comparative, the studies also serve as an initial attempt at comparing the forms and effects of Chinese hydropower investment abroad. The studies focus on the collaboration between Sinohydro and the respective government departments and regulatory bodies in order to identify the processes over which the governments retain control, the areas of responsibility delegated to the company and the mechanisms for ensuring its compliance with national legislation and international standards.

References

Acharya A (2004) How ideas spread: whose norms matter? norm localisation and institutional change in Asian regionalism. Int Org 58:239–275

Arnstein SA (1969) A ladder of citizen participation. J Am Inst Planners 35(4):216–224

Bosshard P (2010) The dam industry, the World Commission on Dams and the Hydropower Sustainability Assessment Forum (HSAF) process. Water Altern 3(2):58–70

Briscoe J (2010) Overreach and response: the politics of the WCD and its aftermath. Water Altern 3(2):399–415

Checkel JT (1997) International norms and domestic politics: bridging the rationalist-constructivist divide. Eur J Int Relat 3(4):473–495

Conca K (2002) The World Commission on Dams and trends in global environmental governance. Polit Life Sci 21(1):67–70

Cortell AW, Davis JW (2000) Understanding the domestic impact of international norms: a research agenda. Int Stud Rev 2(1):65–87

Dingwerth K (2005) The democratic legitimacy of public-private rule-making: what can we learn from the World Commission on Dams? Glob Gov 11(1):65–83

Dore J, Lebel L (2010) Gaining public acceptance: a critical strategic priority of the world commission on dams. Water Altern 3(2):124–141
Dubash NK (2009) Global norms through global deliberation? reflections on the World Commission on Dams. Glob Gov 15:219–238
Dubash NK, Dupar M, Kothari S, Lissu T (2001) A watershed in global governance? an independent assessment of the World Commission on Dams. World Resources Institute, Washington
EU (2004) Directive 2004/101/EC of the European Parliament and of the Council of 27 October 2004 amending Directive 2003/87/EC establishing a scheme for greenhouse gas emission allowance trading within the community, in respect of the Kyoto Protocol's project mechanisms (Text with EEA relevance). http://eur-lex.europa.eu/LexUriServ/LexUriServ.do?uri=OJ:L:2004:338:0018:0018:EN:PDF. Accessed 29 July 2013
Fiorino DJ (1990) Citizen participation and environmental risk: a survey of institutional mechanisms. Sci Technol Human Values 15(2):226–243
Finnemore M, Sikkink K (1998) International norm dynamics and political change. Int Org 52(4):887–917
Geddes B (2003) Paradigms and sand castles theory building and research design in comparative politics. The University of Michigan Press, Michigan
Goldstein J, Keohane R (eds) (1993) Ideas and foreign policy: beliefs, institutions, and political change. Cornell University Press, Ithaca
Guardian (2011) New rights challenge to Belo Monte Dam in Brazil, 12 April. http://www.guardian.co.uk/environment/2011/apr/12/belo-monte-dam-work-suspended. Accessed 29 July 2013
IAP2 (2007) http://www.iap2.org/associations/4748/files/IAP2%20Spectrum_vertical.pdf
IHA (International Hydropower Association) (2010) Hydropower Sustainability Assessment Protocol. http://www.hydropower.org/sustainable_hydropower/hsaf_Hydropower_Sustainability_Assessment_Protocol.html. Accessed 29 July 2013
Johnston AI (2007) Social states: China in international institutions, 1980–2000. Princeton University Press, Princeton
Katsumata H (2011) Mimetic adoption and norm diffusion: "Western" security cooperation in Southeast Asia? Rev Int Stud 37:557–576
Keck M (1998) Planafloro in Rondonia: the limits of leverage. In: Fox JA, Brown DL (eds) The struggle for accountability. The World Bank, NGOs, and Grassroots Movements. MIT Press, Massachusetts
Keck M, Sikkink K (1998) Activists beyond borders: advocacy networks in international politics. Cornell University Press, Ithaca
Kelley J (2008) Assessing the complex evolution of norms: the rise of international election monitoring. Int Org 62:221–255
Khagram S (2004) Dams and development transnational struggles for water and power. Oxford University Press, New Delhi
Khagram S, Riker JV, Sikkink K (2002) From Santiago to Seattle: transnational advocacy groups restructuring world politics. In: Khagram S, Riker JV, Sikkink K (eds) Restructuring world politics. Transnational social movements, networks and norms. University of Minnesota Press, Minneapolis
King SC, Feltey KM, Susel BO (1998) The question of participation: toward authentic public participation in public administration. Public Adm Rev 58(4):317–326
Kunreuther H, Slovic P (1996) Science values and risk. Ann Am Acad Polit Soc Sci 545:116–125
Locher H, Hermansen GY, Johannesson GA et al (2010) Initiatives in the hydro sector post-World Commission on Dams- the Hydropower Sustainability Assessment Forum. Water Altern 3(2):43–57
Mahoney J (2007) Qualitative methodology and comparative politics. Comp Polit Stud 40(2):122–144
Park S (2009) The World Bank, dams and the meaning of sustainable development in use. J Int Law Int Relat 5(1):93–124

Phillips T (2011) Brazilian judge orders construction of Amazon Dam to stop, Guardian, 29 September. http://www.guardian.co.uk/environment/2011/sep/29/brazilian-judge-monte-bello-dam. Accessed 29 July 2013
Petts J (1999) Public participation and environmental impact assessment. In: Petts J (ed) Handbook of environmental impact assessment, environmental impact assessment: process, methods and potential, vol 1. Blackwell Science, Cornwall:, pp 45–177
Ragin C (1987) The comparative method moving beyond qualitative and quantitative strategies. University of California Press, Berkeley
Reus-Smit C (1997) The constitutional structure of international society and the nature of fundamental institutions. Int Org 51(4):555–589
Risse T, Sikkink K (1999) The socialization of international human rights norms into domestic practices: introduction. In: Risse T, Ropp SC, Sikkink K (eds) The power of human rights: international norms and domestic change. Cambridge University Press, Cambridge, pp 1–38
Risse-Kappen T (1995) Bringing transnational relations back in: non-state actors, domestic structures and international institutions. Cambridge University Press, Cambridge
Rowe G, Frewer L (2005) A typology of public engagement mechanisms. Sci Technol Human Values 30(2):251–290
Scheumann W (2008) How global norms for large dams reach decision makers: a case study from Turkey. In: Scheumann W, Neubert S, Kipping M (eds) Water politics and development cooperation: local power plays and global governance. Springer, Berlin, pp 55–80
Tarrow SG (2011) Power in movement: social movements and contentious politics, 3rd edn. Cambridge University Press, New York
Warner J (2012) The struggle over Turkey's Ilisu Dam: domestic and international security linkages. International Environmental Agreements, electronic pre-publication March 2012. doi:10.1007/s10784-012-9178-x
Wiener A (2007) Contested meanings of norms: a research framework. Comp Eur Polit 5:1–17
WCD (World Commission on Dams) (2000) Dams and development: a new framework for decision-making. Earthscan, London

Chapter 2
Sustainable Dam Development in Brazil: The Roles of Environmentalism, Participation and Planning

Agnes da Costa

Abstract The chapter explores the reforms of Brazil's environmental and resettlement policies and the influence of domestic and external actors on its dam-related legislation and practices. It analyses two hydropower plant projects: Itaparica, built during the military regime; and Santo Antonio, a public–private project still under construction. The analysis begins with an overview of the relevance of hydropower to Brazil's development plans since the military regime and of the central administration's strategy of developing the country's hydropower potential by maintaining a clean mix of energy sources for the supply of electricity. Domestic opposition to the authoritarian regime had culminated in the promulgation of a highly participation-oriented Constitution in 1988. The chapter then analyses decision-making processes in Brazil's dam-related bureaucracy and the role allotted to civil society since then. By singling out two problem areas—Environmental Impact Assessment and resettlement—the chapter addresses changes in the normative framework and in practice. Case studies then consider these processes, taking Itaparica as a case where planning and participation did not lead to satisfactory implementation in resettlement terms; and Santo Antonio as a possible example for future projects as regards the involvement of affected people at earlier stages of the project cycle.

2.1 Introduction

Brazil differs from the majority of countries in terms of electricity supply in that it relies on renewable energy sources. Although biomass has undergone significant growth in recent years compared to other sources, hydropower remains the key electricity generation technique. In 2007 85.2 % of Brazilian power was generated

A. da Costa (✉)
Brazilian Ministry of Mines and Energy, Brasília, Brazil
e-mail: agnes.dacosta@globo.com

W. Scheumann and O. Hensengerth (eds.), *Evolution of Dam Policies*,
DOI: 10.1007/978-3-642-23403-3_2,

mainly in large hydroelectric power plants, most of which relied on large reservoirs with multi-annual storage capacities. The operation of the transmission and generation facilities by a centralised operation system observes an optimisation rule that takes the storage level at each reservoir and the different hydrological regimes throughout the country into consideration.

Most of the existing reservoirs were built during the military regime (from 1964 to 1985), which means that, given the development and growth policies at that time, decisions to build large dams and flood large areas of land did not necessarily consider the resulting social and environmental impacts. Such decisions were normally taken on a sectoral basis. However, with the re-establishment of democracy and the promulgation of the Federal Constitution (1988; "CF/88"), which requires public participation in decision-making processes, decisions on the construction of dams and the flooding of land are now taken jointly by society, a number of government institutions and especially the population affected.

Furthermore, environmental legislation and standards have developed significantly since the promulgation of the CF/88. As hydropower is still considered necessary to meet growing demand and so to enable economic growth to continue in the country, more new dams are expected to be built but their construction will be subject to a comprehensive Environmental Impact Assessment (EIA), addressing social as well as environmental impacts (i.e. resettlement issues, too) and forming part of a rigorous three-stage environmental licensing process.

As the decision-making on and construction of dams before and after CF/88, during the liberalisation phase and after the reform of the electricity industry in 2003/2004 have different characteristics, the degree to which environmental and social norms were observed differed in each case. As a consequence, two dams have been chosen: the first was planned and built before 1988; the second was planned and built after CF/88.

The next section addresses the strategic role of dams in Brazil. Section 2.3 then conducts a country-level analysis, depicting the decision-making framework for dams in Brazil, identifying actors and networks involved in this process and indicating political opportunities. Section 2.4 presents the two dams in their historical context, while the final section links the two levels of analysis and discusses whether and how the findings relate to and substantiate the three hypotheses.

2.2 The Strategic Role of Hydropower in Brazil

The Brazilian electricity market has been growing at a sustained annual rate of 5 %, requiring the annual addition of 4,500–5,000 Megawatt (MW) of installed generation capacity to maintain economic growth. After the energy rationing in 2001, guaranteed energy supply was one of the main political priorities of the administration headed by President Lula, who was in office between 2002 and 2010.

The 10-Year Energy Plan 2007–2016 (MME 2008a) refers to the development of 90 projects by 2016 (including 19 already under construction), adding more than 36 Gigawatt (GW) of installed capacity to the system. Although the National Energy Plan 2030 (MME 2008b) indicates that hydropower capacity is expected to decrease in comparison to other sources, it should still play a dominant strategic role.

Electricity supply in Brazil is dependent on existing hydropower plants and on large reservoirs with multi-annual storage capacities. The origin of the Brazilian hydropower-based electricity mix is linked to the operation of the state-owned Eletrobras (Centrais Elétricas Brasileiras) and its subsidiaries, but the construction of recent hydropower plants results from both public and private investments and from a stable economy.

Brazil's economically viable hydropower potential is estimated at 260 GW, of which only 30 % is either operational or under construction (World Bank 2008). This means that, if this potential is to be developed at levels similar to those in more developed countries, the subject of sustainable dam construction must be addressed.

The strategic role of hydropower in Brazil and the central government's confidence in the comprehensiveness of the way this issue is being addressed has led officials to call for the recognition of hydropower as a renewable and sustainable source of energy. One commonly recognised[1] event illustrating this political positioning is the role played by the former Minister of Mines and Energy and now President Mrs Dilma Rousseff, head of the Brazilian delegation to the International Conference for Renewable Energies, which took place in Bonn in 2004. In that it included large hydropower plants in the category of renewable energy sources (CEPEL 2004), her intervention was decisive for the outcomes of the Conference. The Political Declaration (Programa Chile Sustentable 2004) agreed by government officials makes no reference to hydropower plants, while the Resolution for the International Parliamentary Forum on Renewable Energies (2004) explicitly refers to hydropower as renewable.

However, the government has encountered strong resistance to the implementation of this strategy. A sign of this political difficulty is the growing number of thermal generation plants contracted to supply the market in the years to come, since the government has been having difficulty obtaining the environmental licences needed before hydropower projects can be handed over to project developers. Yet, at the same time the increasing role of thermal plants has been sharply criticised by society in general (Goldemberg 2009; IRN 2009; Novaes 2009), thus producing a paradoxical behaviour of Brazilian society.

[1] As mentioned in several interviews (I05122008; I26122008; I30012009; I06022008; I12032009b; I13032009; I25032009; I22042009; phone call T11032009).

2.3 Changing Policies and Decision-Making Frameworks for Dams in Brazil

2.3.1 Triggers of Change

Political developments relating to the construction of dams in the periods discussed in this chapter contributed to and were affected by global trends. While a new democracy confident of its participative and decentralized foundations was built in 1988 as a reaction to 20 years of dictatorial military regime, the successor governments were constrained by the changes in the global political economy caused by response to the oil crises of the 1970s and the "Washington Consensus". This is the political and economic context in which the decision-making framework began to evolve.

2.3.1.1 Democratization of the Country: The Federal Constitution of 1988

The worst examples of Brazilian hydropower projects in terms of environmental and social impacts were built during the military regime. It was argued that they were crucial for the country's development, which was based on the growth of the economy and industrial activity. A supply of cheap electricity was considered essential for the success of this strategy.

Brazil faced a slow transition from military to civilian government (from 1974 to 1989) at a time of economic slowdown, which, on one hand, strengthened the opposition to the regime and to its development strategy and, on the other, enabled a positive association of concepts related to democracy, such as participation and decentralization, with those related to environmental protection. The reinforcement of democratic values began with an amnesty in 1979, which prompted the return of activists with new ideas and strategies. This was followed by the publication of a law on the protection of "diffuse interests",[2] which was passed in 1985 (Law 7347). This movement culminated in the promulgation of a participation-oriented Constitution in 1988 (Hochstetler and Keck 2007). The Constitution's longest article is devoted to the protection of the rights of the individual (Article 5).

According to Hochstetler and Keck (2007, p. 13), democratisation was also the origin of the Brazilian concept of socio-environmentalism, since it convinced environmentalists:

> (…) to broaden the social bases of their appeal to be able to influence newly democratic decision-making processes in a context of economic crisis (…) Based on this experience, many environmental movements in Brazil have focused as much on social equity and participation as on protecting the environment.

[2] Diffuse interests or rights are those interests or rights shared by a group, class or an indeterminate category of individuals in the same situation.

The Constitution itself was a result of a participative process. 122 popular amendments were presented to the constituent assembly, 83 of them being adopted. The gains achieved by these popular amendments included the recognition of the rights of the Indian,[3] the entitlement of associations to submit acts of unconstitutionality to the Supreme Court and the creation of the popular initiative of drafting laws (Lopes 2008, pp. 55–58).

The 1988 Constitution altered the framework for decision-making on the construction of dams. Two problem areas covered by CF/88 should be highlighted as possibly having the greatest impact on this framework: (i) the distribution of powers under the federative pact and (ii) the protection of the environment.[4]

Brazil's federal system has three constitutionally designated levels of political authority: the federation (usually called the Nation, the Union or the State), the states and the municipalities. Each has elected executive and legislative branches.

Because of the 20 years of dictatorial military regime, which concentrated many of the decisions and resources at federal level, CF/88 guarantees that responsibilities, resources and powers were shared among the three levels. The powers of the states are known as residual powers, since they are the powers which the Constitution does not explicitly prohibit the states to exercise (Article 25(1)), resulting in vague constitutional mandates, as pointed out by Hochstetler and Keck (2007, p. 14).

As a result of decentralised principles and consequent jurisdictional conflicts, Brazilian Federalism, known as the "federative pact", has been trying to find a way to deal with decision-making on projects or policies that have national repercussions, such as the construction of new hydropower plants. Thus, at national level, this decision-making process must consider the distribution of powers throughout the federation, the many actors who may have conflicting interests and rights of veto in certain subject areas.

The drafting of the legislation specifically concerning the environment, which is also covered by CF/88, began in the 1980s, when shared powers became a principle: Article 10 of Law 6938/81 (which created the National Environmental Policy) and Articles 4 and 5 of Resolution 237 of the National Council on the Environment (CONAMA) of 29 December 1997 form the legal bases for the shared powers of the three levels of government.

CONAMA laid down general rules defining the jurisdiction of the Brazilian Institute for the Environment and Natural Renewable Resources (IBAMA). IBAMA is responsible for the licensing and inspection of potentially or effectively polluting activities that have a national or regional impact, concern two or more states or are located on the border with other countries, on Indian lands or on areas protected by the Union and activities involving nuclear materials. CONAMA

---

[3] A whole chapter of CF/88 (Title VIII, Chapter VIII, Articles 231 and 232) is dedicated to this subject.

[4] Article 5 (LXXIII); Article 23 (VI); Article 24 (VI); Article 24 (VIII); Article 129 (III); Article 170, VI; Article 174, § 3°; Article 186 (II); Article 200 (VIII); Article 220 (II) and the whole of Chapter IV of Title VII (Article 225).

assigned to the environmental agencies of the states, the federal district and to implementing bodies the authority to license and inspect potentially or effectively polluting activities affecting two or more municipalities or forests and other forms of natural permanently protected vegetation. Furthermore, CONAMA made the municipal level responsible for the licensing and inspection of projects with a local impact.

One consequence of federalism as applied to the environment is the considerable variation in subnational environmental policies and practices accompanied by conflicting norms. This may make for greater flexibility and consideration of local specificities but also, in processes involving the construction of hydropower plants, for example, differences in the standard and quality of environmental licensing processes and studies. In some cases, this is solved in practice in a rather informal way, as observed by Hochstetler and Keck (2007), with IBAMA sometimes asked to take over the environmental licensing process that is the responsibility of a state environmental agency.[5] This causes further delays in the issuing of environmental licenses, increasing associated risks and costs.

As for water resources, CF/88 assigns the ownership of rivers that flow through two or more states or along or across borders with neighbouring countries to the Union (cf. Article 20). Article 21 also delegates to the Union responsibility for creating the National System for the Management of Water Resources, i.e. the institution ruling water resources issues in view of the federative pact.[6]

With respect to indigenous communities, Article 231 (1) of CF/88 recognises the indigenous groups' original right to the lands traditionally occupied by them (those used for living, for production activities and for their cultural and physical reproduction). Nevertheless, indigenous lands are federal lands forming part of the Union's exclusive domain, and, as stated above, the exploitation of water resources on these lands is subject to previous legislative authorisation, which entitles indigenous communities to be heard at specific public hearings (Article 231(3)). Considering that much of Brazil's remaining hydropower potential lies in the Amazon region (MME 2008b), where the majority of indigenous groups live, the role of the National Congress in dam decision-making is expected to grow in the years to come.

The protection of indigenous groups' rights is the responsibility of the Ministry of Justice, its associated foundation, the Brazilian National Indian Foundation (FUNAI) and the *Ministério Público.* In civil society, the indigenous organisations that fight for indigenous rights include the Indigenous Missionary Council (CIMI),

---

[5] Also addressed by I30012009 and World Bank (2008, p. 37), which mentions the case of the UHE Itumirim, the Preliminary Licence for which was issued by the environmental agency of the state of Goiás, in view of the vagueness of the term "local/regional environmental impact". Following a question by the *Ministério Público* the judicial authority decided in favour of this Ministry, and the process had to be restarted, this time conducted by IBAMA.

[6] Kelman and Veras (2008) point to the most important constitutional provisions with respect to hydroelectricity and the construction of dams: Article 20 (VII); Article 21 (XXII (b)); Articles 21 and 175; Article 22 (IV); Article 49 (XVI); Article 176.

which is linked to the Catholic Church, the Pro-Yanomami Commission (CCPY) and the Coordination of Indigenous Organisations of the Brazilian Amazon (COIAB). Nevertheless, some claim that indigenous groups are easily manipulated[7] by preservationist non-governmental organisations (NGOs) seeking to block the construction of such infrastructure as hydropower plants, given their constitutional prerogatives. The preservationist NGOs' fight for indigenous rights also attracts international attention in view of the sensitiveness of the indigenous rights issue in the global arena.

The *Ministério Público* is the State institution responsible for ensuring the effectiveness of legislation. Where legislation is violated by State organs, the *Ministério Público* is required to enforce (usually through the judiciary) compliance with the law (Lopes 2000). The *Ministério Público* is frequently referred to as the fourth branch of the Brazilian State because, although legally bound to the executive, it is not legally subordinate to it. As its task is to ensure that public agents respect the law, the members of the *Ministério Público* are entitled to functional independence, which also means that each prosecutor (procurador) is able to bring forward legal actions. The structure of the *Ministério Público* conforms to the distinction made by Brazilian federalism between the Union and the states, meaning that there is a *Ministério Público Federal* (at the level of the Union) and several *Ministérios Públicos Estaduais*, at the level of the states.

Although civil society organisations do not need the intervention of the *Ministério Público*, they choose in most cases to address their claims to the *Ministério Público's* prosecutors, who then file the public civil actions (Lopes 2000; I04032009). One of the reasons for this approach may be that, of all the players involved in the environmental licensing process, the *Ministério Público* has the best educated staff, significant resources and a broad mandate (World Bank 2008; Lopes 2000; I04032009).

The *Ministério Público* does not need to be called upon by other agents to act. Prosecutors may bring a public civil action based on a newspaper article, for example. In this respect its relationship with civil society is qualified in two ways: it is open to demands from civil society, thus constituting an agency that vocalises socially relevant issues as well as giving them political relevance; and it frequently induces demands, favouring specific societal dynamics. The *Ministério Público* is thus an institution whose activities reduce the cost of organising the process of bringing social concerns to the attention of the public sphere and of society in general (Lopes 2000).

Despite the relevance of its role of vocalising social demands and giving them political importance and of acting as a monitor of public decisions and policies, the *Ministério Público* is frequently criticised for the action it takes, especially in the case of the environmental licensing process of hydropower projects. Considering the delays the prosecutors cause in the licensing process, the World Bank (2008)

---

[7] I25032009; I04032009; Costa (1997).

claims that licensing agency employees fear penalties which may be imposed on them by the *Ministério Público* under the Environmental Crimes Act and the Administrative Improbity Act, which hold officials personally and criminally liable even when they act in good faith in complex circumstances. They therefore keep raising new demands concerning the scope of EIA as a means of deferring the need to take a decision (whether or not to issue the environmental licence). Many interviewees confirmed this perception, and the press has sometimes referred to cases where officials have been sued or intimidated by the *Ministério Público*.

The concern of the legislative branch for the environment is generally welcomed by society.[8] One visible consequence is that the environment has been taken up by all the parties of the political spectrum[9] and that the Environmental Parliamentary Front (*Frente Parlamentar Ambientalista*)[10] is composed of 12 (of the 80) senators and 465 (of the 513) deputies.

On the other hand, although Congress members may share environmental concerns, they may also welcome projects that have the potential to attract new investment to their regions. Hydropower plants are under a constitutional obligation to pay financial compensation. This is ruled by the Brazilian Electricity Regulatory Agency (ANEEL), and in 2009, 22 states and 634 municipalities were entitled to financial compensation, which amounted to R$ 1,647 million (this amount comprises also the royalties paid by the binational Itaipu hydropower plant, which are governed by another, similar distributive rule) corresponding to almost 0.1 % of Brazil's gross domestic product in that year. The interest of the legislative branch in these projects must not therefore be overlooked, even if the resources are in fact managed by the executive branches at each level of the federation. Even elected representatives from political parties which are in opposition benefit from having supported a project that increased government revenues.

One important issue to be addressed following the changes caused by CF/88 to the decision-making framework for the construction of hydropower plants is the relationship between social movements and NGOs and the apparatus of state, in view of the behaviour of individuals of organised civil society who constantly migrate from the governmental to the non-governmental sphere and so form a symbiotic relationship between NGOs and social movements on the one hand and the administration on the other, especially in the environmental field. Bernardo (2001, pp. 51–52) mentions that the Ministry of Environment (MMA) is usually regarded as a large NGO right in the middle of the State, an impression corroborated by some interviewees.[11] She argues that the proximity of civil society organisations to the State and to the formulation of environmental policies grew in the 1990s, when the State

[8] I05122008; I06022009.

[9] I05122008.

[10] See http://www.frenteambientalista.com. Accessed 04 July 2010.

[11] I06022009; I12032009b; I29042009; I13032009; I12032009c. On the other hand, because personnel shortages in the State grew during the 1990s, the fact that the State looked for experts in specific fields where this know-how already existed is not a peculiarity of the environmental field.

also backed sustainable development in environmental policies. Until then the concept of command and control had persisted in environmental policies, this being inconsistent with the State's limited capacity to exercise control and impose penalties. Bernardo points out that the approximation of the State and NGOs occurred mainly because of changes to the requirements of the providers of international resources. This type of resource has been critical not only for Brazilian (environmental) NGOs, but also, as Bernardo emphasizes, for the expansion of public environmental institutions, which have historically relied on foreign resources.

International funding became vital for the implementation of environmental policies in Brazil after 1992, when the State began to play a decisive role in the circulation of foreign resources in the environmental field, since large sums had accumulated in programmes in which it acted as intermediary between the funding source and NGOs (Bernardo 2001, p. 52). The leading financial institutions were the World Bank, the Inter-American Development Bank (IADB) and the United Nations Development Programme. Bernardo argues that it was mainly due to these institutions that consultation mechanisms were consolidated and public participation in environmental policies was ensured, these being made conditions for the loans they allocated. Financial institutions realised that integrating NGOs into the implementation of the funded programme was a way of avoiding the inefficiency of the official use of resources since it ensured that they reached the target group and contributed to the achievement of local goals. NGOs thus became instruments of public policies at the target, which the aforementioned financial institutions and the administration were virtually incapable of reaching directly. The partnership with NGOs was also advantageous to the State as a means of satisfying financial needs in the environmental sphere.

With the passage of time, however, the link between the State and NGOs in this field became more complex because of individual relations and combinations of events. In other words, the individuals who deal with environmental policies in Brazil may be found in different positions at different times: they may be working within the administration, for the administration (as consultants), in NGOs as partners of the administration, or in NGOs opposed to certain activities of the administration.

In addition, this "democratic bureaucratic regime" (Lopes 2008) may also help to explain why the public administration is still present in situations that could be left to private actors. The administration has been playing an important role as an intervening and mediating agent in negotiated agreements between project developers and affected people since 2003. The role played by the *Ministério Público*, which is a public institution as well as an intervening and mediating agent, and its use of Protocols for Adjustment of Public Conduct (TACs) also demonstrate that the presence of a public entity is important in giving validity to negotiated solutions. In this regard, the complaint[12] about the administration generally intervening in or coordinating discussions (not encouraging, therefore, the free membership drive—*associativismo*) may be considered pertinent.

---

[12] I13022009a.

2.3.1.2 From Economic Liberalisation and Privatisation to Energy Rationing (1995–2001)

The 1990s were marked by fundamental changes in Brazil's economic policy aimed at establishing a new currency (*Plano Real*, from 1994), opening markets to competition and reducing national indebtedness. The administration also tried to implement the liberal concept of the State: as policy-maker and regulator rather than the owner of economic assets. Consequently, one of the main results in this period was a major transfer of economic assets from the State to the private sector under the National Privatisation Programme (PND) created by Law 8301 in 1990.

Until the privatisation programme of 1995 the main investor in the hydropower industry had been Eletrobras, a semi-public holding company controlled by the Union. As the "Concessions Act" of 1995 satisfied the constitutional requirement that tenders be organized for all new public service concessions, including new hydropower plants, private companies began investing in these projects. They included such international companies as Tractebel (from the Belgian group SUEZ), Endesa and Neoenergia[13] (both of Spanish origin). Some of the private investors are also of Brazilian origin, an example being the CPFL Energy Group. From 2003 to 2009 installed capacity in generation projects (different sources) grew 22,384 MW at a total investment of R$ 31,995.5 million, with private investors accounting for 81 % of this amount.

Other relevant private agents are the machinery and supply industries and contractors companies of national and foreign origin. They dominate in the construction of dams for hydropower generation, and their participation in such projects as shareholders is gaining ground. Other investors whose role has been growing since the 1990s are pension funds and *BNDES Participações*,[14] the investment company of the Brazilian National Bank for Economic and Social Development (BNDES).

BNDES, which is linked to the Ministry of Development, Industry and Foreign Exchange, played a major role in financing privatisations during the National Privatisation Programme and is now the main long-term source of finance for investment in all branches of the economy. The bank is the main financial institution for hydropower plants in Brazil, offering the cheapest loans in the market. BNDES is also engaged in indirect financing through *accredited financial institutions* (almost entirely financial institutions operating in the Brazilian banking sector), which means that private banks as well as other state-owned banks are involved in the financing of new hydropower projects, but usually with resources from BNDES. It is important to realise that many of these banks are signatories to the Equator Principles (Cardoso 2008). Nevertheless, it is possible to identify a few recent projects partially financed by multilateral agencies or regional development banks, such as the Cana Brava and UHE Campos Novos hydropower

[13] Banco do Brasil and Previ (Banco do Brasil's pension fund) are also Neoenergia shareholders.

[14] BNDES Joint-ventures.

plants (UHEs), which relied on resources from the IADB and began operating in 2002 and 2007, respectively.

It was also during the 1990s that the restructuring of the regulatory model of the Brazilian Electricity Supply Industry (ESI) began. In practice, the goals were to replace the previous vertically integrated model, based on natural monopoly concepts, with a new system founded on free price formation, competition and separation of generation, transmission, distribution and commercialisation activities. This first phase of the sectoral reform featured the implementation of the Restructuring the Brazilian Electricity Supply Industry Project (RESEB).[15]

This first market-oriented reform led to a supply crisis that lasted from June 2001 to February 2002. It culminated in an electricity rationing plan, caused by low rainfall, which had reduced reservoir levels (at that time, hydropower plants produced 90 % of the electricity supply), by the depletion of the reservoirs in the 1980s, but mainly by a lack of expansion of the installed capacity of power generation projects.[16] The rationing highlighted the need for the revision of the institutional model of the Brazilian ESI, which was reformulated by President Lula's administration (2003).

2.3.1.3 New Government and New Role for the State in the Brazilian Electricity Supply Industry from 2003

The energy rationing of 2001 and the consequent reduction of economic growth caused changes to policies affecting dams. The New Model of the Electricity Sector of 2003/2004 resulted in the resumption by the State of its planning role, and led the Ministry of Mines and Energy (responsible for this function) to become proactively involved in such related areas as the environment and water resources and to interact more frequently with the Presidential Staff Office (responsible for the coordination and integration of government action) and the Ministry of Environment (with its associated institutions), among others.

As a consequence, since March 2004[17] the electricity policy has changed substantially,[18] primarily to attract the investment needed to guarantee the

---

[15] The most important institutional innovations under the RESEB Project were the creation of the National Council for Energy Policy, with responsibility for proposing national energy policies, of the Brazilian Electricity Regulatory Agency as an independent regulatory entity, of the National Power System Operator, with responsibility for controlling power generation and transmission activities in the interconnected electricity system through a tight pool dispatch system, and of the Wholesale Electric Energy Market, responsible for undertaking all electricity purchase and sale transactions and for promoting the accounting of agents' transactions in the multilateral short-term market under market rules.

[16] For more details on the electricity supply crisis in Brazil, see Melo et al. (2009).

[17] For more details on this new phase of the reform at the Brazilian ESI (and of the reform as a whole), see Correia et al. (2005, 2006), Melo et al. (2007), Melo et al. (2009), Araújo et al. (2008).

[18] With the publication of Law 10848 and Decrees 5081, 5163, 5177 and 5184.

development of the sector. To attract investment in generation, long-term power purchase agreements (PPAs)—15 and 30 years (in which the delivery period of the energy is fixed, as well as the day on which the project must start operating)[19]—have been auctioned with a view to controlling energy contracting by the distribution utilities. The aim of this scheme is to reduce the risks to investors (who can then raise long-term loans for their projects with the backing of their PPAs). The auction by least/lowest price stimulates economic efficiency and in principle gives correct signals for the cost of expanding the system through competition.

The long-term PPA made BNDES the major financier of hydropower plants in Brazil. Interviewees[20] gave several reasons for this development:

- Multilateral financial institutions had been the major financial backers of infrastructure projects implemented by the State in Brazil until the 1970s, but they have refrained from financing new infrastructure projects since then because their new agenda has focused on market reforms and because of pressure from NGOs.[21]
- After a decade of sectoral adjustments and reforms (the 1990s), a period in which investment in new infrastructure was unsuccessfully left to private investors (with great hopes of foreign investment) and in which macroeconomic stabilisation was achieved, the State was ready to reestablish its goals and propose its own agenda.
- The reform of the ESI in 2003 identified and solved a major problem created by the previous reform, which had been based solely on the self-organising capacity of the free market: the need for a predictable cash-flow for hydropower plants, which are highly dependent on long-term capital. The energy sales auction, which resulted in a 30-year Power Purchase Agreement for the energy-selling agent (and, in this case, the developer of the new hydropower plant project), made it possible for BNDES to finance these projects by a "project finance" procedure (adding an alternative to the standard corporate financing)[22] for the least expensive resource on the market.
- BNDES trusts the environmental licensing process. It also controls and monitors the use of resources to satisfy the conditions set out in the environmental licences (concerning compensatory and mitigating environmental and social measures to be taken by the investor) since the bank's employees are considered personally liable for the environmental crimes defined in the environmental legislation.[23]

---

[19] Agents who do not meet these requirements are liable to severe penalties.

[20] The financing of hydropower plants was discussed with a total of 12 interviewees.

[21] For more information, see da Costa (2002), Mallaby (2004).

[22] In a "project finance" procedure, the loan guarantee accepted by the banks is the project cash flow, whereas in the case of "corporate financing" banks ask project developers for corporate guarantees, such as stocks. In view of the size of this type of project, the more project developers become involved in different projects, the fewer corporate guarantees each has to provide.

[23] I13022009b; T23072009 (see Law 9605/1998 and Law 6938/1981).

- BNDES is, moreover, one of the world's largest development banks and one of the few institutions in the world that can afford exposure to the risk inherent in such large individual projects as hydropower plants. Being a development bank, it has operational policies that may favour investments with long maturation periods, unlike commercial banks, which may prefer short-term returns.

As to the administration's call for feasible projects, the main change introduced by this model was that only projects equipped with a preliminary environmental licence might be included in the portfolio of projects to be auctioned. This requirement was introduced to avoid reliance on energy which could not be offered as a consequence of environmental (and social) problems. When President Lula's administration took office in 2003, it inherited a long list[24] of unbuilt hydropower plants and of dissatisfied investors (holding the concession rights) complaining about an unpredictable environmental licensing process.

The projects proposed by the administration also reveal a major change in the planning and study phases for hydropower projects. The previous, more liberal model permitted any agent interested in studying and planning a hydropower project to seek authorisation from ANEEL to carry out inventory and feasibility studies. ANEEL was able to issue study authorisations to more than one agent, having, at the end of the study phase, to choose the best study (the criterion being the best development of hydraulic potential)[25] on which to base the standards for the auction[26] and for future construction.[27] The innovation introduced by the "New Model" was that the Energy Research Company (EPE, linked to the Ministry of Mines and Energy), a specialised technical and public institution created in 2004 with the main objective of conducting the necessary research and planning in the energy sector, has since been permitted to study and propose future hydropower plants. In this context, decision-making on future plants is expected to become more coordinated and integrated from the earliest stages, with the public interest becoming the target function of the decision (which was not necessarily the case when almost all stages of the process were conducted by agents of the industry/investors).

In the former framework, investors played the predominant role, being responsible for studies and for applying for all sorts of permits and licences. The new framework, on the other hand, gives the government a more important role, allowing the investor to act from the basic project phase onwards. While this may make for a more centralised decision-making process in the hands of the government, it also means that the decisions are to be taken in a more coordinated manner, given the involvement of different areas and levels of government.

[24] According to I06022009, this list consisted of 45 projects.

[25] In accordance with Article 5 of Law 9074/1995.

[26] In respect of expected costs, budget and prices.

[27] I06022009. In this case, the cost of all alternative studies is reimbursed to the agents who financed them.

Moreover, this trend may preclude conflicts of interests, since investors usually put a great deal of effort into minimising costs in order to improve project profitability. It also makes the government more accountable for its decisions.

2.3.2 *Specific Decision-Making Processes in Dam-Related Policy Arenas*

In Brazil decisions on the construction of hydropower dams are governed by legislation and policies in three public policy areas: energy, water resources and the environment.

Energy policies are framed in a multi-level policy area whose main institutions have been created since the first market-oriented reform (Fig. 2.1).

In this regard, one important function of the Ministry of Mines and Energy (MME), with the support of the EPE, is to assess options on different planning horizons through the development of plans for the expansion[28] of Brazil's energy and electricity industries, such as the National Energy Plan (PNE), studies of Brazil's future energy mix, the Hydroelectric Inventory of Hydrographic Basins and the 10-Year Energy Expansion Plan (PDEE), which are submitted for public discussion.

As to the participative feature of energy planning in Brazil, provision is made for public consultations, although they are a recent innovation and may have room for improvement. The *Ministério Público*, for example, recommended MME to extend the period of public consultation under PDEE 2009 and to make further public announcements, since the original period was set to begin on Christmas Eve 2008 and to end in January of the following year.[29] Although this recommendation was accepted, the *Ministerio Público* nevertheless promoted a public hearing on the PDEE at its headquarters in Brasília.

It should be realised that the discussion of energy planning still attracts the attention only of organised civil society, including universities,[30] and that Brazilian society as a whole is still not[31]—or is only just becoming[32]—interested in participating in this kind of debate, although it may take an interest in the consequences of the decisions made.[33] Although the final impact of participation in the planning phase may be considered low, the pressure exerted by society has already proved capable of changing such projects as the UHE Belo Monte (auctioned in 2010).

---

[28] For more details, see MME (2007, 2008a, b) and EPE (2006).

[29] I04032009.

[30] I13022009b; I13032009.

[31] I26122008; I04032009; I06022009; I30012009.

[32] I16012009; I12022009.

[33] I05122008.

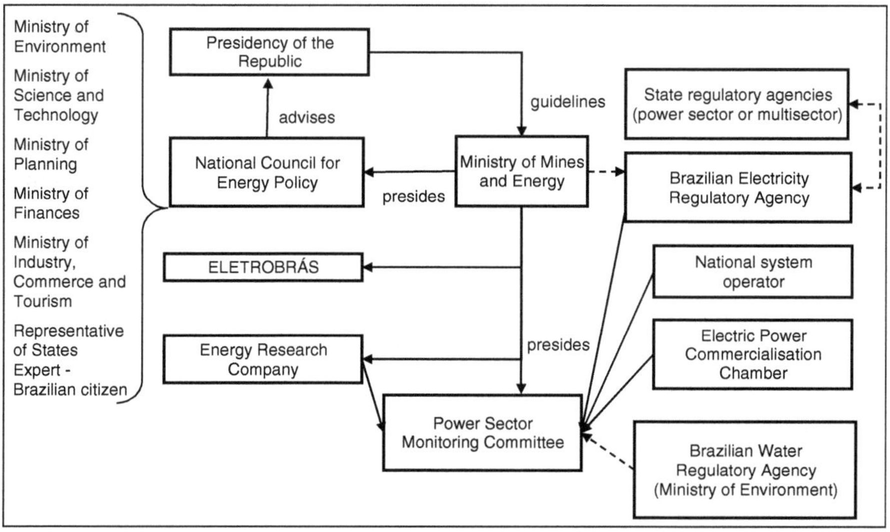

Fig. 2.1 Multi-level policy arena in the electricity sector (2010) (author's own compilation)

The energy sector has more experience of planning than other industries.[34] It began considering the socio-environmental variable in its studies in the 1980s, in response to the questioning of the environmental impacts of hydropower projects implemented in the 1970 and 1980s and to pressure from international bodies, and especially financial institutions.[35] The EPE is also responsible for conducting the studies that lead to the Integrated Environmental Evaluation (AAI)[36] of hydrographic basins, with the focus on hydropower development. This responsibility resulted from a Protocol for Adjustment of Public Conduct issued by the *Ministério Público* as a condition for the continuation of the licensing of the UHE Barra Grande. The EPE developed the first AAI in 2005, for the Uruguay river basin. Since then, a further thirteen river basins have been the subject of an AAI. This is, however, very limited coverage, when it is remembered that Brazil has 12 hydrographic regions and 56 planning units.

An AAI thus enables a basin to be described, its fragilities and potentialities to be identified (temporal and spatial evaluation of cumulative and synergic impacts), directives to the power sector which must be borne in mind in the environmental licensing of new and the management of existing hydro plants to be defined and detailed and complementary programmes to be recommended to other sectors. AAIs are also subject to at least two public hearings, the first being held when potential conflicts are being identified, the second when the preliminary

[34] Previously undertaken by Eletrobras and now by MME/EPE.

[35] For more details, see EPE (2006).

[36] Given the differences between AAI and Strategic Environmental Evaluation, an AAI would be what the World Bank considers to be a sectoral Strategic Environmental Evaluation (I12022009).

conclusions and directives are presented.[37] The AAI of the Uruguay river basin entailed sixteen public consultation seminars.[38]

Taking the water resource and environmental policy areas into consideration, Fig. 2.2 shows that related decisions take place in the same arena.

The management of water resources follows a unique pattern that extends beyond the Brazilian federative system. If river basins are viewed as spatial units (based on the French system),[39] each with its own committee and plan, a new paradigm in terms of decentralisation and participation in decision-making processes emerges. The National System for the Management of Water Resources,[40] provided for in Article 21 of CF/88 and set up under Law 9433 of 8 January 1997,[41] is composed of the following organisations with shared powers:

- The Ministry of Environment and state secretariats are responsible for formulating water resource policies.
- The Brazilian Water Regulatory Agency and state water resource agencies are responsible for implementing the system, for granting rights to use water resources and for monitoring.
- Basin Committees, which are highly participative in composition, are responsible for deciding on water resource plans (which must be drawn up at central, state and basin levels) and on how to charge for the use of water resources.

Finally, it is worth mentioning that it is not yet clear what effects the water policies are having. While the current legislation requires decisions on dams to be based on basin plans, prepared by the appropriate river basin committee, plans have not yet been generally developed.

2.3.2.1 A Three-Stage Environmental Licensing Process

Brazil's environmental legislation dates back to 31 August 1981, when the Brazilian Environmental Policy (PNMA) was adopted[42] and the National Environmental System (SISNAMA) and CONAMA were established. As the Ministério público (2004) points out, Brazilian environmental legislation was formulated in a context of growing environmental concerns in the 1970s and of increasing demands from the World Bank and IADB, for example, for environmental studies on projects which they were financing (including the Tucuruí and Sobradinho

[37] For more details, see EPE (2007).

[38] I12022009.

[39] A sign of international norm diffusion, in this case mainly through environmentalists who worked in the administration.

[40] More information about National System for the Management of Water Resources may be found in Braga et al. (2008).

[41] Known as the "Water Law", which also instituted the National Policy for Water Resources (PNRH).

[42] Law 6938/1981.

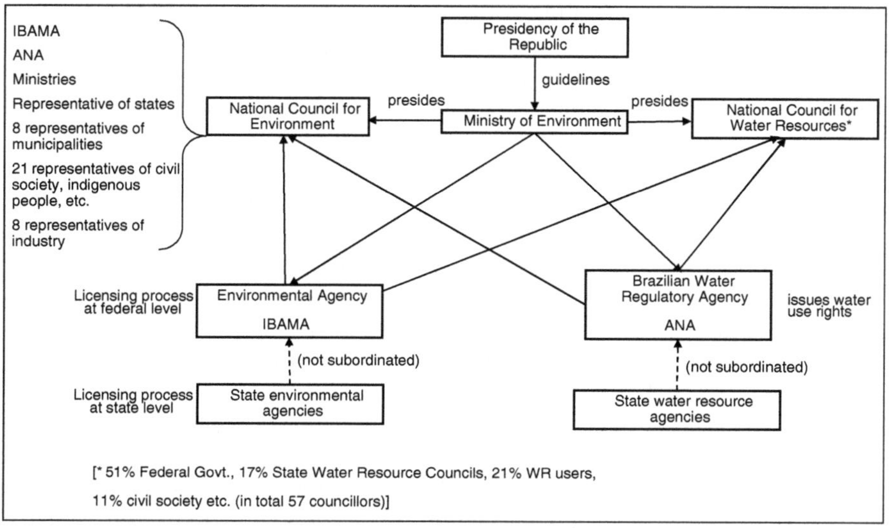

Fig. 2.2 Multi-level policy arena for the environment and water resources (2010) (author's own compilation)

hydropower plants). This was also a time of greater social mobilisation against the social and ecological consequences of major development projects implemented by the military regime, the reaction to these projects constituting a focus of opposition to the regime itself.

The CF/88 maintained this concern. Following the promulgation of the Constitution, IBAMA, the environmental agency at the federal level, was created in 1989.[43] In the same year the National Environmental Policy was updated.[44] It now requires prior environmental licensing together with environmental impact assessments of projects. Environmental licences are granted by the environmental agencies and bodies composing the National Environmental System (at federal, state and municipal levels). But even though IBAMA is required to issue these licences, it is also meant to consider technical assessments provided by municipal or state environmental agencies and by the Brazilian National Indian Foundation and the National Institute of Historical and Artistic Heritage (IPHAN). As a rule, states are entitled to elaborate their own sets of environmental laws and norms but, without contravening this principle, the central administration may also establish general licensing norms through CONAMA, standardizing procedures at national level.

As regards environmental licensing, the World Bank (2008) claims that "Brazil is one of the very few countries (if not the only one) to employ a three-stage process (Preliminary License, Installation License and Operating License), with

43 Law 7735/1989.

44 Law 7804/1989.

separate procedures for granting licenses at all three stages" (Summary Report 2008, p. 19). The Preliminary Licence is issued by the licensing agency for the planning phase of a project's implementation, modification or expansion. The Installation Licence then authorises the beginning of the construction or installation of the project. Finally, the Operating Licence is required before project operation and may need to be renewed.

In the case of projects expected to cause significant environmental harm, such as hydropower plants,[45] the granting of the Preliminary Licence depends on the approval of both the EIA and the corresponding Environmental Impact Report (RIMA, a summarised version of the EIA), which present an environmental diagnosis for the influence area of the project, an analysis of environmental impacts of the project and alternatives, a list of measures designed to mitigate negative impacts and an indication of survey and monitoring programmes.[46] Investors are expected to bear the cost of environmental licensing, the most relevant components of which are environmental studies (EIA; RIMA, etc.), the arrangement of public hearings, the publication in the press of actions related to the licensing process, the implementation of the programmes for which the licences provide (mitigation measures) and environmental compensation.[47]

Prior to EIA/RIMA, terms of reference are usually proposed by the project proponent and must be approved by an environmental licensing agency, which also submits them to the National Institute of Historical and Artistic Heritage and FUNAI for its appraisal.[48] These terms of reference define the scope of investigations to be undertaken and set out other requirements concerning, for example, the qualifications of the technical team responsible for drafting the studies. Although there is no federal legislation obliging the environmental licensing agency to comply with the terms of reference, it normally does so. As regulations on the environmental licensing process vary among the states, the enforcement of terms of reference also varies (MPF 2004).

Brazilian legislation[49] has also institutionalised a public participation mechanism as part of the environmental licensing process. Public hearings may be held whenever the environmental agency or the body responsible for issuing the licences sees fit or when called for by a civil society entity, by the *Ministério Público* or by 50 or more individuals. This means that public hearings usually take place before the issue of the Preliminary Licence, since the complaints and concerns voiced by local communities usually lead to the inclusion in the license of mitigation or compensatory measures to be taken by the project proponent, the

---

45 Resolutions CONAMA 01/86 and 11/86.

46 Resolution CONAMA 237/97.

47 The payment of this compensation is obligatory for projects deemed to be responsible for significant environmental impacts. It is allocated to the financing of conservation areas/units from the integral protection group which may also be created as a result of the licensing process.

48 According to I06022009, although the *Ministério Público* (MPF 2004) argues that these institutions and others (e.g. Cultural Foundation Palmares) should be more involved at this stage.

49 Article 2 of Resolution CONAMA 09/87.

penalty for failure to do so being the withdrawal of the environmental licence. However, as public hearings may be called for at any moment of the decision-making process, they may also be held before the terms of reference are approved by the environmental agency. IBAMA demands announcements of public hearings to be made on the radio and on banners and published in the official press and in major local or regional newspapers. As the location must be easily accessed by the interested parties, more than one public hearing on a single project may be held. Even so, criticism continues to be levelled at the ineffectiveness of the local population's participation due to a lack of information on the project plans and a lack of people with appropriate technical skills (Fearnside 2006).

In 2008 the World Bank undertook a thorough and critical analysis of the environmental licensing process for the construction of hydropower plants, its main observations being:

- Most of the problems associated with environmental licensing in Brazil occur at the first stage of a three-step process owing to a lack of adequate planning at government level; a lack of clarity about which level of government (federal or state) has the legal authority to issue environmental licences; delays in issuing the terms of reference for the environmental impact assessments (EIAs) required by law; the poor quality of the EIAs submitted by project proponents; the subsequent uneven evaluation of the EIAs (by the government); the lack of a suitable dispute resolution system; the absence of comprehensive rules on social compensation for populations affected by hydropower projects; and the shortage of qualified social development specialists within the government's federal environmental agency (World Bank 2008, (I), p. 9).
- The preparation of the terms and the subsequent analysis of the EIA/RIMA are not always carried out in a competent, interdisciplinary fashion or within a reasonable timeframe as a result of insufficient financial and human resources of licensing bodies considering Brazil's growing need for energy (World Bank 2008, (I), pp. 24–25).
- Public prosecutors (Ministério Público) enjoy virtually unlimited autonomy in Brazil. This power has no parallel in any of the other countries examined in the study[50] and plays an important role in the lack of predictability and timeliness of the environmental licensing process. This allows prosecutors to be involved in technical or administrative acts related to the environment, which would otherwise fall under the mandate of the environmental agency (World Bank 2008, (I), p. 10, author's footnotes).
- "The licensing process offers few opportunities for dispute resolution. As a result, disputes related to environmental licensing tend to proliferate and are rarely resolved. This has led to a widespread perception that excessive environmental regulation exists in Brazil. The absence of regular revisions to the environmental licensing rules has led to an increase in political and social

[50] Norway, Colombia, Indonesia, Germany, China and Canada.

disputes; these disputes have created serious complications and delays for infrastructure projects" (World Bank 2008, (I), p. 20).

In this context, Hochstetler and Keck argue that "individual licensing decisions often become crucibles for the airing of giant social conflicts that really should be settled in other ways" (Hochstetler and Keck 2007, pp. 45–46).

This perception was also shared by some interviewees, who emphasised that the process is used to discuss energy planning options rather than the impacts of specific projects (indicating that society is not using the discussion forums created at planning level).[51] One possible reason for this may be that licensing processes attract more attention from society as a whole than the planning phase and that this process suits those interested in a broad ideological discourse in a sphere where the focus should be on technical data and information.[52]

The weakness of the environmental licensing process lies therefore in the lack of implementing capacities and resources of the licensing bodies (at federal and state levels). The considerable variation in subnational environmental policies and practices (along with conflicting norms) is an inevitable consequence of federalism applied to the environmental sphere in a country marked by strong economic disparities among regions.

The main steps in the overall decision-making on the construction of a hydropower plant within the current legal framework are shown in Fig. 2.3.

2.3.2.2 Resettlement

Resettlement policies and practices form part of the environmental licensing process. Discussions on resettlement are initiated through EIA/RIMA, which establishes the area that will be directly affected by the reservoir (comprising both flooded areas and areas that will be isolated by the reservoir),[53] indicating the number of people who will be displaced and possible solutions.[54] Once the Preliminary Licence has been granted, the concessionaire must draw up the Basic Environmental Plan (PBA), in which it must give more details about the specific measures it intends to take.[55] IBAMA encourages concessionaires to resolve conflicts and to begin implementing its resettlement and compensation programmes during the validity of the Installation Licences, 6 months before the reservoir is filled.[56]

The second observation concerns the reasoning behind this procedure: it is a characteristic of the Brazilian environmental culture that it addresses social and

[51] I30012009.

[52] I25032009; I06022009.

[53] I30012009.

[54] I16012009; I30012009.

[55] I30012009; I13022009b.

[56] I30012009.

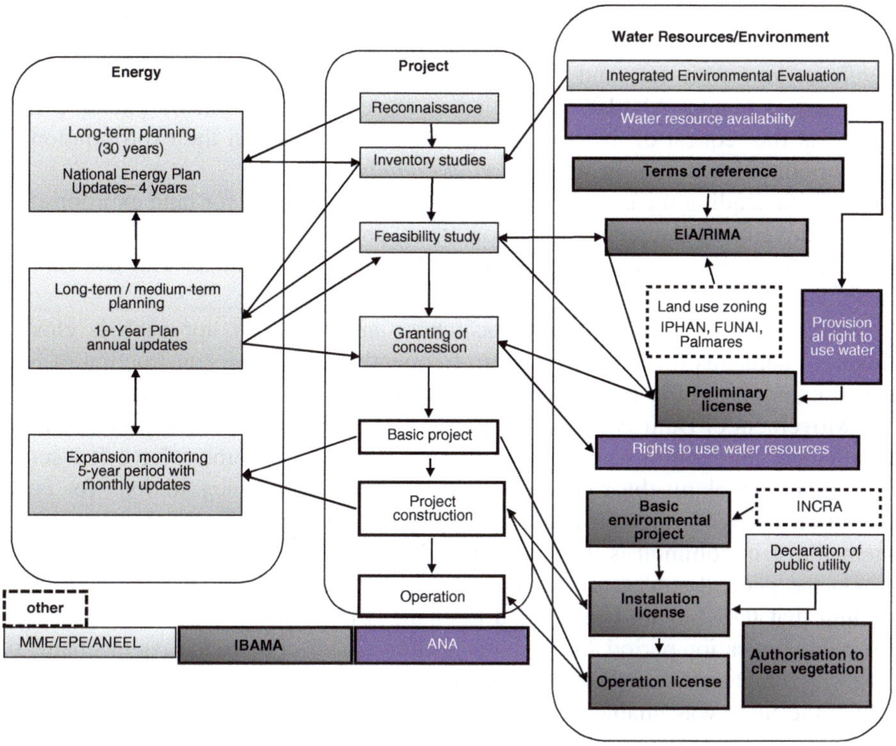

Fig. 2.3 Stages in water-/energy-related decision-making (author's own compilation)

environmental problems together, since they exacerbate each other. This includes a guarantee of the various rights of communities whose livelihoods depend on the environment (Hochstetler and Keck 2007, p. 109). This concept was officially adopted by the Ministry of Environment in 2003, when President Lula chose Marina Silva, a senator from the state of Acre and a symbolic leader of the rubber-tapper movement,[57] to be the head of the MMA.[58]

Nevertheless, it must also be pointed out that Brazil does not have a specific set of laws and norms that address infrastructure-related resettlement, including hydropower plants.

As for other sets of laws and norms relating to resettlement, while CF/88 guarantees property rights, it also gives the State the right to expropriate private

[57] More details about the rubber-tapper movement may be found in Hochstetler and Keck (2007).

[58] I29042009.

property in the public interest. Article 5 (XXIV) called for future legislation[59] to establish the procedures for expropriation in cases of *public need*, *social interest* or *public utility*, with fair and previous pecuniary compensation. In the case of hydropower plants, ANEEL is entitled to issue a Declaration of Public Utility (DUP) at the request of the concessionaire in order to effect the expropriation of land or administrative easement (*servidão administrativa*).

Notwithstanding the constitutional right to compensation for land expropriated, social movements frequently question these processes, since they address only those who can prove ownership of the land rather than all those who live on and from that land.

Some of these social movements, whose anti-capitalist approach is closely linked to demands for agrarian reform, frequently oppose the construction of new dams, examples being the Landless Workers Movement (MST), Via Campesina and Movement of Dam-Affected People (MAB). The lack of information available to affected communities and their lack of political engagement have led social movements to claim the right to act as intermediaries in negotiations for compensation between affected people and project developers.[60] What these movements have in common is the rural land struggle.

On the general subject of the legitimacy of Brazilian NGOs, it is worth mentioning that they have all responded to changes in political opportunities: during the democratisation period and with the reinforcement of participative mechanisms and with NGOs acting in the 1990s as instruments of public policies at local level, which the State was unable to reach efficiently, NGOs were clearly recognised as representatives of society as a whole. But, as time passed and they became better organised, having established a wide network of allies, the presence of the same individuals in different organisations, networks, forums and councils[61] and their ambiguous relationship with the State led to the NGOs' representativeness and legitimacy being questioned, since they were faced with the same problem as the State: it became more difficult for them to keep in touch with the grass-roots movements while representing them in every instance. In other words, people who had their origins in grass-roots movements became—paradoxically—remote from them as their involvement in all the newly created participative bodies grew.

This prompted the NGOs to change their support base: they do not seek to defend a cause as representatives of society, which would then legitimise their action. Their source of legitimacy is now the cause itself: the preservation of a certain species of fish or a certain biome or education in resettlement areas.[62]

---

59 Law 9785/1999; Provisional Measure 2183-56/2001; Law 11977/2009 updating Decree-Law 3365/1941; Law 2786/1956; Law 4686/1956; Decree-Law 856/1969; Law 6071/1974; Laws 6306 and 6602/1978.

60 I06022009; I12032009a.

61 This problem of over-representation was also addressed by I22042009.

62 This reflection can be attributed to the discussions with I29042009.

Possibly because of the legacy of the pre-2002 administration (more than 40 hydropower projects approved but not constructed) or again because of common origins in the democratisation period, the Workers' Party administration began engaging in regular round table discussions with the MAB[63] to address this movement's claims and to set up an interministerial working group (GTI) in October 2003 to analyse those claims and propose solutions to problems of people affected by dams. The GTI is recognised (Casa Civil da Presidência da República 2004) as the first attempt ever in the history of Brazil's electricity industry to include this matter in the federal administration's official agenda. The GTI's report referred to the basic need to define who, in addition to the land-owners, is entitled to compensation for the effects of resettlement programmes, thus demonstrating that in Brazil everyone living in the area and from its resources is considered to be affected by a dam. The work of the GTI culminated in the publication of a presidential Decree[64] which established an official social and economic cadastre for the identification, the qualification and the public registry of dam affected people and created an inter-ministerial committee to work on the means to implement the referred cadastre. This committee will probably deal with the problem of establishing a starting point for social auditing procedures, since a growing number of dispossessed people are attracted to such areas in the hope of eventually benefiting from any resettlement, compensation or social programme or even from recently created job opportunities.

While there was clearly an absence of a specific set of norms relating exclusively to resettlement, it is nevertheless possible to detect a change in the way this issue has been addressed since the adoption of CF/88 and by President Lula's administration. Interviewees attributed the evolution of resettlement practices to several factors, such as a reaction to (and lessons learnt from) past experience,[65] the process of democratisation,[66] the role played since then by NGOs and social movements,[67] by the media[68] and by the *Ministério Público*,[69] and the reinforcement of environmental legislation and EIA/RIMA since CF/88,[70] since these procedures are addressed during the environmental licensing process.

The interviews revealed a wide range of instruments commonly used under resettlement policies to assist affected people and of goals pursued, although it was also said that concessionaires preferred pecuniary compensation (usually in the form of letters of credit), which causes severe social damage (such as the loss of valued neighbourly relations) and is therefore discouraged by federal

[63] Installed in April 2004 (Casa Civil da Presidência da República 2004).

[64] Decree 7342, of October, 27th 2010.

[65] I12032009b.

[66] I30012009; I25032009.

[67] I05122008; I30012009; I16012009; I13022009b; I25032009.

[68] I16012009; I30012009.

[69] I05122008; I30012009.

[70] I16012009.

administrations and by the *Ministério Público*.[71] While compensation is still paid in the form of letters of credit, concessionaires are also expected to offer the options of collective resettlement, individual plots of land and land in areas close to cities, as people or families affected choose.[72] Although it has become common practice to celebrate agreements, concessionaires are not (and could not be) encouraged to negotiate with social movements: they are meant to deal directly with affected people, they being entitled to select (or not to select) a representative to engage in the negotiations. Social movements tend to feel neglected when negotiations take place between the concessionaire and individual families or groups of people.[73]

Living conditions in the resettlement areas are better than in the previous living areas,[74] since they have better infrastructure (access, public transport, sanitation, connection to the electricity grid, education and health care systems).[75] If they are in rural regions (as they usually are), there must be access to technical assistance,[76] since resettlement programmes must ensure not only the maintenance of livelihoods[77] but also economic inclusion.[78] Consequently, concessionaires are also expected to offer income programmes and replacement jobs.[79] In the latter case, they are required to provide a temporary income until the new land starts producing[80] and compensation for lost profits.[81] The continuation of neighbourly relations should be ensured.[82]

However, people who do not have property rights are also eligible to participate in resettlement and compensation programmes.[83] They include people with links to the affected area, such as employees, sharecroppers and tenants,[84] who are known as "children of the land" (*filhos da terra*)[85]; and even those earning a temporary income from the work at the construction site (such as people selling snacks in the area) are eligible.[86] The environmental agency is, however, expected

[71] I06022009; I04032009.

[72] I30012009; I13022009b.

[73] I30012009; I06022009; I12032009a.

[74] I12022009.

[75] I05122008; I16012009; I26122008.

[76] I16012009.

[77] I06022009; I12022009.

[78] I12032009b.

[79] I05122008.

[80] I16012009; I13022009b.

[81] I30012009; I06022009.

[82] I12022009; I13022009b.

[83] I05122008; I26122008; I12022009; I13022009b; I12032009b.

[84] I30012009.

[85] I06022009.

[86] I30012009.

to set a minimum time for people to be considered bound to the land and so entitled to the benefits.[87]

Even people whose income previously came from informal or illegal[88] activities are eligible for resettlement and compensation. Two possible reasons were cited by the interviewees[89]: these programmes are not intended to be discriminatory,[90] but are designed to provide for social compensation rather than the mere disbursement of financial resources.[91] It may be inferred that the administrations are adopting this lenient position because it is thought that these people had no choice before the dam but to engage in informal or illegal activities, a situation which should be changed with the economic resources created by the construction of the hydropower plant.

The selection rules on the acquisition of new land require the concessionaire to offer the affected people more than one option to choose from.[92] The land is to be close to the original location,[93] have similar qualities,[94] be productive, be reasonably flat,[95] have good infrastructure and ensure easy access.[96] How much compensation is paid depends on the value of the assets (comprising not only the land but also machinery and equipment)[97] and revenue lost.[98]

The renewal of licences and the issue of the next ones are considered to be opportunities for the environmental agency to perform its task of determining whether the agreed social compensation measures have been taken.[99] The withdrawal of an environmental licence or its non-renewal is the penalty imposed on concessionaires if they fail to take the agreed measures.[100]

Recent efforts to cope with the social aspects of dam construction are also recognised by the World Bank (2008, (I), p. 27, author's footnote):

> In discussions with government agencies we noted that the social problems associated with hydropower projects, which tended to outweigh the purely environmental issues, were duly recognized. Considerable interest was shown in seeking solutions. It should be mentioned that the many social demands arising during the licensing stages of hydropower developments are essentially social ones. Many of these are often unrelated to the project

[87] I16012009.

[88] Like illegal mining or sand removal.

[89] I05122008; I12022009; I16012009; I30012009; I06022009.

[90] I12022009.

[91] I05122008.

[92] I13022009b; I12032009b.

[93] I05122008; I16012009; I26122008.

[94] I16012009; I12032009b.

[95] I30012009.

[96] I16012009; I30012009.

[97] I16012009; I12022009; I13022009b.

[98] I16012009; I06022009; I12022009.

[99] I05122008; I26122008.

[100] I12022009; I13022009b.

> and generally precede the project itself. The burden of costs to satisfy these demands cannot be borne entirely by the project proponent. The Government could play a more effective role in the licensing process as a partner and meet demands of this type. Based on Conta 10,[101] the present study has estimated that the costs involved in addressing the social issues linked to environmental licensing are of the order of US$ 94 per installed kW on average.

From inventory data the World Bank concluded that the figure observed for social and physical environmental costs in relation to total plant costs (both measured in terms of installed kW) may be "estimated at an average of 12 %. Outlays for social aspects, such as population resettlement, support to communities and municipalities and supplying infrastructures, represents the major part of these costs (about 80 %), whereas the costs associated with the physical environment represent only 2 % of the total cost" (World Bank 2008, p. 26).

2.4 Case Studies

Now that the current regulatory framework, the agents and networks involved in the process and the structural and policy changes in recent decades have been presented, this chapter will analyse individual dam projects where these changes can be observed. Table 2.1 provides an overview of the project characteristics.

2.4.1 Dam Building During the Military Dictatorial Regime and the Newly Implemented Democracy (1974–1988): The UHE Itaparica

The best known projects built during this period, with the exception of the binational UHE Itaipu on Brazil's southern border with Paraguay, were undertaken in the northern and northeastern regions of the country. One of the many projects is presented: the UHE Itaparica,[102] a plant built in the northeastern state of Pernambuco on the River São Francisco. The plant is known for its controversial and expensive resettlement experience. Its extensive reservoir affected several municipalities in the densely populated São Francisco valley. As many people were displaced, the resettlement programme became a challenge for the military regime.

The UHE Itaparica was planned to increase electricity supply in the northeast (Salomão 2007), since links to other systems were poor and the strategy was to

---

[101] Conta 10 is a component of the budget structure used by the EPE and Eletrobras to estimate the cost of future hydropower plants.

[102] Its name changed to UHE Luiz Gonzaga, as a tribute to a regional and renowned musician.

Table 2.1 The dam projects investigated at a glance (author's own compilation)

Historical moment	Military dictatorial regime	Newly regulated environment
Hydropower plant	Itaparica	Santo Antônio
Region	Nordeste	Norte
State	Pernambuco	Rondônia
River	São Francisco	Madeira
Year of the concession contract	1945	2008
Equity holders	Chesf	Furnas, Odebrecht, Andrade Gutierrez, Cemig e investment fund
Origin	Public	Public and private
Planning studies	Enenorde, plano 90	Eneram
Year of the study	1973	1973
Beginning of construction	1979	2008
Operation (1st turbine)	1988	2012
Installed capacity (MW)	1480	3150
Reservoir area (km^2)	828	271
Reservoir volume (hm^3)	10782	2075
Maximum maximorum level (m)[a]	305	70
Total investment	US$ 1622 million	R$ 13500 million
Major equipment suppliers	Gie, Siemens, Voith	Alstom Hydro, Bardella, Areva, Siemens, Va Tech, Voith Siemens Hydro
Financial institutions	International Bank for Reconstruction and Development, Eletrobras	BNDES, Santander, Banco do Brasil, Banco do Espírito Santo, Banco do Nordeste, Banco da Amazônia, Itaú BBA e Caixa Econômica Federal
Displaced population	41,200 (incl. 1,200 indigenous persons)	578 non-Indian families[b]
Environmental studies	EIA/RIMA not required, several studies undertaken	ToR, EIA/RIMA, Strategic Environmental Evaluation, four public hearings, 64 participative meetings
Licensing bodies: licences	IBAMA: LO	IBAMA: LP, LI, LO

[a] above sea level

[b] resettled population as to 2009 (negotiations with affected population is still in progress)

ToR: Term of Reference

LI: Installation Licence

LO: Operation Licence

LP: Preliminary Licence

develop hydropower projects at a time of high oil prices. It would help to regularise the flows of water to the Paulo Afonso hydropower plants located on the river downstream. The 1,480 MW hydropower plant was built by Chesf, another subsidiary of Eletrobras, and was financed by the International Bank for Reconstruction and Development (of the World Bank Group) and by Eletrobras. Its construction began in July 1979 and it went into operation only in June 1988.

The Itaparica dam has a reservoir area of 828 km^2,[103] having flooded three municipalities in the state of Bahia (Chorrochó, Rodelas and Glória) and four in the state of Pernambuco (Belém do São Francisco, Floresta, Petrolândia and Itacuruba) and displaced some 7,000 families (about 40,000 persons), including 200 families of Tuxá Indians (about 1,200 people) (Salomão 2007).

The news of the construction of a new dam in the densely populated region of Itaparica was spread by people contracted by Chesf to pave the way for the project and to identify all those to be resettled and by *pólo sindical* (an organisation representing different municipal rural workers' unions) (Scott 2006). The need to organise was a response to previous experience of resettlement in the area of the UHE Sobradinho (Lima 2007; Scott 2006). *Pólo sindical* was conversant with documents guiding the actions of the State and was well advised by people and institutions with experience of defending the rights of excluded people. The strength of *pólo sindical* and of other workers' unions during the negotiations with Chesf since 1986 is confirmed by Scott (2006), the entire resettlement process being a result of an agreement signed in 1986 between Chesf and the affected people, represented by *pólo sindical.*

Like the UHE Sobradinho resettlement programme and as the population of the UHE Itaparica area lived mainly from subsistence agriculture (Lima 2007), the resettlement programme for the UHE Itaparica focused on rural resettlement and on keeping the resettled people working in agriculture (Scott 2006). This was Chesf's response to the affected people's motto "Land for Land".

Lima (2007) argues that the São Francisco Valley Development Company, responsible for irrigation systems and technical assistance, and Chesf designed a self-management model entirely new to the resettled people, who were used to traditional agriculture. In other words, although instruments and technical assistance were generally available, the people did not know how to operate irrigation systems, leaving them feeling strange, vulnerable and incapable and having to relearn cultivation methods. Besides this, the use of agrochemicals and pesticides, common in modern farming at that time, caused new health problems (Scott 2006). The different completion times of the irrigation systems and the agro-villages gave rise to new social and economic differences among resettled families (Lima 2007).

With the construction of the villages and the resettlement programme delayed and indeed paralysed, Chesf created an additional mechanism to give the resettled families an income until they started producing on their own: the Temporary

---

[103] See http://www.chesf.gov.br/portal/page/portal/chesf_portal/paginas/sistema_chesf/sistema_chesf_geracao/conteiner_geracao?p_name=_Luiz_Gonzaga. Accessed 18 August 2010.

Maintenance Allowance (VMT). They are still receiving the VMT, since the construction of the agro-villages was not completed on the ground that they would not be economically viable. As a result, many resettled families are entirely dependent on Chesf (some families have benefited from such alternative income as pensions and wages), having failed to reestablish their farming activities (Lima 2007). Chesf, on the other hand, has not offered them the opportunity of developing any other economic activity in which the VMT could be invested to guarantee the continuation of their social position as workers (Scott 2006; World Bank 1998).

The World Bank's Independent Evaluation Group described the programme as a "costly misadventure": the cost of the programme presents a ratio of resettlement costs to dam and power plant costs of almost 1:1. It concluded that "the outcome would have been less likely if the Bank and CHESF had designed an effective compensation package," since "few of the resettled families were technically qualified or experienced commercial farmers," the handling of the VMT being the most delicate issue within the overall solution because it "(...) has been converted from a short-term palliative into a permanent component of families' household economies" (World Bank 1998, pp. 1–3).

Given the outcomes of the various resettlement programmes, Lima (2007) believes that the UHE Itaparica programme represented an advance in qualitative terms mainly as a result of social mobilisation by the workers' unions, which succeeded in obtaining compensation, a resettlement programme and VMTs for the resettled population. "Polo sindical's remarkably professional performance as a leader of an unorganised peasantry" was also highlighted by the World Bank (1998). Another recognised gain was the access to land granted to all farmers: landowners, employees, sharecroppers and tenants. Landowners were also entitled to compensation for flooded and lost assets. Lima accepts that the construction of the hydropower project created employment opportunities, that the new city has better infrastructure and access to public services and that there are now more dynamic commercial and farming activities. But he also emphasises that this development was not accomplished without affecting lives formerly based on friendship, neighbourhood and family ties, even if the planned resettlement had provided for the preservation of some community ties. Scott (2006), on the other hand, refers to the continuing dependence of resettled people on Chesf and its VMT, recalling that, before the dam, there had been almost no unemployment in the region and that, after the dam, these workers lost their production basis.

As regards the resettlement of the Tuxá, Salomão (2007) claims that the *Ministério Público* played an important role in defending their rights. In 1991, after FUNAI had denounced Chesf for failing to comply with targets and deadlines set in an agreement signed in 1987, the *Ministério Público* ensured that Chesf signed a new agreement in 1994, the first to involve the Tuxá community directly. In 2001, the *Ministério Público* also succeeded in granting new Tuxá families the right to the VMT. In 2006 it organised a meeting with the three Tuxá communities (Rodelas, Ibotirama and Inanjá) and Chesf, with the aim of ordering Chesf to allocate R$ 1.4 billion to Tuxá resettlement for the acquisition of land, for the introduction of irrigation systems and for their operation, maintenance and any technical assistance needed. By the end of the negotiations it had been decided that the irrigation system project should not be implemented and that the amount set aside for it should be shared among all 442 Tuxá families, who would receive approximately R$ 110,000 each. This sum would be paid by Chesf in three (annual) instalments and the payment of the VMT would not be suspended until 2 years after the payment of the last instalment. Salomão (2007) concludes by recognising that, although these communities may be living in wealthier conditions since the construction of the dam, they miss several aspects of their former lives.

The case of the UHE Itaparica illustrates that even in an authoritarian political regime resettlement was considered. Also worthy of note is that resettlement is a continuous process which the project developer adjusts as time passes and circumstances change. It is not a watertight solution. On the other hand, projects like the UHE Itaparica that were affected by changing institutional frameworks, unfavourable macroeconomic policies and budgetary constraints are subject to various contingencies, for which no provision was made in this case, resulting in delays and failure to take planned and agreed measures. Consequently, as social liabilities become evident and as the public questioning of the companies' failure to comply with commitments becomes unavoidable and is even backed by the *Ministério Público,* companies have to allocate considerable economic and technical resources to deal with plans that have not been implemented.

2.4.2 Dam-Building in a Newly Regulated Environment (from 2003): Santo Antônio Hydropower Plant

The second case to be analysed is the 3,150 MW UHE Santo Antônio hydropower plant. The decision to construct the plant was made after President Lula took office. The act marking this fact was the publication of Resolution 4 of the National Council for Energy Policy (CNPE) on 28 September 2007. This resolution categorised the UHE Santo Antônio as an electricity generation project which was to be regarded as a priority[104] when it came to granting a concession after a public tendering procedure because of its strategic importance and public interest.

The planning of this project began in the 1970s with the studies of the hydropower potential of the Amazon region, Eneram, which identified the River Madeira as a possible site for hydropower projects (Switkes 2001). The Madeira is a transboundary river which is formed by a wide network of smaller rivers that flow through Brazilian, Bolivian and Peruvian territory (such as the Rivers Beni, Mamoré, Guaporé, Madre de Dios and Orthon) and eventually flow into the River

[104] This categorisation became possible with paragraph VI in the Article 2 of Law 9478/1997 under Law 10848/2004.

Amazon (Leme 2005a). But it was only in 2001, in the middle of the liberalisation period of the Brazilian ESI's recent reforms, that Furnas and Odebrecht, a construction company, decided to undertake a joint study of the River Madeira's hydropower potential. The inventory studies of the river were initiated in 2001 with the ANEEL's authorisation and were concluded in November 2002 (Leme 2005a). The environmental licensing process then started in August 2003.

EIA[105]/RIMA[106] were carried out by Leme Engenharia Ltda. from 2003 to 2005, a period in which 156 on-site observation campaigns took place, guaranteeing the coverage of a complete hydrological cycle. The drafting of the studies entailed the participation of and contributions from several local and regional research institutes and organisations, such as the Federal University of Rondônia (palaeontology, hydrobiogeochemistry, vegetation, insects, medium-sized and large mammals, fish and fisheries, and social and economic aspects), the Amazonian National Research Institute (INPA) (water quality, small mammals, reptiles and amphibians, large fish, birds, aquatic mammals, studies of vectors of medical interest), the Mineral Resources Research Company (geology, geomorphology, mineral resources, hydrogeology), the *Museu Paraense Emílio Goeldi* (pre-historic and historic archaeology), the Tropical Pathologies Research Institute (evaluation of the health and condition of the riverside population) and the NGO Research Center for the Cuniã Traditional Populations (CPPT) (interaction with riverside communities, communication of all the actions to be implemented from the study to the operation phase) (Leme 2005a).

Following the scoping of the Terms of Reference, the EIA was designed to cover three characteristic areas: direct influence, indirect influence and regional coverage. The analysis of the regional level addressed cumulative effects and possible synergies between the project and other activities in the same river basin, with specific regard to its Brazilian part. Besides meeting the requirements laid down in the terms the project proponent also carried out a Strategic Environmental Evaluation, which focused on political, commercial and environmental potentials throughout the Amazon region, addressing development and integration issues which might help to develop this region in the middle of South America. The concept of the Strategic Environmental Evaluation of the Madeira was consistent with the initiatives taken by the South American Regional Infrastructure Integration Initiative (IIRSA)[107] and was welcomed by the licensing authority and the *Ministério Público*.

Before the Preliminary Licence was issued, the project proponent held four public hearings in the region in November 2006 in accordance with IBAMA's instructions. But this was not the first time the people were asked to participate.

[105] The EIA is a very detailed document, rich in information, having more then 1,500 pages plus annexes, maps, figures, etc. and an additional 321 pages describing environmental (including social) programmes.

[106] The RIMA is designed as a shorter, summary version of the EIA. The one carried out for this project is 82 pages long.

[107] T18092009.

They were involved by the project proponents before the terms of reference for the environmental studies were validated. During the EIA/RIMA phase Furnas and Odebrecht, on the other hand, held 64[108] meetings of local communities and affected people in order to disclose information on the project and the process and to learn about the people's expectations, anxieties and demands for participation. The information gathered in this process was passed to IBAMA and contributed to the definition of the mitigation measures associated with the Preliminary Licence,[109] the appropriate documents being passed to local administrations, since the project proponents saw this information as a record of social demands that might help in the design and planning of public policies.[110] Besides arranging these participative meetings, the project proponents presented the project and the process at a further 82 events[111] (including meetings, seminars and conferences) attended by a number of institutions, such as those of the Brazilian and Bolivian public administrations (at federal, state and municipal level, and in the three different branches), banking and international institutions, NGOs and universities.

Resettlement issues to be covered by a Social Compensation Programme were addressed in the EIA/RIMA and discussed at the public hearings. As a result, the Preliminary Licence was granted on the basis of the EIA/RIMA on condition that the project developer addressed mitigation and compensatory measures in greater detail, setting out, for example, programmes and projects that reconcile the demand for and supply of public services; mitigation measures for families that do not own land in the influence area of the reservoir, but whose economic activities would be affected; the support of resettlement initiatives related to the agrarian reform and initiatives for small farmers and riparian communities in the area influenced by the reservoir designed to develop environmentally sustainable activities; an action plan for the control of malaria; the consideration of the recommendations made by FUNAI in the Programme of Support for Indigenous Communities; and support for the revision of Porto Velho's urban plan. The Preliminary Licence contained a total of 36 requirements to be met by the project developer in addition to those referred to in the EIA/RIMA (IBAMA 2007).

The discussion of the environmental impacts became so intense that the MME commissioned an independent study with support from the World Bank. The study was carried out by an international sedimentation expert,[112] Sultan Alam, who recommended some changes to the project to minimise the problem of environmental impacts. In this context, it should be noted that the role of multilateral

---

[108] T180909. This experience inspired the production of a documentary film called "Vozes do Madeira" (The Madeira's Voices) available at http://www.cemig.com.br/cemig2008/content/campanhaspublicitarias/empreendimento_santo_antonio.asp. Accessed 19 September 2009.

[109] I14092009; T18092009.

[110] T18092009.

[111] The last of these events, which are listed in Table A.II.1 on pages II-13–15 Tomo A, Volume 1 of EIA (Leme 2005b), was held in September 2004.

[112] Mitigation measures for this issue also form part of the 36 additional requirements attached to the Preliminary License (IBAMA 2007).

agencies in Brazil has shifted in recent years. The former director of the World Bank for Brazil, John Briscoe published an article in a Brazilian newspaper on 16 August 2008 pointing out that the World Bank was designing a new model for engagement in the Amazonian region and indicating that its lack of enthusiasm for complex projects in sensitive biomes had often led to blockages in the process, preventing resources, knowledge and sustainable practices from reaching places where they were most needed as an alternative to predatory and uncontrolled development (Briscoe 2008). This may mean that the World Bank is focusing its know-how on improving the implementation of the development agenda and priorities set by developing countries.

Following the issue of the Preliminary Licence and the auction, *Santo Antônio Energia* faced the task of designing the Basic Environmental Plan, detailing all the programmes and mitigation measures it intended to implement. Although BNDES was to provide a large part of the financing needed for the project, the fact that Banif and Santander were participating as equity holders led the company to consider the Equator Principles[113] and the International Finance Corporation's environmental guidelines while drawing up the PBA.[114]

The UHE Santo Antônio's PBA was approved by IBAMA in August 2008, when the installation licence was also issued. The PBA, another document which counted on the participation of the same local and regional institutions that collaborated in the drafting of the EIA/RIMA, presents some 24 environmental programmes (and other subprogrammes), which may be grouped in four types: monitoring programmes, preservation and conservation programmes, impact reduction programmes and compensation programmes.[115]

The UHE Santo Antônio helps to illustrate how intricate dam decision-making has become, involving several public and private agents, with their different priorities and preferences. This is a case where the process has contributed to better outcomes (compared to previous experiences), considering the years of planning and of project discussion in a number of arenas. The strategy of contacting and informing the local population at earlier stages of the process has helped the project proponent to advance the licensing process, which may form a bottleneck in the decision-making process.

The role of multilateral institutions also seems to have changed: the fact that they keep withdrawing from the financing of projects no longer constrains the process, since infrastructure projects have been successfully backed by BNDES. On the other hand, the knowledge and expertise of multilateral institutions have helped the country's development strategies, priorities and agenda to achieve better results.

---

113 The consideration of these principles is also mentioned in the documentary film, "The Madeira's Voices" available at http://www.cemig.com.br/cemig2008/content/campanhaspublicitarias/empreendimento_santo_antonio.asp. Accessed 19 September 2009.

114 T180909.

115 For more information, see http://www.santoantonioenergia.com.br. Accessed 19 September 2009.

The UHE Santo Antônio is also a good example of how environmental legislation and practices have been consolidated since the 1980s, while the economic stabilization achieved in the 1990s and the institutional reforms of the ESI since the beginning of this century have succeeded in attracting another type of foreign investor: banks and private equity funds. As the UHE Santo Antônio has shown, these agents may form a new group of international norm carriers.

2.5 Summary and Conclusions

This chapter has sought to analyse the main factors that persuade national decision-makers to internalise and apply international standards of sustainable dam development against the background of domestic political structures and political processes. Previous sections have covered the evolution of decision-making on hydropower since the advent of the present Federal Constitution in 1988, identifying political opportunities (CF/88, the liberalisation of the ESI in the 1990s and the reform of the ESI in 2003/2004), specific processes (and access points) and the agents involved (their support or criticisms and their role as norm carriers) with a view to explaining the possible internalisation of international standards.

It cannot be denied that the Brazilian government is committed to a wide variety of international agreements and involved in international forums where fundamental international norms, including those related to sustainable dam development, are discussed (good examples in this context are the International Conference for Renewable Energies that took place in Bonn in 2004 and the Convention on Biological Diversity, the United Nations Framework on Climate Change and Agenda 21). The Brazilian government is convinced that, especially because of their democratic and participative features, national legislation and practices are no less rigorous than anywhere else where hydropower plants are still being built and that new hydropower projects have led to the construction of sustainable dams. The confidence in the country's decision-making framework and dam-building procedures may justify the lack of interest in the World Comission on Dams's (WCD) recommendations (ANA 2001, 2003). Nevertheless, the role of the WCD as a catalyst for improving sustainable dam development for the sake of human beings is not ignored (ANA 2001).

The Brazilian Regulatory Water Agency (ANA) was concerned about the right of veto claimed by the WCD for stakeholders, arguing that, in Brazil, this would infringe the principles of Law 9433 of 1997, which instituted the National Policy of Water Resources, and would also be incompatible with the participatory decision-making process in force.

In 2002 the Ministry of Environment issued a joint ministerial order (MMA 2002), signed by the Ministers of Mines and Energy and of Foreign Relations, setting up a working group composed of representatives of the two ministries to draw up a statement by the Brazilian government on its position on the WCD report. The statement written by the working group was sent to the Minister of

Environment in March 2003 (ANA 2003) and acknowledged the contribution of the report in systematising values and strategic priorities to be considered in decision-making on dams and in catalysing the global debate on the role of dams in the development process, pointing out, however, that the case studies on which the report was based were not representative of existing dams, most of which were old. It also noted that the proposals made in the WCD report had been incorporated in Brazilian legislation long before the WCD had come into existence. It also addressed the issue of option assessment:

> An additional error of the Report is the absence of a systematic comparison of the environmental and socio-economic costs of dams as compared to the available alternatives for energy generation: nuclear plants and thermoelectric plants based on gas or coal (...). (Brazil proposes that a comparison should be made of the) socio-economic and environmental sustainability of dams with feasible alternatives that would provide the population with the same multiple services: electricity, flood protection, navigation, domestic and industrial water supply as well as food production (ANA 2003, pp. 8–9)

Despite this political strategy of not accepting the WCD's recommendations as strict directives, it is possible to find comments on them (primarily as food for thought rather than a manual to be followed) in official publications of the energy sector (e.g. MME 2007, pp. 19, 623, 638). Although little reference is made to the WCD report in official documents and statements, some discursive effects[116] in Brazil can be discerned: it has influenced the discourse of the anti-dam movements,[117] although other Brazilian social movements make little mention of these recommendations[118]; and it has influenced the discourse of international NGOs active in Brazil and of governments of the developed world against large hydropower projects as a renewable source of energy.[119]

Notwithstanding the influences of globalisation and other international trends, the government's embeddedness in international institutions does not therefore seem to be a determinant cause of the internalisation of international standards,: Brazil is rather a case in which recent changes in the environmental field (and to related procedures) are more attributable to a general process of democratisation and internal policy learning.

---

[116] Some interviewees also claimed that the WCD recommendations had had a wider effect in that it had reached a normative and even a material level. Something of a normative effect was detected in the planning of the energy sector, where the WCD recommendations had been used as a reference in the adoption of premises, criteria and procedures, although there is no agreement on all the recommendations (I12022009). Some best practices have been adopted by the electricity sector, but this had already been a trend before the WCD published its recommendations and was not necessarily due to its work (I12022009; I13032009). A material effect was mentioned by one interviewee, who believed that it was due to the WCD report that investment in new hydropower plants had risen, as the growing number of hydropower projects under construction worldwide showed (see World Bank 2009).

[117] I16012009; I30012009; I06022009; I13032009; I25032009.

[118] I13022009a; I12032009b.

[119] I06022009.

All governments from the 1990s onwards have appreciated the importance of private national and foreign capital and have therefore been committed to creating a credible investment environment that goes beyond a guarantee of stable macroeconomic bases. This has necessitated the reform of the ESI on two occasions: in the mid-1990s and in 2003/2004. The "reform of the reform" addressed several issues related to the business of attracting investment (such as the design of auctions and related long-term PPAs; compliance with contracts concluded within a previous framework; a guarantee of funding from BNDES on competitive terms), but where the specific issue of sustainable dams is concerned, what was most important was a guarantee of the environmental (including social) feasibility of any hydropower project before it was handed over to a private and[120]/or public project developer.[121] This was a necessary change and recognition that Brazil's environmental licensing process is complex, rife with opportunities for vetoes and therefore unpredictable, a process that does not necessarily attract private investment, according to the World Bank's analysis (2008).

As this chapter and Hochstetler and Keck (2007) argue, social mobilisation and environmental awareness (which endorses the value of sustainable dams) are two closely linked phenomena in Brazil, each of which has reinforced the other's goals since the democratisation of the 1980s. The Movement of Dam-Affected People is a clear example of a product of this process, as the analysis by Rothman and Oliver (2002) shows. This has created fertile soil for the dissemination of the sustainable dam discourse, whatever its origins.

As for the internalisation of international standards, some pathways can be distinguished which eventually had normative, discursive, regulatory and material effects. These pathways and factors are:

1. the constant relationship between national social and environmental movements and NGOs and their international counterparts since that time,
2. added to international funding for their activities and especially for some activities of the State in the environmental sphere in the 1980 and 1990s,
3. combined with the continuous transit of individuals (an explicit example of norm carriers) from the environmental sphere to different positions according to the moment (working within the administration or for the administration—as consultants or partners of the administration in NGOs—or in NGOs opposed to a certain administration), and[122]
4. added to the growing power and legitimacy of the *Ministério Público*, because of its responsibility for protecting diffuse rights and its symbiotic relationship with organised civil society.[123]

[120] Through special purpose companies.

[121] Through the obtention of preliminary license (the most critical license) prior to the energy auctions.

[122] As presented by Bernardo (2001).

[123] As addressed by Lopes (2000).

The fact that no hydropower plant is built unless the requirements of the environmental licensing process are satisfied and its procedures are completed, shows that this confluence of phenomena has irrevocably affected decision-making on the construction of new hydropower plants.

The requirements of the environmental licensing process must be satisfied and the procedures associated with it must be completed not only because they are legally mandatory, but particularly because of the access points and opportunities for vetoes which this process entails: as the process involves innumerable institutions, agents and actors at the three levels of the federation and the three branches of the political system, which do not have the same interests, priorities and values and as all decisions (at the various stages of the environmental licensing process and of the life cycle of the project itself) may be questioned by the *Ministério Público* or by social movements and NGOs in the form of associations and then taken to court, the less visible the process is, the greater the possibility of its outcome being unpredictable. In this regard, it has become easier and quicker to act in accordance with all the rules governing the process, satisfying all its requirements, although this does not prevent some institutions with a contrary view from obstructing the process, since the deadlines for decisions at each stage are not clearly defined.

As evidence of how these access points may also work as opportunities for exercising vetoes, of the 40 or so in 2003 still unbuilt hydropower projects (despite the granting of concession rights prior to 2003), 13 remain unbuilt in 2010 either because of questions raised during the environmental licensing process or because the environmental agency declared them environmentally non-viable.

As a consequence, the coordinating function of the Casa Civil has been reinforced at least at the level of the federal executive branch. On the other hand, the creation of the EPE and the delegation to it of the task of undertaking feasibility studies and EIAs/RIMAs of future hydropower plants may be seen as a first attempt to guarantee that all requirements are met.

One common argument for questioning the process is that affected people have not been informed or able to participate as required by law. Therefore, the participation of the affected communities since the early stages of the licensing process for the UHE Santo Antônio, in which they also relied on the cooperation of a local NGO to facilitate this contact, may serve as an example of what needs to be done in the future to avoid questioning the process.

Regardless of the pressures created by the development of the national environmental institutional framework, the internalisation of international standards also occurred in the energy sector (including the ESI): reliance of domestic dam developers—until recent years, state-owned companies—on foreign financial services induced domestic decision-makers to implement international standards. This also accounts for the creation of units within Eletrobras dedicated to the environmental field and of specific methodologies to address this issue in the budgets of hydropower plants (*Conta 10*). The long planning experience of the sector, together with its need to comply with requirements laid down by international financiers (multilateral agents, export credit agencies and foreign banks), has moulded a planning policy that considers socio-environmental and participative issues, as this chapter has shown.

The case of the UHE Itaparica analysed above indicates this direct influence. The second case, the UHE Santo Antônio is, on the other hand, emblematic: the Equator Principles were taken into account because the foreign banks acting as equity investors in the project were signatories to them.

Consequently, five observations on international norm diffusion through foreign financial services and expertise are made:

1. Recent Brazilian administrations have put an enormous effort into creating an investment environment that succeeds in attracting private capital. In this respect, investors are free to apply the best practices they choose, provided that they are not inconsistent with national legislation and norms. The initiative of the International Hydropower Association may serve as an example of this trend, as long as its recommendations bind industry agents and not governments. Private investors may now play the role that international financial institutions played in the past as international norm carriers.
2. The relevance of international financing has waned in Brazil's recent history owing to the macroeconomic stability that has been achieved and to the growing role and size of BNDES (Santo Antônio received significant funding from BNDES).
3. International financial institutions, on the other hand, seem to be a better target for social movements and NGOs. A likely reason for this is that every incident involving these institutions may act as an international shop window for local problems, giving them an international dimension. This means that these institutions may also be regarded as another channel through which social mobilisation succeeds in influencing practices. This finding has implications for past and future projects.
4. However, the internalisation of international standards may also occur through the expertise of the international financial institutions that support studies on the construction of new hydropower plants. It should again be pointed out that this is not a prerogative of financial institutions, foreign research institutes having always been involved in such studies, as the UHE Itaparica demonstrates. Nevertheless, these cases also serve to illustrate that the involvement of international agents in the past did not suffice to guarantee the construction of sustainable dams.
5. The World Bank's recent endorsement of the MME in studies on the UHE Santo Antônio illustrates that the role of development cooperation agencies may be boosted and may also contribute to much better and more balanced results if they accept the strategies of the countries they intend to support, a view highlighted by Briscoe (2008). This is because what singles out these institutions is that they concentrate within themselves the most qualified portfolio of specialists in development policies. This know-how is much needed in developing countries, where the professionalization and qualification of the State is still a work in progress.

To conclude, while international influence on environmental awareness in Brazil, legislation and practices is undeniable, the social content of Brazilian environmentalism and its political implications are also a feature of the collective identity that is typical of Brazil. The country's recent democratisation and economic history have formed a State that is conscious of its considerable strength, which legitimises its actions: it is supported by a political system that extols democratic and participative values.

Acknowledgments There is a large number of people whom I had the pleasure to interview and therefore contributed incommensurably to the end result of this chapter and to the conclusions I came to. I would like to first express my deep appreciation to them—thank you for your time, availability and willingness to share with me your experiences and opinions. I would also like to thank Marisete Pereira, Marcio Zimmermann, Dr. Elbia Melo and Dr. Maritta Koch-Weser for their constant encouragement and intellectual contributions in developing this text. Finally, a big thanks goes to Dr. Albert Cordeiro Geber de Melo who has supported my first presentation on this issue.

References

ANA (Agência Nacional das Águas) (2001) Ofício n° 031, de 12 de fevereiro de 2001

ANA (Agência Nacional das Águas) (2003) Ofício n° 002/AE, de 10 de março de 2003

Araújo J et al (2008) Reform of the reforms in Brazil: problems and solutions. In: Sioshansi F (ed) Competitive electricity markets: design, implementation, performance. Elsevier, Oxford, pp 543–572

Bernardo M (2001) Políticas públicas e sociedade civil. In: Bursztyn M (ed) A difícil sustentabilidade—Política energética e conflitos ambientais. Editora Garamond, Rio de Janeiro

Braga B et al (2008) Pacto federativo e gestão de águas. Estudos Avançados. USP 22(63):17–42

Briscoe J (2008) Solução Planejada. O Globo 16 Aug 2008, Caderno Opinião, Rio de Janeiro. http://web.worldbank.org. Accessed 17 Sept 2009

Cardoso F (2008) Princípios do Equador devem incentivar a responsabilidade social dos bancos. ENVOLVERDE Revista Digital. http://envolverde.ig.com.br/materia.php?cod=48602&edt=. Accessed 22 July 2009

Casa Civil da Presidência da República (2004) Relatório Final—Grupo de Trabalho Interministerial—Atingidos por Barragens, May 2004

CEPEL (Centro de Pesquisas de Energia Elétrica) (2004) Energias Renováveis—Conferência de Bonn e PROINFA. CRESESB Informe, Rio de Janeiro, November 2004, Year IX, 9, pp 4–5

Correia T et al (2005) Contra-reforma institucional da indústria elétrica brasileira e novas perspectivas de mercado: VI Congresso de Regulação, Associação Brasileira das Agências de Regulação (ABAR), May 2005, Manaus, Brazil

Correia T et al (2006) Trajetória das reformas institucionais da indústria elétrica brasileira e novas perspectivas de mercado. Revista Economia/ANPEC 7(3):307–627

Costa N (1997) ONGs coordenam evento indígena Internacional. Alerta em Rede, 5 Oct 1997. http://www.alerta.inf.br/Geral/1104.html. Accessed 25 July 2009

Da Costa A (2002) O Banco mundial na década de 90: política de empréstimo para energia e o papel do setor privado. Federal University of Rio de Janeiro, Rio de Janeiro

EPE (Empresa de Pesquisa Energética) (2006) A questão socioambiental no planejamento da expansão da oferta de energia elétrica, Rio de Janeiro

EPE (Empresa de Pesquisa Energética) (2007) Avaliação ambiental integrada dos aproveitamentos hidrelétricos na Bacia do Rio Tocantins, relatório final, sumário executivo, July 2007. http://www.epe.gov.br. Accessed July 2008

Fearnside P (2006) Dams in the amazon: belo monte and Brazil's hydroelectric development of the Xingu river basin. Environ Manage 8(1):16–27

Goldemberg J (2009) Energia—o Brasil na contramão? O Estado de São Paulo, 16 Feb. 2009, São Paulo

Hochstetler K, Keck M (2007) Greening Brazil—environmental activism in state and society. Duke University Press, Durham

IBAMA (Instituto Brasileiro do Meio Ambiente e dos Recursos Naturais) (2007) Licença Prévia N° 251/2007. http://www.ibama.gov.br/licenciamento. Accessed 26 May 2010

International Parliamentary Forum on Renewable Energies (2004) renewable energies—the challenge for the 21st century, resolution for the international parliamentary forum on renewable energies, Bonn, 2 June 2004. http://www.ipf-renewables2004.de/IPF_Resolution_en.pdf. Accessed 4 July 2009

IRN (International Rivers Network) (2009) Letter to world bank by brazilian ngos re $1.3 billion environment loan. http://www.internationalrivers.org/en/latin-america/brazilian-dams/letter-world-bank-brazilian-ngos-re-13-billion-environment-loan. Accessed 6 July 2009

Kelman J, Veras L (2008) A Constituição Federal de 1988 e o setor elétrico brasileiro (CF/88 and the Brazilian electricity sector). Revista Justiça e Cidadania—Especial 20 anos da Constituição 99:62–66

Leme E (2005a) Usinas hidrelétricas de Santo Antônio e Jirau—RIMA. http://www.ibama.gov.br. Accessed Sept 2009

Leme E (2005b) Usinas hidrelétricas de Santo Antônio e Jirau—EIA. http://www.ibama.gov.br. Accessed Sept. 2009

Lima S (2007) Barragem de Itaparica: vinte anos após o programa de reassentamento—(des)envolvimento e (des)estruturação de vidas: II encontro ciências sociais e barragens, 19–22 November 2007, Salvador, Bahia. http://www.ecsb2007.ufba.br. Accessed May 2009

Lopes J (2000) Democracia e cidadania: O novo ministério público brasileiro. Lumen Juris, Rio de Janeiro

Lopes J (2008) A carta da democracia. Lumen Juris, Rio de Janeiro

Mallaby S (2004) The world's banker: a story of failed states, financial crises, and the wealth and poverty of nations. The Penguin Press, New York

Melo E et al (2007) A perspective of the Brazilian electricity sector restructuring: from privatization to the new model framework: Conference on Developments in Economic Theory and Policy, Bilbao, 5–6 July 2007

Melo E, Neves E, da Costa A (2009) The new governance structure of the Brazilian electricity industry: how is it possible to introduce market mechanisms? In: 32nd IAEE International Conference: Energy, economy, environment: the global view, San Francisco, 21–24 June 2009

MMA (Ministério do Meio Ambiente) (2002) Portaria Interministerial n° 005, de 22 de novembro de 2002

MME (Ministério de Minas e Energia) (2007) Manual de inventário hidrelétrico de bacias hidrográficas. E-papers, Rio de Janeiro

MME (Ministério de Minas e Energia) (2008a) Plano decenal de expansão de energia 2007–2016. MME, EPE, Brasília. http://www.mme.gov.br. Accessed 2 July 2009

MME (Ministério de Minas e Energia) (2008b) Plano nacional de energia 2030. MME, EPE, Brasília. http://www.mme.gov.br. Accessed 2 July 2009

MPF (Ministério Público) (2004) Deficiências em estudos de impacto ambiental: síntese de uma experiência. Escola Superior do Ministério Público da União, Brasília

Novaes W (2009) Expansão de energia trafega na contramão. O Estado de São Paulo, 06 March 2009, Espaço Aberto, A-2, São Paulo

Programa Chile Sustentable (2004) Seminario latinoamericano sobre energias renovables—Hacia la conferencia mundial de Bonn—2004, LOM Ediciones, Chile. http://www.fonamperu.org/general/energia/documentos/sem_conosurpa.pdf. Accessed 4 July 2009

Rothman F, Oliver P (2002) From local to global: The anti-dam movement in southern Brazil. In: Smith J, Johnston H (eds) Globalization and Resistance: transnational dimensions of social movements. Rowman and Littlefield Publishers, Lanham, Hd, pp 115–149. http://www.ssc.wisc.edu/~oliver/PROTESTS/ArticleCopies/Dams_moby.pdf. Accessed 25 July 2009

Salomão R (2007) A construção da hidroelétrica de Itaparica e de Paulo Afonso e o reassentamento dos Índios Tuxá de Rodelas: II Encontro Ciências Sociais e Barragens, pp 19–22 November 2007, Salvador, Bahia. http://www.ecsb2007.ufba.br. Accessed May 2009

Scott P (2006) Re-assentamento, saúde e insegurança em Itaparica: um modelo de vulnerabilidade em projetos de desenvolvimento. Saúde e Sociedade 5 (3):74–89. http://www.abep.org.br. Accessed Nov 2008

Switkes G (2001) Considerações sobre a precipitada corrida do Ibama para avaliar o licenciamento ambiental do complexo hidrelétrico-hidroviário do rio Madeira. http://www.riosvivos.org.br/arquivos/1268161633.doc. Accessed 19 Sept 2009

World Bank (1998) Recent experience with involuntary resettlement: Brazil—Itaparica. http://lnweb90.worldbank.org/oed/oeddoclib.nsf/DocUNIDViewForJavaSearch/D514BFC6E53762BC852568150065D70C?opendocument. Accessed 22 May 2010

World Bank (2008) Licenciamento ambiental de empreendimentos hidrelétricos no Brasil: uma contribuição para o debate. Relatório N° 40995-BR. http//:www.mme.gov.br. Accessed 2 July 2009

World Bank (2009) Directions in hydropower. http://www.siteresources.worldbank.org/INTWAT/Resources/Directions_in_Hydropower_FINAL.pdf. Accessed 1 May 2010

Chapter 3
Between Local and Global Norms: Hydropower Policy Reform in China

Oliver Hensengerth

Abstract The chapter explores reforms of China's environmental and resettlement policies and the influence of domestic and external actors on Chinese dam-related legislation. It also analyses the impact of these reforms on two dam projects: the Nu River Project and the Xiaolangdi Multipurpose Dam Project. The analysis starts with an overview of the strategic role of hydropower in the economic development plans of central and local government. This forms the context for domestic conflicts that ensue between economic planners and people affected by construction projects. The chapter then analyses decision-making processes in the Chinese dam bureaucracy and the role allotted to civil society. By singling out two issue areas—Environmental Impact Assessment and resettlement—the chapter examines legal changes and the reasons for them. The case studies then consider these processes, taking the Nu River Project as an example of using Environmental Impact Assessment and the Xiaolangdi Multipurpose Dam Project to illustrate resettlement.

3.1 Introduction

In 2001 the report of the World Commission on Dams (WCD) was rejected by the Chinese government on the grounds that it was not represented on the Commission. However, the central government is not averse to international cooperation. China's domestic legislation on resettlement and Environmental Impact Assessment has been influenced from two sides: domestic policy learning processes and international debates. Moreover, China must comply with international standards when borrowing from multilateral development banks (the World Bank and the Asian Development Bank) for the construction of dams, as in the case of the Xiaolangdi

O. Hensengerth (✉)
Department of Social Sciences and Languages, Northumbria University, Newcastle, UK
e-mail: Oliver.Hensengerth@northumbria.ac.uk

W. Scheumann and O. Hensengerth (eds.), *Evolution of Dam Policies*,
DOI: 10.1007/978-3-642-23403-3_3,

dam on the Yellow River. Some ministries are more open than others to the idea of improving environmental protection. The classic conflict is between the economics related ministries and the Ministry of Environmental Protection.

This raises two issues. First, as has been pointed out for many years, the Chinese government is not a unitary block: its ministries pursue different strategies and goals. The classic studies here are Lampton (1987), Lieberthal and Oksenberg (1988) and Lieberthal and Lampton (1992). The key concept in these studies is fragmented authoritarianism, meaning a structure through which individual units in central and local government acquire bargaining power, enabling them to exert more leverage than should otherwise be the case in an authoritarian state.[1]

Second, the rejection of the recommendations of the WCD was no more than a surface reaction, hiding the far-reaching reforms under way in the environmental and resettlement policy fields.

This chapter seeks to explain changes in dam-related regulations and practices in China and the agents of those changes. The WCD serves as the reference point, given that it was a world-wide forum that elaborated guidelines for social and environmental sustainability to be followed during the dam-planning process.

The present chapter explores reforms of national environmental and resettlement policies and their impact on individual dam projects. It begins with an overview of the strategic role of hydropower in the economic development plans of central and local government. This sets the context for the domestic conflicts that ensue between economic planners and people affected by construction projects. The chapter then analyses decision-making processes in the Chinese dam bureaucracy and the role allotted to civil society. By singling out two issue areas—Environmental Impact Assessment and resettlement—the chapter examines legal changes and the reasons for them. The case studies then consider these processes, taking the Nu River Project as an example of using Environmental Impact Assessment and the Xiaolangdi Multipurpose Dam Project to illustrate resettlement.

The cases chosen are relevant to the development of China's hydropower potential. The Nu River Project will demonstrate the domestic situation by showcasing alignments of national environmental and economic actors and their power to influence governmental decision-making processes. The Xiaolangdi project is partly funded by the World Bank and an example of international involvement in China's dam-building sectors. In addition, both the Nu River and the Xiaolangdi Project qualify as large dams as defined by the International Commission of Large Dams (ICOLD).[2]

The author conducted semi-structured interviews in China between October 2008 and January 2009 with researchers, members of Chinese non-governmental

---

[1] For an excellent application of the concept of fragmented authoritarianism to the decision-making processes for the Three Gorges Project see Heggelund (2004).

[2] Large dams are "those having a height of 15 m from the foundation or, if the height is between 5 and 15 m, having a reservoir capacity of more than 3 million cubic metres". The definition is used by the WCD and the Asian Development Bank (ADB) and the World Bank takes it as the basis for its Operational Policy 4.37: Safety of Dams (World Bank 2001b, paragraph 3).

organisations, policy advisors and government officials. Further interviews were conducted with representatives of foreign companies providing environmental consulting services in China. Given the delicate political nature of the issue in China, all interviewees were assured of anonymity. Data from all the interviews cited in the text are therefore encoded.

3.2 The Strategic Role of Dams for China's Economic and Social Development

In 2004 China generated 238 billion kilowatt hours (kWh) from hydroelectric stations. This represented 15.8 % of total electricity generation (Energy Information Administration 2006). With ample hydropower potential on the Nu, Jinsha, and Lancang (Mekong) Rivers, Yunnan Province is the focus of China's hydropower expansion plans. In the late 1990s, Yunnan was said to have a hydropower potential of 90 gigawatts (GW), or 23.3 % of China's total hydropower potential (People's Daily 2000).

Plans drawn up by the National Development and Reform Commission (NDRC), the Chinese government's chief planning body, feature hydropower as the main component of the renewables target. The focus is on the development of the water resources of south-western China. It is planned to increase hydropower capacity from around 129,000 megawatt (MW) in 2006–300,000 MW by the end of 2020 (Stanway 2007b).[3] Hydropower performs several functions, perhaps the most important being the development of the landlocked western areas, as set out in the Western China Development Strategy launched in 1999. The policy covers the provinces of Gansu, Guizhou, Qinghai, Shaanxi, Sichuan and Yunnan, the autonomous regions of Guangxi, Inner Mongolia, Ningxia, Tibet and Xinjiang and the municipality of Chongqing (Cui 2006, p. 246).

The development of hydropower is also part of China's 2006 Renewable Energy Law. China's economic rise has put a strain on domestic energy resources. Since 1989 its Gross Domestic Product (GDP) has risen by an annual average of 9.5 % (calculated from the ADB Key Indicators, 2007). Until 1993 China was self-sufficient in oil. Since 2003 China has been the world's largest oil consumer after the United States, and since 2004 the third-largest oil importer after the United States and Japan (Downs 2006). China has also lost its self-sufficiency in natural gas and has to import coal to satisfy peak demand.[4]

The importance of hydropower in China's energy policy does not preclude Chinese interest in related reform processes. However, the economic planners on the National Development and Reform Commission argue that strict adherence to

[3] For a brief overview of historical calculations see Cui (2006, p. 24), Wei and Fan (2006, p. 9).

[4] For detailed figures on imports and exports see the continuously updated country balance sheets of the International Energy Agency at http://www.iea.org. Accessed 19 June 2013.

WCD recommendations would obstruct the rapid build-up of hydropower capacity. Dam financiers and professional organisations involved in China are muted. The World Bank, which financed the Xiaolangdi resettlement, said in its statement for the WCD report:

> The World Bank's conclusion on the guidelines is best summarized by the Chair of the WCD, who has explained that our guidelines offer guidance—not a regulatory framework. They are not laws to be obeyed rigidly (World Bank 2001a).

The International Hydropower Association (IHA), which represents the hydropower industry, says that it "draws on WCD Core Values and Strategic Priorities, along with other existing principles and policies, in its work to develop a practical assessment tool for hydropower sustainability" (IHA 2009, pp. 2–3). Consequently, IHA uses the WCD as a reference point, but does not adopt its guidelines.

However, the International Hydropower Association (IHA) initiated a political process in the form of the Hydropower Sustainability Assessment Forum (HSAF), which revised the 2004 IHA Hydropower Sustainability Assessment Protocol. The Chinese members were Zhou Shichun of the General Institute of Water Resources and Hydropower Planning and Design at the China Hydropower Engineering Consulting Group Corporation and Yu Xuezhong of the China Institute of Water Resources and Hydropower Research, which is affiliated to the Ministry of Water Resources. The National Research Center for Resettlement at Hohai University in Nanjing was also involved in the IHA-HSAF, though not as a permanent member.[5] A draft protocol was published in 2009. The new finalized protocol was published in November 2010.

3.3 Decision-Making Frameworks and Their Change Over Time

China's dam-building activities are governed by a highly complex bureaucracy and a party-state that is in the process of redefining its relationship with society. Within this complex setting, the following decision-making processes can be outlined.

3.3.1 Dam Decision-Making

Starting with governmental decision-making, central government ministries have counterparts on all lower levels of government, where they are known as bureaus. The Ministry of Water Resources corresponds to local Water Resources Bureaus on the five levels of government: province, prefecture, county, township and

[5] The Center was also asked by the WCD to compile the China country report. It declined, however, because the Chinese government did not welcome the WCD process.

village. On the township and village levels, the bureaus are known as stations. Traditionally, the bureaus functioned as implementing agencies for the central ministries. During decentralisation, local government has been given considerable decision-making authority, raising it above the status of mere implementing agencies. Accordingly, the Ministry of Water Resources, for instance, does not have direct administrative relations with the bureaus.[6] The power relationship is between the bureaus and local government because the bureaus depend on local government for funding. Local government exerts leverage over the bureaus through the allocation of taxes (Economy 2004, p. 92).

Whether or not dams are constructed is decided by the National Development and Reform Commission on the basis of its assessment of China's economic development and energy needs. Any project must also have the consent of the Ministries of Agriculture, Land and Resources, Environmental Protection, and Water Resources. Together with the National Development and Reform Commission, these ministries have to approve the pre-feasibility and feasibility studies, the design plans and the Project Application Report.[7] Mertha (Table 3.1) reports the following approval benchmarks.

Projects larger than 300 MW require the approval of the National Development and Reform Commission (Makkonen 2005, p. 10 citing Magee no year, publication after 2005). If a dam project is situated on a river that crosses county or prefecture borders, the approval of the administratively higher water resources bureau is needed (Mertha 2008, p. 41). If the project is on a river that crosses provincial or national borders, central government's approval is required.[8] In the case of international rivers, the provincial level is excluded from decision-making and may only provide input (Magee 2006a, p. 250, b, pp. 38–40) on environmental and social conditions in the site area. Since the reform of the water sector in 2002, Article 15 of the revised Water Law of 2002 requires project developers to submit their plans for approval to one of the seven basin commissions under the Ministry of Water Resources rather than to the National Development and Reform Commission (Magee 2006a; for a detailed analysis of the Water Law and the changes that were made with regard to previous legislation see Wouters et al. 2004).

This means that the planning of any hydropower station begins at the level of the seven basin commissions. They draw up the comprehensive basin plans, including the theoretical and technically feasible hydropower potential. Next, five state-owned energy companies (China Huaneng, China Datang, China Huadian, Guodian Power and China Power Investment)—or project-specific organisations created by central government for trans-provincial rivers, such as the Xiaolangdi Water Resources Construction and Management Bureau for the Yellow River—have the right to construct hydropower stations in the various basins (Magee 2006a; for the relationship between the comprehensive basin plans and the

---

[6] Interview B17112008.

[7] Interviews K12122008.

[8] Magee (2006a) and interviews K12122008.

Table 3.1 Decision-making authority of local government (author's own compilation adapted from Mertha 2008, p. 41)

Government level	Reservoir volume	Wall height (m)
Provincial water resources bureau	10 million cubic metres	more than 50
Prefecture water resources bureau	1–10 million cubic metres	30–50
County water resources bureau	100,000–1,000,000 cubic metres	15–30[a]

[a] There appears to be a typographical error in Mertha's book, as it says that the county bureau approves dams with a height of 50–30 m. This is corrected, as he presumably means 15–30 m

environmental protection plans and the problem of institutional responsibility between the Ministry of Water Resources and the Ministry of Environmental Protection see Ongley and Wang 2004).

The five state-owned companies were the result of a large-scale reorganisation of the energy sector in 2003 that dismantled the State Power Corporation. Their plans for developing hydropower must comply with the comprehensive basin plans, compliance being checked by the commissions. Once a basin commission approves a hydropower development plan, it sends its decision to the National Development and Reform Commission for further approval. Project developers sometimes circumvent the basin commissions and go directly to the National Development and Reform Commission (Magee 2006a, pp. 136–137, 171, 172 note 113, 249–250). Consequently, the initial steps in the approval process—in particular, the initial project idea—proceed not from the comprehensive plan but rather from the comprehensive hydropower plan drawn up by the hydropower development company in cooperation with one of the survey and design institutes (Magee 2006a, p. 250).

Once a basin commission (or, if it was bypassed by the energy companies, the National Development and Reform Commission) has decided that a dam should be built, one of the five energy companies having the right to develop the river basin in question "solicit bids" from the China Hydropower Engineering Consulting Group Corporation and its eight design institutes (Magee 2006a, p. 250). The design institutes develop detailed plans for individual dams:

> including installed capacity, approximate dam site, type of turbines, and other technical features. Precise dam sites and installed capacities may vary somewhat at this point, but the big picture provided by the comprehensive plan is supposed to be followed (Magee 2006a, p. 250).

At this point, the energy companies commission appropriate agencies to carry out pre-feasibility and feasibility studies. The (pre-) feasibility studies must contain data on site selection, technological and economic feasibility (e.g. demand-sales projection), resettlement design and environmental appraisals (including evidence of public consultation). If both studies and the design studies are approved, the operator can compile the Project Application Report, which includes the final EIA Report (NDRC 2007a, b; McElwee 2008, p. 4). The Environmental Impact Assessment (EIA) is conducted by a certified EIA agency, such as a private consulting company. The resettlement plan is drawn up by the design institute in cooperation with relevant governmental and other experts.

Each component of the pre-feasibility and feasibility studies and the Project Application Report is examined by the relevant governmental expert departments. For projects that have to be approved by central government, the EIA is passed on to the Ministry of Environmental Protection. The Ministry of Water Resources has to approve the resettlement plan. Both Ministries notify the National Development and Reform Commission of their decisions. Only then should the National Development and Reform Commission approve the Project Application Report. For projects on international rivers, the National Development and Reform Commission sends the project documents to the State Council. If the State Council approves, it forwards them to the National People's Congress. Once rubber-stamped by the National People's Congress, construction can begin.

Figure 3.1 summarises these procedures. The upper half depicts procedures governed by the Water Law, the lower half those subject to EIA and resettlement legislation. It is only in the lower half that public participation becomes relevant, that is, after dams have been approved. No provision is made for public participation in overall decision-making on dams and water management plans.

Non-governmental organisations (NGOs) and the people affected cannot therefore begin to influence dam decision-making before EIA and resettlement planning. Importantly, Plummer and Taylor (2004, p. 37)[9] caution against overstating the scope for action by NGOs in China, because the official:

> objectives of community participation in development projects in China have all been instrumental in nature [...] to improve the efficiency of investment in rural development and the rate of progress in poverty reduction.

Ho and Edmonds (2007) argue that Chinese civic organisations exist in a state-society relationship that is best described as "embedded social activism", their environmental activism taking the form of "embedded environmentalism". Being embedded means that in China's semi-authoritarian political system with its limited scope for action, civic organisations and individuals rely on a diffuse and informal network with other individuals, groups, loose networks and organisations that enables them to bridge the divide between party-state and society and therefore re-negotiate their relationship with the party-state, resulting in incremental change. Yang and Calhoun (2007) emphasise the use of the mass media to influence the environmental debate in China and to redefine the relationship between the party-state and society, through the emergence of a "green public sphere".[10] Like Plummer and Taylor (2004), Economy (2004, pp. 21 and 129–175) argues that Chinese leaders have allowed the growth of NGOs and "aggressive media attention" to environmental degradation for an instrumental purpose: to make up for the weakness of the Environmental Protection Bureaus in local government. The strengthening of the societal level is therefore designed to keep a check on local government.

---

[9] See also Taylor (2004).

[10] See also Harris (2006, p. 11).

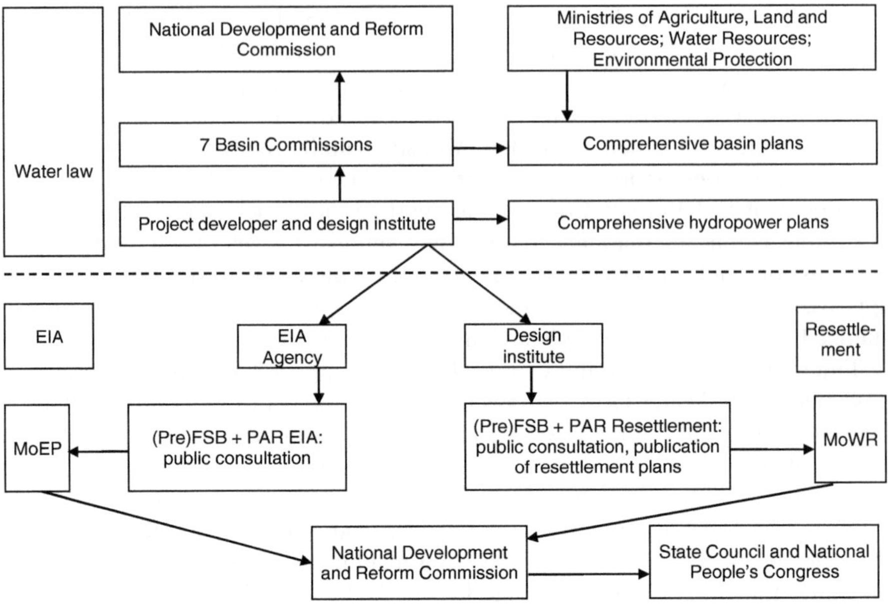

Fig. 3.1 Decision-making for hydropower development (author's own compilation adapted from Magee 2006a; McElwee 2008; EIA Law; Land Administration Law; NDRC 2007a, b)

In this context, the government published the Guidelines for Full Implementation of the Rule of Law in 2004. In March 2008 the State Environmental Protection Agency was upgraded to the Ministry of Environmental Protection. In May 2008 the State Council issued the Government Information Disclosure Regulations, followed by the Environmental Impact Disclosure Measures of the Ministry of Environmental Protection. All three documents allow individuals access to information. More specifically, the Environmental Impact Disclosure Measures require enterprises to submit pollution data to environmental agencies, which then have to release this information on a timely and regular basis (Qin 2008; Brewer 2008, p. 19; Powell et al. 2008, pp. 11, 36–38).[11]

3.3.2 Environmental Impact Assessment

The development of EIA in China is both the result of domestic learning processes and international examples studied by Chinese experts. The first mention of EIA was in Article 6 of the 1979 Environmental Protection Law. This development coincided with the reform processes that began in December 1978. The 1979 Law stipulated that a:

[11] For specific regulations governing NGO registration see Gough (2004, p. 14).

> report on environmental impact must be submitted when construction, transformation and extension projects are to be undertaken, and the design and construction may not be initiated until approval by the department in charge of environmental impact and other relevant authorities (Article 6, quoted in Qin 2008, p. 2).[12]

The first details on EIA procedure and content then came in the 1981 Measures on Administering Environmental Protection in Infrastructure Construction Projects. This was modified in the 1986 Measures on Administering Environmental Protection in Construction Projects, when the requirement to conduct an EIA was extended from infrastructure projects to all projects with potential environmental impacts. The 1986 regulations were set out in a ministerial decree issued by the National Environmental Protection Agency and modelled on the United States National Environmental Policy Act.

Its coming into force meant that the Chinese government was attempting to coordinate environmental policy and economic policy and that it regarded EIA as an effective means of achieving this.[13] Modelling it on the United States National Environmental Policy Act meant that the Chinese leadership was prepared to learn from foreign countries' experience.[14] The 1986 EIA applied to single projects only, not to plans and programmes. Thus EIA was not used to appraise regional or zoning plans, only projects within a plan (e.g. a building).[15]

While a number of further regulations were passed after 1986 (see Table 3.2), the sea change came with the 2003 EIA Law (formally the Law of the People's Republic of China on the Environmental Impact Assessment), which requires Strategic Environmental Assessments (SEAs) for all regional and special programmes or plans involving construction, such as industrial development zones or water conservancy projects, and EIAs for all construction projects that either stand alone or form part of a regional or special programme (Articles 3, 7, 8, 16, 18). However, the environmental authorities can do no more than comment on SEAs, proposing changes with a view to mitigating impacts. They cannot reject SEAs (Articles 14, 15). The 2009 Regulations on Planning Environmental Impact Assessment attempt to strengthen SEA implementation (Qin 2009; Ministry of Environmental Protection 2009; for a detailed review of post-2003 EIA regulations and the relationship between EIAs and SEAs see the in-depth study of Zhu and Lam 2009).

When EIAs are carried out, the general public has the right to participate "in appropriate ways," that is, by holding "demonstration meetings or hearings or by any other means." The EIA document must indicate whether the public's views have been considered and give the reasons if they have been rejected (Articles 5, 11, 21). The "other means" are often surveys or questionnaires distributed by the

---

[12] If not otherwise mentioned, the following historical overview follows Qin (2008, pp. 2–3).

[13] Interviews W03122008, S17122008. In addition, Chernobyl was mentioned as a potential catalyst, with a nuclear station in operation in Dayawan: interviews W03122008.

[14] Interviews W03122008.

[15] Interviews W03122008.

Table 3.2 Changes in Environmental Impact Assessment regulation over time (Qin 2008, 2009; Ministry of Environmental Protection 2009; NDRC 2005, 2011; Cameron and Luo 2012)

Year	Legislation	Content/change from previous legislation
1979	Environmental Protection Law	Report on environmental impacts of construction projects
1981	Measures on Administering Environmental Protection in Infrastructure Construction Projects	First details on content and administrative procedures for environmental impact assessments
1986	Measures on Administering Environmental Protection in Construction Projects	EIA requirement expanded from infrastructure to all construction projects
1989	Revised Environmental Protection Law	EIA report must propose procedures for mitigating environmental impact of construction projects
1996	Decision on Several Issues in Environmental Protection	First State Council Regulation on EIA: authorities must consider economic, social and environmental impacts "before making major decisions on regional and resources development, urban development, industrial development, restructuring of sectors and [...] other issues of economic production and social development"
1998	Regulations on the Administration of Environmental Protection in Construction Projects	Expansion of scope, content, administrative procedures and legal liabilities for EIA
2003	Environmental Impact Assessment Law	Requirements for SEAs and EIAs
2005	Notice on Strengthening Environmental Protection during Hydropower Construction	Site preparation EIA ("santong yiping" EIA) introduced, which is separate from project EIA. Therefore, site preparation can commence without approval of the project EIA
2006	Provisional Measures for Public Participation in EIA	First details on scale of public consultation
2009	Regulations on Planning Environmental Impact Assessment	Standardisation of SEA administrative procedures
2011	Provisional Measures for the Evaluation of River Hydropower Plans and Environmental Impact Statements	Specifies criteria for the examination of EIA reports hydropower plans
2012	Notice on further Strengthening Environmental Protection during Hydropower Construction	EIA law and related legislation including those for public notification of projects and public participation are made applicable to hydropower EIAs. The Notice also stresses the importance to include the site preparation EIA into the project EIA

EIA consultant to people potentially affected. When reviewing the EIA Report, the Ministry of Environmental Protection may decide to hold further public consultations, such as public hearings (McElwee 2008, p. 4).

However, as public disclosure of the entire EIA report is not necessary, only summary EIAs are published. In addition, the EIA Law cannot govern the quality and formal setting of public consultations, and it does not define what public consultations are (meetings for discussion, face-to-face interviews, etc.). Nor does it specify the groups or number of people affected who are to be consulted. For example, it does not state how many questionnaires must be circulated and does not therefore say anything about the number of people affected that have to be consulted. The problem persists in the 2006 *Provisional Measures for Public Participation in EIA* (see below).[16]

There are three levels of EIA according to the scale of the expected environmental impact. A certified EIA consultant advises the project developer which level applies to his project. The lowest level is Environmental Impact Registration of projects with minimal impact. The Ministry of Environmental Protection has developed a form for this, which has to be completed by the project owner. For the second level there is an Environmental Impact Assessment Form for projects with a "slight" impact. This form requires a brief analysis of the environmental impact and must be completed by a certified EIA consultant. The third level requires a full EIA Report for projects having a "major" impact. No standard forms for this level exist. Reports must be completed by certified EIA consultants and also call for "a comprehensive assessment of the project's environmental impact" (McElwee 2008, pp. 1–2; EIA Law, Article 16).

The EIA Report must include a description of the project; a review of the construction plans; an engineering analysis to determine its environmental impact during construction and operation; and a review of the production process and pollution control options. The EIA Law does not require the production process to be described in detail, and indeed some processes or their components may be proprietary. However, the EIA Report must include a clean production analysis[17] and a risk assessment (especially if hazardous substances are involved). The projected environmental impacts are compared with national and local environmental standards, such as mass load limits for pollutants. Further factors may include energy efficiency and water use. If the environmental impacts exceed national standards and limits, the EIA Report must include mitigation measures (McElwee 2008, pp. 3–4; EIA Law, Article 17).

EIA Reports or Forms are submitted to the Ministry of Environmental Protection for review. If another department is responsible for an issue area affected by the project (e.g. housing), EIA Reports or Forms are first submitted to that department for preliminary examination. The Ministry's approval department makes a decision within 60 days of receiving the EIA Report, within 30 days of

[16] Interview S17122008.

[17] Clean production analysis means a "review of the production process to determine if the same product can be made with fewer resources and the generation of fewer wastes" (McElwee 2008, p. 3).

receiving the EIA Form and within 15 days of receiving the Environmental Impact Registration (EIA Law, Article 22).

EIAs for the following projects are reviewed and approved by the Ministry of Environmental Protection: special nuclear and top secret construction projects, construction projects situated in more than one province, and all construction projects subject to final approval by the State Council (EIA Law, Article 23). The last category includes, for instance, projects situated on international rivers or border-forming rivers, such as large hydropower stations. Projects requiring an investment of US$100 million or more also require State Council approval (McElwee 2008, p. 4). All other project EIAs may be approved by provincial authorities. At the discretion of these entities, lower-level authorities may approve project EIAs (EIA Law, Article 23).

Four cases spurred the debate on EIA in China[18]: first, the Nu River Project in 2003, the first project to be halted and subjected to environmental investigation under the EIA Law. The project is described in detail in the case study below. Second, riots in Hanyuan in Sichuan Province in October 2004 in connection with the Baopu hydropower project and the inadequacy of compensation for expropriated land.[19] Third, the dispute of March 2005 over the Lake lining project in the old Summer Palace, the Yuanmingyuan. The dispute attracted national attention and was widely debated in the Chinese media.[20] Fourth, the plan in 2005 of the Taiwanese company Tenglong Aromatic PX to build a chemical factory near a residential area in Xiamen's Haicang district.[21]

Triggered by these events, which revealed the growing social unrest and the greater ability of citizens to organise themselves suddenly and spontaneously, the central government drew up a bye-law on public participation in EIA in February 2006, the Provisional Measures for Public Participation in EIA. These measures are more detailed than the sections of the EIA Law concerning participation and stipulate the following:

- Whom to consult: citizens, corporations and other institutions according to their geographical distribution, profession, expertise and the degree to which they are affected
- When to involve the public: public notification of the project when the EIA is initiated; public notification detailing project scope and effects during compilation of the EIA Report; release of EIA outline and indication of approval or rejection during examination and approval of the EIA Report;
- How to participate: there are five participation channels: expert comments, discussions, demonstration meetings, hearings and questionnaires. Detailed rules are laid down on the organisation of hearings, starting with preparation,

[18] Interviews B20122008, W03122008.

[19] For details see Lee (2007).

[20] For details see Qin and Li (2005), Powell et al. (2008, p. 25).

[21] For details see Powell et al. (2008, p. 27).

selection of participants, procedures, publication of reports, media reports and how to deal with responses from the public;
- Access to information: developers are required to release information within 7 days of contracting an EIA agency. During compilation, developers or EIA agencies are to notify the public of the project, its potential impacts on the environment, a summary EIA and schedules and forms of public consultation. Methods used are hearings, solicitation of comments, public forums, expert forums, etc. Information can be disclosed in three ways: public notices in the local media, dissemination of publicly accessible information and other means;
- After being notified, the public may submit opinions to developers, EIA agencies or environmental departments for examination, approval and re-approval of EIA Reports. These opinions must be archived. If the public feel that developers or EIA agencies have failed to address their concerns, they may call on the environmental authorities to resolve the situation (Qin 2008, pp. 7–13).

The *Provisional Measures* leave it to the developer and EIA agencies to define the "major items subject to public consultation," "range and major issues subject to public consultation" and the public groups qualified to participate. The Disclosure Regulations and Disclosure Measures may at least mitigate this problem by giving the public the right of access to information, but to enforce all these mechanisms, government commitment is essential (Qin 2008, pp. 13–14).

In cases where government and EIA agencies are perceived to have bypassed public involvement or to have involved the public inadequately,[22] the public have found ways to remedy the situation, aided by policy entrepreneurs, journalists, environmental NGOs or NGOs that provide legal assistance. Increasingly, NGOs are taking to alliance-building (Gough 2004, p. 18) with media and/or politicians. Only rarely do NGOs cooperate with each other. Yet the predominant approach is still to "maintain a low profile [...], a non-confrontational approach, and to establish a good relationship with the government" (Gough 2004, p. 16; International Rivers Network 2005; Stanway 2005, 2007a; Osborne 2006, pp. 58–59; Hirsch and Jensen 2006, p. 60). Organised popular protest with or without the organisational means of NGOs is targeted at different levels of government: municipal and provincial, but rarely central (Mertha and Lowry 2006; Powell et al. 2008, case studies 1–20, 24–41).

Future EIA reforms under discussion include a strengthening of participation, and particularly of the rules laid down in the Provisional Measures for Public Participation in EIA. In addition, the 2012 *Notice on further Strengthening Environmental Protection during Hydropower Construction* considers the need to integrate the site preparation EIA into the project EIA (Ministry of Environmental Protection 2012; Cameron and Luo 2012). The site preparation EIA (the so-called santong-yiping—"3 connections and 1 levelling"—EIA) that was introduced in 2005 in the *Notice on Strengthening Environmental Protection during Hydropower Construction* (NDRC 2005) concerns the creation of access to roads, water

[22] Interviews W03122008.

and electricity (the 3 connections) and the levelling of the construction site in order to begin construction. The project EIA then is necessary to build the actual dam. The fact that the site preparation EIA is separate from the project EIA means that the project developer can begin preparation of the dam site once the site preparation EIA is approved. Preparation of the dam site is therefore not conditional on approval of the project EIA (NDRC 2005; Cameron and Luo 2012).

Enforcement of regulations therefore plays a prominent role, with particular emphasis on the reinforcement role of the local Environmental Protection Bureaus. An important model for Chinese decision-makers considering the reform of participation is the Aarhus Convention, known formally as the United Nations Economic Commission for Europe (UNECE) Convention on Access to Information, Public Participation in Decision-Making and Access to Justice in Environmental Matters. Other models originate from the European Union: EIA Directive 85/337/EEC, Public Participation Directive 2003/35/EC and the chemicals directive REACH, which served as the model for China's own chemicals directive.[23]

The content of these regulations is transmitted to Chinese experts and decision-makers through comments and translations by Chinese scholars and through government delegations (kaochatuan, literally: study groups, often mocked as "tourist groups") to European countries and the United States. In addition, such intergovernmental forums as the China-EU summits and ministerial-level meetings between the Ministry of Environmental Protection and its EU counterparts help with the transposition of these regulations into the Chinese law-making process.[24]

According to a Beijing-based NGO, China's cooperation in international bodies stems from a general interest on the part of the leadership and such companies as PetroChina in being acquainted with international norms not only in the dam construction sector but also as they relate to exports and extractive industries.[25] It should be borne in mind, however, that most of Beijing's policy-makers have no interest in engaging with traditional donor countries in ideological discussions on good governance. On the contrary, they are pragmatic and want to see quick results through the application of new and innovative technologies.

3.3.3 Resettlement Planning

The other big issue in dam construction in China is resettlement. Resettlement is governed by a number of laws and regulations, most importantly the 2004 Land Administration Law and the 2006 Rules of Land Compensation and People Resettlement in Medium and Large Hydraulic and Hydroelectric Projects.

[23] Interviews W03122008 and S17122008 with environmental law researchers and government advisors.

[24] Interviews W03122008.

[25] Interview B27112008.

Chapter 5 of the Land Administration Law contains regulations on land expropriation and compensation. Article 46 states that local government from the county level upwards are responsible for carrying out expropriation after due notification. Expropriations are part of general land use plans drawn up by all levels of government, including the township level, although lower-level plans must comply with the plans of the next higher level of government. They must also comply with the national and social development plans (Articles 17, 18, 21) drawn up by the National Development and Reform Commission. Article 47 of the Land Administration Law specifies four types of compensation:

1. land compensation payments,
2. resettlement fees,
3. compensation payments for ground attachments and young crops,
4. if vegetable fields in suburban areas are expropriated, the units using the land after expropriation pay into a new vegetable field development fund.

The most important recent reform has been affected under the 2006 Rules of Land Compensation and People Resettlement in Medium and Large Hydraulic and Hydroelectric Projects. These rules and the Land Administration Law have established the following resettlement practice: displaced people must be provided with a level of livelihood similar to or higher than that before their displacement; resettlement plans must include economic development plans, not just cash payments for lost land and resources; resettlement plans should create jobs; and they must include plans for reclaiming new farmland for displaced rural communities in order to avoid the overcrowding of resettlement areas (Brewer 2008, p. 19).

However, the 2006 rules are too general: for example, they contain no provisions for indirect effects of flooding (that is, people who are partly affected), as when peasants' fields are flooded, but not their houses. Compensation for such indirect effects must be negotiated with local government on a case-by-case basis.[26]

While people are generally resettled near their area of origin, with the focus on agricultural resettlement, exceptions are made where there is a shortage of farmland and land is prone to erosion (State Council 2006a, Articles 7–15). People are then moved to other provinces or cities. Resettling people outside their area has raised assimilation problems and resulted in conflicts with the resident population, since the new arrivals are equipped with compensation packages (Heggelund 2006, pp. 12–13).

The development of resettlement regulation began with the reform period in 1978/79. People resettled before the reform of the land use system suffered from poverty and a lack of public infrastructure.[27] The government recognised this

[26] Interviews N15122008 with resettlement researchers and government advisors.

[27] Interview N23102008 with a resettlement researcher. The reform of the land use system introduced user rights for each family under the household responsibility system. Yet officially the land belongs to the state, and in villages it belongs to the collective and is therefore administered by the village committee.

problem at an early stage. The first resettlement regulations, under which the Ministry of Finance and the Ministry of Power established a reservoir maintenance fund, were adopted in 1981. The owners of hydropower stations under the Ministry of Power were required to contribute 1 Renminbi per thousand kWh to this fund. The money was then used to compensate relocatees for lost livelihoods and means of production and to pay for the maintenance of reservoirs, drinking water supply, and irrigation and transport structures (Ministry of Power and Ministry of Finance 1981; Bizer and Ragsdale 1997; van Wicklin 1999, p. 239).

To assess the social impact of resettlement before the reform period, the State Council asked the Ministry of Water Resources in 1984 to conduct a survey of people who were resettled before 1980. This showed that one–third were affected by poverty (by that time 10 million people had been resettled since 1949). As a consequence, the State Council published the Circular on Transmitting the Report of Paying Close Attention to Reservoir Resettlement in 1986. This Circular established a reservoir construction fund for those who had been resettled before 1985. The fund was initially endowed with 240 million Renminbi a year. The money was drawn from hydropower revenues, water tariffs and local and central government budgets (State Council 1986; van Wicklin 1999, p. 239).[28] Again in 1986, resettlement regulations were set out in Chapter 5 of a new Land Administration Law. In addition, the Ministry of Water Resources established a resettlement office and a coordinating committee, on which the Ministries of Finance, Agriculture and Power were also represented. The committee was responsible for the construction of roads, schools, water supply and public infrastructure, and it supported relocatees in farming (land development, planting of fruit trees, agricultural techniques and non-farming skills). On the local level, provinces, counties and cities also set up resettlement offices, and the funds from the reservoir construction fund were distributed down the administrative chain.[29]

In 1991, the State Council issued Rules of Land Compensation and People Resettlement in Medium and Large Hydraulic and Hydroelectric Projects, which were updated in 2006 (State Council 2006a). According to the Rules, relocatees would be supported for 5–10 years after resettlement was completed. They thus provided not only for compensation, but also for the restoration of incomes. These provisions can be seen as a follow-up to the 1986 Land Administration Law (especially Chapter 5).[30]

In the reform processes that followed, external influence through the Asian Development Bank (ADB) and the World Bank became important. In December 1996, at China's request, the ADB extended its Technical Assistance (TA) 2735-PRC entitled Capacity Building for Natural Resource Legislation (ADB 1996). The result of the ADB's TA was the revision of the 1986 Land Administration Law, which became effective in January 1999. The revision had four goals: first, to

[28] Interview N23102008.

[29] Interview N23102008.

[30] Interviews N23102008 and N02122008.

increase compensation and resettlement fees and to pay them directly to the victims of expropriation rather than through the rural collective. This means that the provincial resettlement office pays the compensation funds to the local resettlement office in the resettlement area. Then the local resettlement office directly allocates land and pays compensation directly to the resettled villagers; second, to enhance consultation on resettlement plans: local government must now publicise plans and consult with local people; third, to improve transparency and accountability in the use of compensation fees; fourth, to improve contract adjustment and dispute settlement procedures. This includes a provision enabling local government decisions to be challenged in a lawsuit to be filed within 3 days of having received the decision (ADB 1998a, p. 26; note 2; Chen et al. 2004, p. 213).

This was followed in March 1998 by ADB Regional TA (ADB RETA) 5781: Review of National Resettlement Policies and Experience with Involuntary Resettlement Projects (ADB 1998b). The lead agency for implementing the Regional Technical Assistance (RETA) was the Environmental Protection and Natural Resources Conservation Committee of the National People's Congress. Then, in September 2000, came ADB RETA 5935: National Resettlement Policy Enhancement and Capacity Building (ADB 2000), the lead agency being the Ministry of Land and Resources.

The aim of both RETAs was to update Chapter 5 of the 1999 Land Administration Law. The China country report for both RETAs was compiled by the National Research Center for Resettlement at Hohai University. For the first RETA the report proposed "standardizing resettlement entitlements and procedures between sectors, provinces, and urban–rural areas." This was put into effect by a State Council decree in 1999. For the second RETA the report proposed that the term "requisition" in Chapter 5 of the Land Administration Law be replaced with the term "expropriation" (Shi et al. 2001; ADB 2002, pp. 5 and 10). The aim here was to stop the illegal expropriation of land by local government and to consolidate the ownership of land in the hands of the state and the collective (Ho 2005, p. 26; note 34).

The Ministry of Land and Resources, the State Development and Planning Commission, the State Economic and Trade Commission, the Ministries of Commerce, Water Resources, Railways, Transport, and Agriculture, the Resettlement Bureau of the Three Gorges Construction Committee of the State Council and the National Electric Power Corporation were involved in the second RETA (ADB 2002, p. 7).

ADB RETA 5935 not only revised the Land Administration Law in August 2004, but also led to a change in China's constitution in March 2004. The latter was based on a proposal from the Ministry of Land and Resources to the State Council, which led the State Council to submit legislation to the National People's Congress.[31] The changes consisted in instituting a distinction between expropriation and requisition. Accordingly, the constitution was amended and the term

[31] Interview N02122008.

"expropriation" added to the third sentence of Article 10.[32] The sentence was transferred verbatim to the 2004 Land Administration Law (Article 2). Following the recommendations of the RETA 5935 country report, the term "requisition" was almost entirely deleted from the revised 2004 Land Administration Law and replaced with the term "expropriation", the exception being in Article 2.

In 1996, the National Development and Reform Commission and the Ministries of Finance, Power and Water Resources set up a post-relocation support fund for relocatees from the area of large and medium-sized dams that had gone into operation between 1986 and 1995 and dams that had been approved and on which work had started after 1996 (NDRC 1996). In 2006 the 1991 Rules of Land Compensation and People Resettlement in Medium and Large Hydraulic and Hydroelectric Projects were updated (State Council 2006a). Specifically, monitoring and evaluation were added to the resettlement rules, as were better protection for the cultures of ethnic minorities. These new elements were derived directly from World Bank and ADB procedures, which had hitherto been applied in China only to projects financed by those banks.[33] From a new post-relocation support fund consisting of revenues generated by electricity sales, relocatees receive 600 Renminbi a year for 20 years. As there are currently 22.8 million registered relocatees, the fund amounts to 13.68 billion Renminbi.[34] The aim of the current review of the 2004 Land Administration Law is to establish dynamic compensation criteria with a view to improving implementation and the compensation paid to farmers whose land has been expropriated (Qin 2009) (see Table 3.3).

Planning for resettlement is carried out by the General Institute of Water Resources and Hydropower Planning and Design, the national-level design institute that is affiliated to the China Hydropower Engineering Consulting Group Corporation. The Ministry of Construction and its local bureaus are in charge of urban house removal, i.e. relocation and reconstruction of buildings. The land management sector represented by the Ministry of Land and Resources and its local bureaus are responsible for policy formulation and enforcement of national laws relating to expropriation (Chen et al. 2004, p. 204). On a basin level, resettlement bureaus exist for four of the seven basin commissions under the Ministry of Water Resources: the Yellow River, Changjiang, Hai and Pearl Rivers (Zhang 2007). The resettlement bureaus are responsible for approving resettlement plans, monitoring and supervision of resettlement implementation and the review of resettlement completion (Zhang 2007).

Once a resettlement plan is approved by the resettlement bureaus, implementation is the responsibility of county governments and their resettlement offices.

[32] "The state may, in the public interest, appropriate or requisition land for its use in accordance with the law, while making compensations." Further constitutional changes were made at the same time. For details, see China Daily (2003).

[33] Interview N15122008. See also Chen et al. (2004, p. 204).

[34] Interview N23102008, and State Council (2006b, Article 2, 5–10).

Table 3.3 Changes in resettlement regulation over time (Ministry of Power and Ministry of Finance 1981; Bizer and Ragsdale 1977; State Council 2006a; ADB 1998a, 2002; Shi et al. 2001; NDRC 1996; Qin 2009)

Year	Legislation	Content/change from previous legislation
1981	Reservoir maintenance fund	RMB0.001 per kWh: cash compensation, infrastructure (drinking water, transport)
1986	Reservoir construction fund	RMB0.004 per kWh = RMB240 million per year for people resettled before 1985
1986	First Land Administration Law contains resettlement regulations	Establishment of resettlement offices
1991	Rules of Land Compensation and People Resettlement in Medium and Large Hydroelectric Projects	Post-relocation support and income generation (5–10 years)
1996	Post-relocation support fund	RMB0.005 per kWh for projects in service and approved 1985–1995
1998	Land Administration Law amended through ADB Technical Assistance	More transparency and compensation
2004	Land Administration Law amended through ADB Technical Assistance	"Requisition" replaced with "expropriation"
2006	1991 Rules updated	Monitoring and evaluation; post-relocation support fund: RMB600 per year for 20 years
Under	consideration	Amendment of Land Administration Law
Dynamic	compensation criteria to improve implementation and compensation	

Monitoring and evaluation are carried out by such external agencies as research institutions, consulting companies and the department of design and planning at the Chinese Academy of Social Sciences (Chen et al. 2004, p. 204).[35]

During resettlement planning and implementation and post-relocation support, the communities affected must be involved in the process. The rural communities are represented by village committees, which are involved in the implementation phase in relation to land adjustment, location of housing, management of collective property, land compensation, employment and other areas. Urban communities are represented by residents' committees, which "assist" district and county governments and resettlement agencies with socio-economic surveys and the relocation process (Chen et al. 2004, p. 203).

As previously mentioned, the project developer must submit a pre-feasibility study and a feasibility study prior to approval, and both must contain plans for

[35] In fact, land acquisition and resettlement work are the responsibility of the project developer. However, since land is owned collectively and no market for land exchange exists, this work is contracted to local government bodies (Chen et al. 2004, p. 203).

resettlement. The study requirements and the resettlement specifications they include are set out in the 2006 Rules of Land Compensation and People Resettlement in Medium and Large Hydraulic and Hydroelectric Projects (State Council 2006a). The technical details are specified by the National Development and Reform Commission in a publication series entitled Standards of the People's Republic of China for the Power Industry. The most basic standards in this series are the Code for Preparation of Hydroelectric Project Feasibility Study Report and the Specification of Resettlement Planning and Designing for Hydroelectric Project (NDRC 2007a, b). Both standards were announced in 2003 and renewed in July 2007.

Local government from the county level upwards in the originating and host areas must be consulted on the resettlement plans (State Council 2006a, Article 6). The resettlement plans for the pre-feasibility and feasibility studies must contain details on the economy, society and environmental capacity of the originating and host areas as well as details on the relocatees themselves, such as their sources of income, location and agricultural production levels so that their current and future living standards may be assessed. The studies must also detail post-relocation support measures. Whenever necessary, the plans must be established on the basis of hearings conducted with the targeted resettlement population.

3.4 Dam Case Studies: The Nu River Project and the Xiaolangdi Multipurpose Dam Project

In the following, the study applies the national-level analysis to case studies in order to assess the effects of national legislation on dam projects. The first case applies EIA to the Nu River Project, the second considers resettlement in the context of the Xiaolangdi Multipurpose Dam Project. The Nu River and Xiaolangdi Multipurpose Dam Projects are part of the Western China Development Strategy and therefore cornerstones of China's plans for energy development. The Nu River Project showcases the roles of domestic reform through domestic learning processes and of NGOs and other domestic non-state actors (such as media outlets) in influencing governmental decision-making. The Xiaolangdi Multipurpose Project is an example of the role played by international actors (in this case, the World Bank) in China's dam-building programme. Table 3.4 briefly outlines the main project characteristics.

[47] For a detailed description of all thirteen dams see Brown and Xu (2010).

[48] The Liuku dam is the first of the thirteen dams to be built. It is also the smallest. The largest is the Songta dam with a wall height of 307 m and a reservoir volume of 6312 million cubic metres (Brown and Xu 2010).

Table 3.4 Summary of project characteristics (Brown and Xu 2010; Information Center of Xiaolangdi Multipurpose Project Construction and Administration Bureau at the Yellow River Conservancy Commission no year a)

Project	No of dams	Purpose	Dam height (m)	Reservoir volume	Installed capacity in MW	Average annual output (kWh)	No of re-locatees	Start of construction
Nu[47]	13	Hydro-power	35.5 (Liuku dam[48])	8.1 MCM (Liuku dam)	21,320	131.4 billion	50,000	June 2008 (Liuku dam)
Xiaolangdi	1	Multi-purpose	154	12.65 BCM	18,000	5.1 billion	200,000	September 1991; fully operational since 2001

3.4.1 Case Study 1: Environmental Impact Assessment of the Nu River Project

As the Nu is an international river, its development is a central government responsibility. The preliminary feasibility studies for this project were conducted as early as the 1970s; a development plan for the Nu River was presented in 1989, and, the National Development and Reform Commission adopted the Nu River Project in 1999 on the basis of its assessment of China's energy needs (Brown and Xu 2010). The General Institute of Water Resources and Hydropower Planning and Design then invited bids and awarded planning to the Beijing Survey and Design Institute and the East China Survey and Design Institute, who both suggested two reservoirs and thirteen dams (Mertha 2008, p. 117). While the two institutes drew up the general plan, they subcontracted the technical design of the dam at Liuku—the first dam to be built—to the Kunming Hydropower Survey and Design Institute (Magee 2006a, pp. 136–137, 171, 172 note 113).

It must be emphasised here that planning started well ahead of the reorganisation of the power sector in 2003, the revision of the Water Law of 2002 and the passing of the EIA Law in 2003. As mentioned above, five energy companies have had development rights for China's river basins since 2003. Since 2002 the basin commissions have been in charge of dam planning. For the Nu River, the commission in charge would have been the Changjiang Water Resources Commission. However, Magee reports that no comprehensive basin plan exists for the Nu, and the Commission was bypassed in that Huadian, the company in charge of Nu River development, referred the development plan directly to the National Development and Reform Commission (Magee 2006a, p. 250). In addition, the EIA Law would subject the Nu River Project to Strategic Environmental Assessment and project EIA studies (Qin 2008, p. 6).

The "real action" in developing the Nu began when the central government announced the reform of the energy sector in December 2002, splitting the State Power Corporation into five energy companies, among them the China Huadian

Corporation, which acquired the right to develop the Nu River (Dore and Yu 2004, p. 17). On 14 March 2003, before the official announcement of the project and before the State Council had given its approval, Huadian and the provincial government of Yunnan signed a letter of intent: the power station at Liuku would be the first to be built, construction to begin on 20 September 2003 (Mertha 2008, p. 117; Brown and Xu 2010 citing Zhou 2003). Engineering design for the dams at Maji, Bijiang, Yabiluo, Lushui, Saige and Yansangshu would also start in 2003 (Brown and Xu 2010 citing Beijing News 25 November 2003). By June 2003, it was reported that the Kunming Hydropower Survey and Design Institute "has completed the supplementary report of pre-feasibility study for Liuku project, and is conducting feasibility study on its installed capacity" (Living River Siam 2003).

In July 2003 the two design institutes submitted a report entitled *Nu River Middle and Lower Reaches Hydraulic Planning Report* to the National Development and Reform Commission. Mertha states that the report "was 'very simple' and contained no provisions on the impact of the Nu River Project on the environment" (Mertha 2008, p. 121). From 12 to 14 August 2003, the National Development and Reform Commission convened a meeting of around 140 people from government and research institutes, with opponents in the minority.[36] As usually happens at meetings of the National Development and Reform Commission, the planning report was approved "in principle",[37] a mere two weeks before the EIA Law was to enter into force in September.

Yunnan's Environmental Protection Bureau also supported the project, and Huadian tried to rush the State Council into approving the project before the EIA Law could come into effect. By then, however, Wang Yongchen and her NGO, Green Earth Volunteers, had got wind of the project and begun organising opposition. Encouraged by Wang, Mu Guangfeng of the State Environmental Protection Agency voiced opposition and moved to form a coalition of scientists and NGO activists, who managed to delay the project significantly (Mertha 2008, pp. 120, 121–122, 126). In September 2003, the State Environmental Protection Agency convened a panel of experts. In the same month, it announced "serious reservations" about, among other things, alterations to the World Heritage Site, the protection of the unexplored biodiversity, the loss of the wild rice gene pool, geological instability and the potential for poverty alleviation that might also be achieved through ecotourism (Dore and Yu 2004, p. 17). In late September, provincial and prefecture governments also convened panels of experts, who presented contrary opinions (Dore and Yu 2004, p. 17).

On 1 October 2003, Yu Xiaogang and his NGO, Green Watershed, conducted a survey of the Nu valley (Mertha 2008, p. 120). On 25 October 2003, Wang Yongchen's Green Earth Volunteers organised a petition, which was signed by 62 people from the media, arts, science and environmentalist communities at the second meeting of the China Environment and Promotion Society, and this was

[36] Interview in B14112008.

[37] Interview in K12122008.

publicised by the media (Mertha 2008, p. 119). In November 2003, Green Earth Volunteers and other environmental organisations attended the third meeting of the China-United States Environment Forum and succeeded in having the Nu River Project debated. The meeting let to a "diffusion of opposition to the NRP [Nu River Project] throughout China's NGO network" (Mertha 2008, p. 120).

To boost the project, Xie Yi (Chairman of Nujiang Prefecture's Communist Party Committee) and Ou Zhiming (Governor of Nujiang Prefecture) met officials of the National Development and Reform Commission, the Ministry of Water Resources, the State Environmental Protection Agency, the Water Resources Planning Academy, the Ministry of Transport and the State Ethical Affairs Commission in Beijing on 22 September 2003. At this meeting, Xiaoshaba at the Liuku dam was identified as the first village whose inhabitants would be resettled (Brown and Xu 2010 citing Beijing News, 25 November 2003).

In early 2004, Shen Xiaohui, a senior researcher at the State Forestry Bureau, submitted a letter to the National People's Congress and the Chinese People's Political Consultative Conference (CPPCC) with the help of Liang Congjie, a CPPCC member and head of the NGO Friends of Nature (Mertha 2008, p. 121, citing Deng 2005). This established a link between researchers, NGO activists and politicians. Shen Xiaohui prepared two proposals for the forthcoming CPPCC meeting, where they were presented by Liang Congjie (Mertha 2008, p. 126). Having attracted political attention in Beijing, the project came to a standstill. Huadian was forced to conduct an EIA and submit it to the State Environmental Protection Agency (Mertha 2008, p. 122, citing Cheung 2004).

On 18 February 2004 Prime Minister Wen Jiabao put the project on hold on the grounds that EIA procedures had not been properly followed. In February 2004 twenty journalists, environmental activists, scholars and other experts led by Wang Yongchen visited the thirteen proposed dam sites in the guise of tourists and subsequently exhibited photographs in Beijing in March. One week before the exhibition, they launched a bilingual website on the Nu River Project, www.nujiang.ngo.cn (Mertha 2008, pp. 125–126).

In May 2004, Green Watershed's Yu Xiaogang filmed the resettlement areas of the Manwan and Xiaowan power stations on the Mekong River, together with community leaders from Xiaoshaba village. VCDs and DVDs showing the squalor of the resettlement communities were made and subsequently distributed by peasant activists throughout Yunnan and also circulated among NGOs in Beijing (Mertha 2008, pp. 127–130).[38]

From 27 to 29 October 2004 the United Nations Symposium on Hydropower and Sustainable Development was held in Beijing. To the dismay of Chinese officials attending the conference, Yu Xiaogang was accompanied by two peasant activists, one of whom gave a speech to the audience. The speech was also included in the conference proceedings and therefore made official (Mertha 2008,

[38] It may be mentioned here that Beijing-based NGOs are lobbyists rather than activists, whereas NGOs in the field are more activist (Interview K12122008).

pp. 130–133). In December 2004 the National Development and Reform Commission joined the opposition to the project, and its Vice-Director, Zhang Guobao, called for plans with less environmental impact (Mertha 2008, p. 133).

The State Environmental Protection Agency's Vice-Director, Pan Yue, then announced public hearings under the 2003 EIA Law, while in March 2005 Wang Shucheng, Minister of Water Resources, announced at the session of the National People's Congress that work on four dams would go ahead. In June 2005 a meeting of pro-dam experts in Yunnan submitted a report to Prime Minister Wen Jianbao. The State Council responded by saying that the report had to be revised, but without detailing what should be changed (Mertha 2008, pp. 133–134).

On 25 August 2005 61 organisations and 99 individuals signed an open letter entitled *Petition for Disclosing the EIA Report on the Nujiang River Hydropower Project in accordance with Legal Requirements.*[39] The number of signatories increased exponentially, reaching a total of 84 organisations and 331 individuals by 27 September 2005 (Qin 2008, p. 6; Haggert 2005a, b).

On 13 November 2005 the National Development and Reform Commission convened a panel of experts in Beijing, "but key members of the opposition did not receive their invitations until 2 days before the panel was to begin." The panel members were shown EIAs carried out by the East China Survey and Design Institute [which is part of the China Hydropower Engineering Consulting Group Corporation] and Guodian Corporation's Beijing Survey and Design Institute, but only briefly, giving them no opportunity for a detailed study (Mertha 2008, pp. 139–140 citing Deng 2005 and Bezlova 2005).

On 11 January 2006 the National Development and Reform Commission published a decision that four of the eleven dams would be built. At the time, the EIA Report had been completed and approved by the State Environmental Protection Agency. Conclusions from the EIA Report leaked out, confirming that, in order to limit environmental damage, the number of dams was to be decreased from thirteen to four (Maji, Liuku, Yabiluo and Saige). This would also reduce the number of people to be resettled (Yardley 2006). Not many details are known since the EIA Law expressly allows EIA Reports to be withheld if they are relevant to security. As the Nu is an international river, the Ministry of Water Resources and the State Secrets Bureau argued that this was the case (Yardley 2006).

Significantly, the *Wen Wei Po* newspaper, which is based in Hong Kong but has links to the Chinese Communist Party, reported that four dams would be developed first, because they were less controversial, while the others could be built after improved planning and investigations concerning soil preservation, geological impacts and protection of cultural relics had been completed (Wen Wei Po 2006).

[39] *Tiqing yifa gongshi Nujiang shuidian huanping baogao* (Call for public disclosure of Nujiang hydropower development's EIA report in accordance with the law). The report can be found at http://www.green-web.org/infocenter/show.php?id=17595. Accessed 19 June 2013. For an English translation, see http://old.probeinternational.org/catalog/contentfullstory.php?contentId=2726&catid=24. Accessed 19 June 2013.

On 18 March 2008 the National Development and Reform Commission released its Plan on the Development of Renewable Energy during the 11th Five-Year Plan Period following State Council approval. The document announced that construction of the dams at Liuku and Saige would begin before 2010. However, construction of the Liuku dam started as early as June 2008, and with a few exceptions the inhabitants of Xiaoshaba village were resettled in 2007 (Brown and Xu 2010 citing Shi 2008). In January 2011, Shi Lishan of China's National Energy Administration pointed out that damming on the Nu River would form part of the energy targets of the 12th Five-Year Plan and China's emissions reduction targets for 2020 (Si 2011). On 23 January 2013, China's State Council published the 12th Five-Year Plan for Energy Development, outlining plans to begin the construction of more than 50 hydropower dams before the year 2015. For the Nu River, the plan singles out the dams at Liuku, Maji, Yabiluo and Saige (State Council 2013).

Overall, then, the Nu River case presents a mixed picture. While the secretive and *in camera* decision-making processes of the Chinese political-administrative system are still prevalent, the system has also become porous. In an attempt to maintain its legitimacy, the party-state establishes decision-making mechanisms for non-governmental organisations and individuals. While this is not designed to change the authoritarian relationship between state and society by introducing democratic procedures, civil society organisations and individuals have shown that the openings in the political system can be exploited to organise effective protests without the risk of a repressive state response.

3.4.2 Case Study 2: Resettlement at the Xiaolangdi Multipurpose Dam Project

The Xiaolangdi hydropower station is situated on the Yellow River (Huanghe) at Jiyuan in the Province of Henan. It is a World Bank-financed multipurpose project for flood control, prevention of ice jams in the lower reaches of the Yellow River, siltation control, the provision of water for irrigation and consumption in cities and by industries and power generation to replace thermal power in the Central China Power Grid (World Bank 1997). In 2000, when the WCD published its report, Xiaolangdi became operational and resettlement was in full swing. While Xiaolangdi was reviewed critically by some Chinese researchers, the criticism was not linked to the recommendations in the WCD report.

The World Bank approved three components: the Xiaolangdi Multipurpose Project, Xiaolangdi Multipurpose Project Stage II and the Xiaolangdi Resettlement Project for Reservoir Phases I and II. Resettlement was thus treated as a separate component, with a budget of its own.[40] Resettlement at Xiaolangdi was divided

[40] For detailed information on Xiaolangdi financing, see the World Bank pages on Xiaolangdi at http://web.worldbank.org/external/projects/main?menuPK=224076&pagePK=218616&piPK=217470&theSitePK=40941&query=Xiaolangdi. Accessed 19 June 2013.

into four phases, to be completed between 1992 and 2011: Damsite phase resettlement (dam construction and service areas) and Reservoir Phase resettlement I, II and III (during project planning, phase III resettlement, initially planned to take place between 2010 and 2011, was moved forward to the period 2002–2004) (World Bank 2007, pp. 2, 6). A total of 184,080 people had to be moved, up from an initially estimated number of 180,662 (World Bank 2007, pp. 2, 6). The 130 km long reservoir flooded “7.25 million square meters of housing, 381,870 mu[41] of land including 181,709 irrigated land, 12 mini hydrostations, 658 km of canals, 688 km of roads, 548 km of communication lines and 109 cultural relics sites.” (Information Center of Xiaolangdi Multipurpose Project Construction and Administration Bureau at the Yellow River Conservancy Commission no year b).

In addition, the resettlement programme required the relocation of 789 manufacturing and mining companies. The flooding affected eight counties with 29 towns and 174 villages. Resettlement affected a population of 545,000 people in the host area, including 397 villages and 59 townships in the counties of Yima, Mengzhou, Wenxian, Yuanyang, Zhongmou and Kaifeng (Chen et al. 2004, p. 207); Information Center of Xiaolangdi Multipurpose Project Construction and Administration Bureau at the Yellow River Conservancy Commission no year b). After the Three Gorges Project, it is the second largest project in China in terms of population relocation (Shi et al. 2006, p. 41). The resettlement project meant to include the construction of all infrastructure for the resettled population, “including housing for 276 villages and ten towns”; it was to develop “11,100 hectares of land with 7,000 to be irrigated partly by the reservoir”; “252 industries and mines will be relocated, 84 industries will be established for the employment of 20,500 resettlers” (van Wicklin 1999, p. 238).

When borrowing from the World Bank, China must comply with its standards. The first World Bank policy on resettlement was published in 1980 as Operational Manual Statement (OMS) 2.33, entitled Social Issues Associated with Involuntary Resettlement in Bank-Financed Projects. The policy was updated in 1990 as Operational Directive 4.30 and reissued in December 2001 as Operational Policy and Bank Procedures 4.12.[42] One of the World Bank’s major goals is to ensure that benefits are shared between the operator and the communities affected, including relocatees and host communities. To promote benefit-sharing, rehabilitation packages should be linked to the benefits generated by the project (van Wicklin 1999, pp. 234–235). In its 1990 revision of resettlement regulations, the World Bank included the principle of benefit-sharing in its Operational Directive 4.30 (van Wicklin 1999, p. 236).

The agency responsible for building and operating the Xiaolangdi dam is the Xiaolangdi Water Resources Construction and Management Bureau, which reports

[41] 1 mu is equal to 666 2/3 square metres.

[42] For the World Bank’s involuntary resettlement website, see http://go.worldbank.org/MRNITY6XN0. Accessed 19 June 2013. For an overview of all World Bank Safeguard Policies, see http://go.worldbank.org/WTA1ODE7T0. Accessed 19 June 2013.

to the Ministry of Water Resources. Established in September 1991, it is the project owner for Xiaolangdi. It coordinated the resettlement and land acquisition activities of the provincial Project Management Offices of Henan and Shanxi Provinces. It has five subsidiaries: Xiaolangdi Engineering Consulting Corporation to supervise construction, Xiaolangdi Water Resources and Hydropower Engineering Corporation for construction, Xiaolangdi Resettlement Bureau to manage resettlement, Xiaolangdi Tourism Corporation and Xiaolangdi Hydropower Plant. The general manager reports to the Ministry of Water Resources. Consequently, the latter has overall responsibility. A Central Project Leading Group under the State Council was formed to oversee project management, resolve policy issues and solve interdepartmental/interagency problems. The leading group consists of a vice-minister of the Ministry of Water Resources and directors from the Ministry of Finance, the National Development and Reform Commission and vice-governors of Henan and Shanxi Provinces (Information Center of Xiaolangdi Multipurpose Project Construction and Administration Bureau at the Yellow River Conservancy Commission no year c; World Bank 1997).

Resettlement necessitated by the project is managed by the Xiaolangdi Resettlement Bureau. It is supervised by the Resettlement Bureau of the Ministry of Water Resources. The Xiaolangdi Resettlement Bureau is responsible for general management, the organisation of the overall resettlement plan and the individual projects forming part of it, supervision, monitoring, approval, the disbursement of funds, auditing and the training of relocatees. The Provinces of Henan and Shanxi and the resettlement bureaus of the cities and counties directly affected by the flooding form a Leading Small Group, which is responsible for the resettlement work in the area under its jurisdiction (Chen et al. 2004, p. 213; Information Center of Xiaolangdi Multipurpose Project Construction and Administration Bureau at the Yellow River Conservancy Commission no year b, c).

The implementation of resettlement is then the responsibility of the counties affected in the Provinces of Henan and Shanxi. The funds are disbursed on a lumpsum basis under the direction of the Ministry of Water Resources and the supervision of the project owner: the Ministry transfers the funds to the provincial resettlement office, which passes them on to the local resettlement offices. They then allocate the payments directly to the villages. The resettlement planning and implementation process can be described as follows[43]:

The project owner entrusted (1) the Planning and Design Institute of the Yellow River Conservancy Commission with the task of drawing up the resettlement plan; (2) the Resettlement Bureau of the Yellow River Conservancy Commission supervised resettlement; (3) the Office for Resettlement Supervision of the Beijing Water Resources and Hydropower Institute was made responsible for monitoring and evaluation; (4) at the same time, the World Bank set up a panel of environment

[43] The following account follows the project developer's description of resettlement: Information Center of Xiaolangdi Multipurpose Project Construction and Administration Bureau at the Yellow River Conservancy Commission (no year c), Chen et al. (2004, pp. 208–210), Shi et al. (2006).

and resettlement experts as an advisory body. The loan agreement between the Chinese government and the World Bank stipulated that the project owner would put the Office for Resettlement Monitoring and Evaluation of the Beijing Water Resources and Hydropower Institute in charge of monitoring the resettlement process and the progress of post-relocation rehabilitation. The Office also evaluated test runs undertaken to prepare enterprises for relocation. It liaised regularly with the World Bank and the relevant departments and bureaus on the various levels of government.

Work on the initial design for Xiaolangdi resettlement began in 1986 and included an initial socio-economic survey. In 1994, the Design Institute of the Yellow River Conservancy Commission together with the provincial resettlement office, the land acquisition section of the local resettlement office and village representatives conducted an in-depth socio-economic survey in each county and for each household in order to study the future loss of collective land and property during Reservoir Phases I, II, and III. The results provided the basis for the planning and design of the resettlement areas and the levels of compensation. The survey covered the amount of land resources, income levels and local customs.

The results of the survey led the authorities to conclude that most of the future relocatees engaged in farming would not be prepared to subsist in a market economy, had poor technical skills and lacked the ability to take up non-agricultural lines of work. The Design Institute therefore emphasised the primacy of agricultural resettlement (Shi et al. 2006, p. 43). In the survey, observers referred to the case of Yunlin village in Gucheng town, Yuanqu County, where the investigators joined with the Village Committee in appraising each house and its associated land and each enterprise and infrastructure, including roads and communication links. When the Village Committee published the survey in 1997, villagers detected mistakes and omissions. For example, brick-wood houses had been registered as earth-wood houses. Five households were not registered at all. However, observers attributed the errors to changes that had occurred in the village between the survey and its publication. The villagers notified the Village Committee, and the Village Committee notified the town and county resettlement offices. In 2000 the Design Institute of the Yellow River Conservancy Commission and those responsible in the county and town administrations corrected the errors, and compensation was first paid to the county resettlement office, which then distributed the money to the relocatees (Chen et al. 2004, p. 208). The apparent thoroughness of the investigation teams accorded with the World Bank's seventeen resettlement criteria for Xiaolangdi, which state inter alia that resettled communities should accept the resettlement plans and that there should be compensation for flooded infrastructure so that the "previous level of service is maintained or improved" (World Bank 2007, pp. 43–44).

The Design Institute of the Yellow River Conservancy Commission, representatives of Village Committees from originating and host areas and representatives of the local resettlement offices were responsible for the selection and planning of the resettlement locations. Villagers were able to object to the sites chosen, but where they disagreed among themselves, the local branch of the Communist Party and the Village Committee could decide for them (Chen et al. 2004, p. 209).

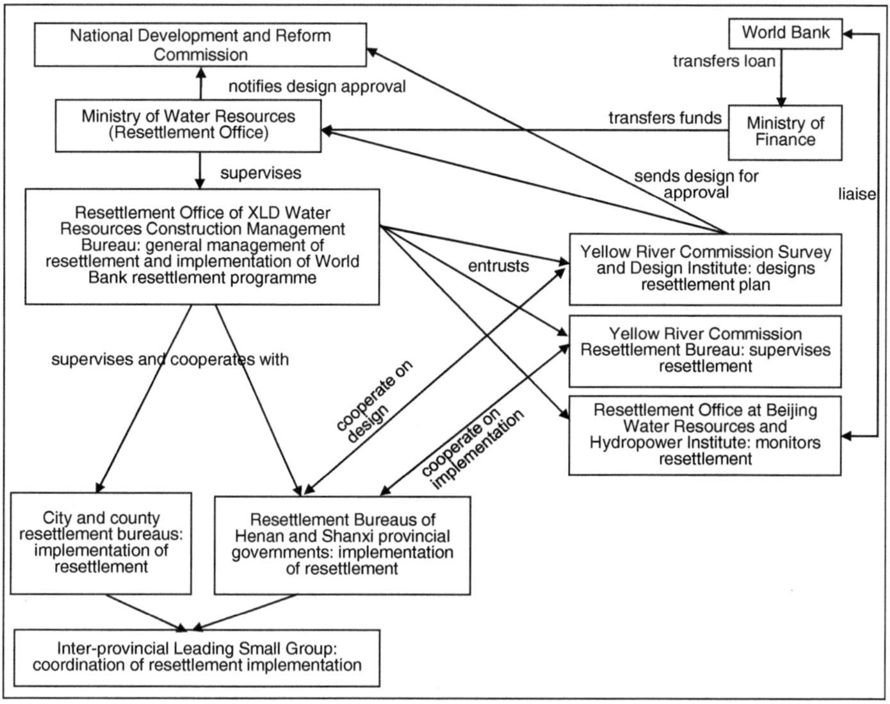

Fig. 3.2 Xiaolangdi resettlement management (author's own compilation adapted from Information Center of Xiaolangdi Multipurpose Project Construction and Administration Bureau at the Yellow River Conservancy Commission no year c)

After the resettlement sites were selected, a residential site plan was drawn up by the same institutions plus the host villages and local government departments. Land was then allocated for general use, production and residential use, with villagers ideally being assembled by county and village governments so that they might have a say in the land they were allocated. Where compensation was concerned, county-level resettlement teams issued guidelines stipulating the procedures to be followed. Chen et al. (2004, p. 210) report that in Yuanqu County, for example, the guidelines provided for compensation of up to 50,000 Renminbi to be decided at meetings of village representatives. Compensation payments above that threshold were to be decided first by village representatives, who were required to present a project proposal to the Village Committee. The proposal then had to be approved by the township government and the county resettlement office (see Fig. 3.2, which summarises the decision-making processes).

The host area for the relocatees was known as Wenmengtan.[44] It lies 20 km downstream from the dam site and is adjacent to Xin'an County. Xin'an County

[44] The name is an amalgam of the names of Wen County in the municipal limits of Mengzhou City, plus the Chinese name for the sandy beaches of the Yellow River (*tan*) at that place.

was initially chosen as the main host area and was to absorb 47,000 people despite its already limited environmental capacity and lack of adequate drinking water for 60,000–70,000 of the original inhabitants. Given the limitations of Xin'an, it was decided to develop Wenmengtan into a further host area in order to provide each Xiaolangdi resettler with an average of 667 square metres of arable land. The allocation size was calculated according to land productivity, per capita grain consumption and other factors established during the socio-economic survey (Shi et al. 2006, pp. 45–46).

In 1990, to advance the plan, the Resettlement Bureau and Leading Small Group for the Coordination of Xiaolangdi Engineering of the Province of Henan and the Planning and Design Institute of the Yellow River Conservancy Commission put forward a plan for river works and soil improvement in the Wenmengtan area. To this end, the Yellow River Conservancy Commission's Design Institute, the Design Institute of the Henan Yellow River Affairs Bureau and the Resettlement Bureau of the Province of Henan began work on the initial design of the Wenmengtan resettlement plan. In 1993 the State Planning Commission approved this initial design. Construction began in 1994. In December 2003 the river engineering works and soil improvement measures in the Wenmengtan area passed the completion check of the Ministry of Water Resources.

Despite the World Bank's involvement, the people directly affected were not involved in the identification of needs and problems (Plummer and Taylor 2004, pp. 43, 51–52. For a critical review of the resettlement process see also McDonald and Webber 2002; Webber and McDonald 2004), but they were able to express their opinions during the planning stage on the suitability of new farmland areas proposed by the Design Institute. However, in case of disagreement or lengthy debate among the villagers, the Village Committee was allowed to make the decision. Errors in registration by the Design Institute resulted in (initial) losses of compensation (Chen et al. 2004, pp. 209–210), and relocatees were "largely unaware of the benefits of participation, and remained passive" (Chen et al. 2004, p. 213). At that time, during the design stage, villagers were merely notified of the results of the design and how this would affect them, with no chance of further interaction; resettlement then proceeded without further consultation (Plummer and Taylor 2004, pp. 43, 51–52). This often resulted in villagers being dissatisfied with the quality of the land allocated to them and the distance it lay from the new settlement (Chen et al. 2004, p. 213). During post-relocation, however, some Village Committees provided further support: for instance, the villagers relocated from Hedi village in Yuanqu county cultivated the new farmland collectively for one year, after which the Village Committee set up an allocation team to evaluate the farmland and divide it into plots on the basis of productivity, soil type, soil quality, water source and distance from the settlement. The plots were then allocated to the villagers (Chen et al. 2004, p. 210).

3.5 Findings and Conclusions

In China's strategic hydropower development setting, the World Commission on Dams' relevance is difficult to discern since it exercises its influence at best indirectly through international organisations such as the World Bank and the ADB, industry representatives such as the IHA, and NGOs. Although past and current resettlement and EIA/SEA reforms have increased the opportunities for NGOs and individuals to make themselves heard, Chinese strategic decision-makers, particularly those in the Ministries of Industry and Water Resources and the powerful National Development and Reform Commission, eschew strategic guidelines that would interfere with top-down decision-making, such as the WCD principle of Gaining Public Acceptance. The adoption of participatory standards is designed to improve transparency of state decision-making at the lower levels of government and help to enforce national legislation where local government refuses to do so. This increases regime legitimacy. The aim of enhancing legislation on participation is not the introduction of democracy on any level of government.

The introduction and reform of resettlement and EIA are therefore most importantly the result of domestic policy learning and the invigoration of the societal level. The former is clear from the fact that reforms in the resettlement and environment spheres began with the onset of the reform period in 1978/79 (environmental reforms certainly being a very protracted process). Looking back at the 1950s, Bizer and Ragsdale (1997) argue that China had resettlement policies in place, but the radicalisation of the political realm since the mid-1950s made their implementation impossible. Rather, resettlement in the context of hydropower construction was seen as a personal sacrifice that relocatees had to make for the sake of the country's development.[45] This changed after 1979. As for the invigoration of the societal level after 1979, this development should be viewed with caution. Rather than allowing the populace greater freedom of expression, the growing opportunities for action by individuals and NGOs are the result of a conscious effort by central government to control local government. The rationale is therefore instrumental.

However, these societal processes increasingly take on self-enforcing dynamics that are difficult for the central government to control. This is partly because some NGOs manage to form alliances with media representatives and members of such political bodies as the National People's Congress and the Chinese People's Political Consultative Conference. Furthermore, such new technologies as the Internet make it easier for individuals to become informed and to protest spontaneously. The government reacts to these dynamics, because it has realised that growing popular dissatisfaction with corruption and environmental pollution has the potential to disturb political stability.

---

[45] Interview N15122008.

It is important to note that, in all this, Chinese NGOs and individuals tend to focus on pressuring central and local government to enforce national legislation. Should the Chinese government be disinclined, for instance, to adopt certain international modes of participation, Chinese NGOs usually abstain from such recommendations given the NGOs' precarious existence. As a rule, protests by NGOs or individual citizens have a bearing on the enforcement of existing legislation rather than the initiation of new laws. Yet NGOs also often set up pilot projects which, if successful, are later adopted by local government (Zawahri and Hensengerth 2012). The Nu River is perhaps the clearest example of NGOs succeeding not only in drawing national attention to environmental protection, but also in forcing government agencies and project developers to engage in an environmental appraisal process.

In addition to the strong domestic reform process, there are international influences. Of direct relevance to dam-building, they are discernible in ADB technical assistance programmes, World Bank financing and China's continued involvement in the IHA-HSAF process until it ended in February 2009 with the publication of the IHA Draft Sustainability Assessment Protocol. Indeed, Qin (2009, p. 2) argues that the 2006 amendments to the Rules of Land Compensation and People Resettlement in Medium and Large Hydroelectric Projects "have incorporated the core requirements of the World Bank Guidelines on Non-voluntary Resettlement." The ADB and World Bank offer practical examples of international influence on Chinese law, and the IHA-HSAF process is an example of direct Chinese participation in norm-building.

Where WCD recommendations have been incorporated into Chinese law, they have most likely been transmitted through World Bank and ADB procedures. Given that the World Bank itself has adopted a cautious attitude towards the WCD, the effect of those recommendations would be difficult to measure. It is more a matter of reforms of the World Bank and ADB resettlement rules finding their way into Chinese legislation. Determining how these reforms are triggered and whether WCD norms have been integrated into World Bank and ADB resettlement guidelines would require not only a study of WCD influence on the World Bank and ADB, but a thoroughgoing examination of the reform processes of the two banks, the persons involved, their networks and their professional background. This goes beyond the scope of this study, but will be fertile ground for future research.

IHA is another international organisation that may be influential, first, because it refers to WCD recommendations, and second, because the Chinese government is involved in the HSAF process. The IHA Draft Hydropower Sustainability Assessment Protocol of February 2009 is perhaps the most transparent way of identifying WCD influence: attached to the draft protocol is a detailed analysis showing which sections of the draft protocol correspond to the WCD's seven Strategic Priorities. As the finalized protocol, published in November 2010, is now in a stage of institutional consolidation, future analysis will have to ascertain how influential the protocol is in China.

Other international influences come from the Aarhus Convention and model legislation in the EU and the United States, and are transmitted through the EU-China Basin Development Programme, Chinese study groups and a range of bi- and multilateral meetings.

Apart from institutions of direct relevance to Chinese dam legislation, other international norm setters are of importance, but their actual influence requires further examination. In January 2008, for example, the Ministry of Environmental Protection announced cooperation with the International Finance Corporation (IFC) in implementing the Equator Principles in China with a view to spreading a concept known as Green Credit. The Equator Principles were developed by a group of ten banks. They are based on World Bank and IFC policies and require compliance with IFC Performance Standards. While the Equator Principles are certainly important for dam project finance, only non-Chinese banks have so far adopted them in respect of dams, an example being the British bank HSBC in its 2005 Freshwater Lending Guideline. In November 2008, the China Industrial Bank was the first Chinese financial institution to announce that it would adopt the Equator Principles. In addition, such Chinese companies as the China Development Bank and Hanenergy have signed up to the UN Global Compact, a UN initiative to which companies can subscribe that monitors companies' (voluntary) adherence to ten principles in the four areas of human rights, labour, environment and anti-corruption. Other Chinese companies are certified under ISO 14001, among them Sinohydro Engineering Bureau No 1 (Corporate Social Responsibility and Sustainability News for China 2008; Brewer 2008, p. 21). A question that remains is whether this is a mere façade or a generic interest in corporate social responsibility. In fact, it can be argued that Chinese companies subscribe to these principles and initiatives because they are investing abroad and wish to be perceived as responsible international players.

Central government is facing a dilemma: it has to cope with increasing energy demand to keep the economy going and so maintain domestic stability by creating jobs for school and university graduates (Information Office of the State Council of the People's Republic of China 2004, Part VI). At the same time, there is growing popular demand for environmental protection. President Hu Jintao's and Prime Minister Wen Jiabao's priority of economic development is therefore increasingly geared to awareness of the social and environmental fallout. Figure 3.3 depicts the forces and actors that influence governmental dam decision-making.

Figure 3.3 shows central government influenced by domestic and international processes. Internationally, the WCD arguably has some influence on World Bank safeguards, ADB regulations and IFC Performance Standards. When China borrows from these three institutions, it has to adhere to their regulations, which will ideally take into account WCD norms.

There is also an exchange of information and expertise between IHA, World Bank and Equator Bank/IFC representatives.[46] Domestically, there are two

[46] Interview L151209.

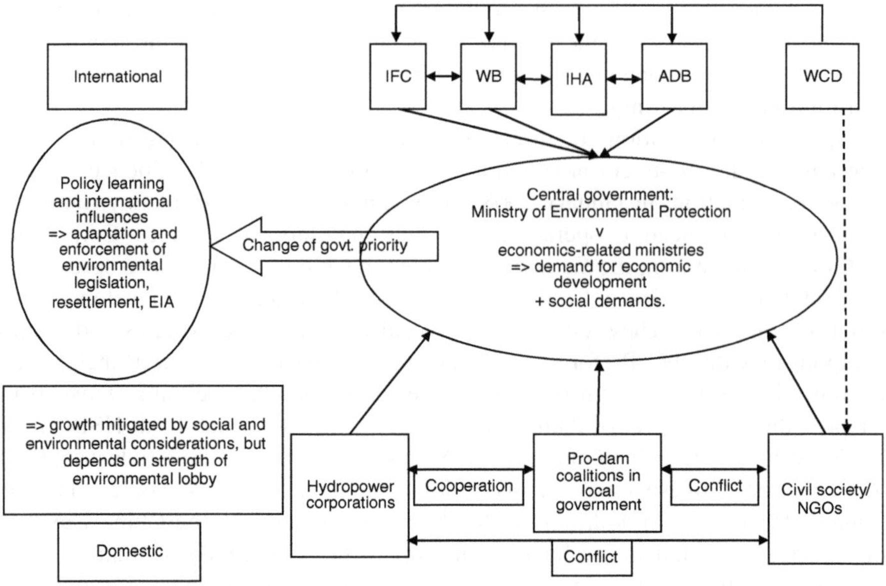

Fig. 3.3 Influences on governmental dam decision-making in China (author's own compilation)

influences: a progressive influence from a civil society seeking the opening of the political system (even if such an agenda is not officially put forward); and a conservative influence from local government and hydropower corporations, which often cooperate in an attempt to circumvent national environmental and participation legislation. As a consequence, progressive and conservative actors are in conflict with each other. It can be assumed that the WCD has an influence on civil society but not ascertained, as this was not investigated as part of the study.

As response to these influences, the government adopts progressive legislation depending on the credibility of international monitoring mechanisms and the power alignment of domestic actors. Among the domestic actors, NGOs and individual citizens, encouraged by a sense in the leadership of the urgency of the need to clean up the environment, have increased the scope of their action against the inertia of conservative actors, as the case of the Nu River shows. Yet the Nu River case also highlights the central problem in these relationships: different actors have different agendas. Conflicts are often perceived as zero-sum games. Whether or not this blocks domestic decision-making processes depends, first, on the power asymmetries and, second, on the conflict management of the stakeholders involved.

Consequently, rather than the adoption of international norms by Chinese policy-makers being attributed to outside pressure, it should be emphasised that a growing number of Beijing's politicians have taken note of the threat to their hold on power from popular anger about the degradation of the environment. China's participation in international institutions and its adoption of participatory procedures depend on their usefulness in legitimising the regime. Chinese policy-makers are not interested in ideological debates with the international community. They are pragmatic. Their interest is in achieving rapid results able to spur on certain domestic reforms, to drive the economy and to secure energy supplies. This should be borne in mind by those who attempt to engage China in international developmental processes.

Acknowledgments I am indebted to Prof Shi Guoqing and Prof Chen Shaojun at the National Research Centre for Resettlement at Hohai University for their invaluable help in assisting this study and for hosting me for 3 months at their Centre. I am also especially grateful to: Jörg Hartmann, WWF Dam Initiative Leader; Prof. Zhang Shiqiu, Institute of Environment and Economy, Beijing University; and Zhang Hui, Ministry of Environmental Protection.

References

ADB (Asian Development Bank) (1996) Capacity building for natural resource legislation: technical assistance 2735-PRC, Manila

ADB (Asian Development Bank) (1998a) Report and recommendation of the president to the board of directors on a proposed loan and technical assistance grant to the People's Republic of China for the Yunnan Dachaoshan Power Transmission project, Manila

ADB (Asian Development Bank) (1998b) Review of national resettlement policies and experience with involuntary resettlement projects: regional technical assistance 5781, Manila

ADB (Asian Development Bank) (2000) National resettlement policy enhancement and capacity building: regional technical assistance 5935, Manila

ADB (Asian Development Bank) (2002) People's Republic of China: national resettlement policy enhancement and capacity building: country report for the regional workshop at ADB headquarters in Manila, 18–20 Feb 2002, prepared by the National Research Center for Resettlement at Hohai University, Nanjing, for the Ministry of Land and Resources

Beijing News (2003) Nujiang 13 ji shuiba reng yao shangma [13 Nujiang dams will go ahead], 25 Nov. http://news.thebeijingnews.com/0547/2003/%2D1125/002@1436.htm. Accessed 19 June 2013

Bezlova A (2005) Let public see secret mega-dam plans. Inter Press Service and Global Information Network, 26 Oct 2005

Bizer J, Ragsdale T (1997) The development of resettlement policy in China: two case studies in Sichuan and Hebei provinces, s. l., Multinational Environmental and Social Assessment Services

Brewer N (2008) The new great walls: a guide to China's overseas dam industry. International Rivers Network, Berkeley. http://www.internationalrivers.org/files/attached-files/intlrivers_newgreatwalls_2012_0.pdf. Accessed 19 June 2013

Brown P, Xu Y (2010) Hydropower development and resettlement policy on China's Nu river. J Contemp China 19(66):777–797

Cameron A, Luo W (2012) An environmental impact assessment for hydropower development in China. Vermont J Environ Law 14(2):275–302

Chen S, Shi G, Zhu W et al (2004) Participatory resettlement: key issues and processes in resettlement in Xiaolangdi and Hexi. In: Plummer J, Taylor JG (eds) Community participation in China: issues and processes for capacity building. Earthscan, London, pp 200–229

Cheung R (2004) Assessment of dam project urged. In: South China Morning Post 12 March 2004

China Daily (2003) Constitutional amendments endorsed, 14 March 2003. http://www.chinadaily.com.cn/english/doc/2004-03/14/content_314719.htm. Accessed 19 June 2013

Corporate Social Responsibility and Sustainability News for China (2008) China's Industrial Bank to adopt Equator principles, 6 Nov. 2008. http://www.chinacsr.com/en/2008/11/06/3544-chinas-industrial-bank-to-adopt-equator-principles/. Accessed 19 June 2013

Cui M (2006) (General Editor), Zhongguo nengyuan fazhan baogao: Nengyuan lanpi shu [The Energy Development Report of China: Blue Book of Energy], Beijing: Shehui kexue wenxian chubanshe

Deng J (2005) Environmental protection's new power is growing. In: Nanfang Zhoumo [Southern Weekend], 27 Dec 2005

Dore J, Yu X (2004) Yunnan hydropower expansion: update on China's energy industry reforms k the Nu, Lancang & Jinsha Hydropower Dams, Kunming, PR of China: Chiang Mai University's Unit for Social & Environmental Research and Green Watershed (Working Paper) March 2004

Downs E (2006) Energy security series: China. Brookings Institution, Washington, DC

Economy E (2004) The River runs black: the environmental challenge to China's future. Cornell University Press, London

Energy Information Administration (2006) Country analysis briefs: China. http://www.eia.gov/countries/analysisbriefs/China/china.pdf. Accessed 19 June 2013)

Gough K (2004) Emerging civil society in China: an overall assessment of conditions and possibilities available to civil society and its organisations to act in China. SIDA, Stockholm. http://www.sida.se/Publications/Import/pdf/sv/Emerging-Civil-Society-in-China-An-overall-assessment-of-conditions-and-possibilities-available-to-civil-society-and-its-organisations-to-act-in-China_1719.pdf. Accessed 19 June 2013

Haggert K (2005a) Open letter puts pressure on Beijing over secretive dam plans. In: Three Gorges Probe 7 Sep 2005. http://journal.probeinternational.org/2005/09/07/open-letter-puts-pressure-beijing-over-secretive-dam-plans/. Accessed 19 June 2013

Haggert K (2005b) Nu river campaign gathers steam. In: Three Gorges Probe 27 Sep 2005. http://journal.probeinternational.org/2005/09/27/nu-river-campaign-gathers-steam/. Accessed 19 June 2013

Harris PG (2006) Environmental perspectives and behavior in China: synopsis and bibliography. Environ Behav 38(5):5–21

Heggelund G (2004) Environment and resettlement politics in China: the three gorges project. Ashgate, Aldershot

Heggelund G (2006) Resettlement programmes and environmental capacity in the three gorges dam project. Dev Change 37(1)179–199 (page numbers quoted in the text above follow the post-print version. http://www.fni.no/doc&pdf/gmh-dc-2006.PDF. Accessed 19 June 2013

Hirsch P et al (2006) National interests and transboundary water governance in the Mekong. Australian Mekong Resource Centre, University of Sydney, Danish International Development Assistance, Sydney

Ho P (2005) Institutions in transition: land ownership, property rights, and social conflict in China. Oxford University Press, Oxford

Ho P, Edmonds RL (2007) Perspectives of time and change: rethinking embedded environmental activism in China. China Inf 21(2):331–344

IHA (International Hydropower Association) (2009) Hydropower Sustainablity Assessment Forum: progress update and invitation to provide feedback, Jan 2009. http://www.hydropower.org/sustainable_hydropower/HSAF-Phase_1_Consultation/HSAF-5_page_consultation_document.pdf. Accessed 19 June 2013

Information Center of Xiaolangdi Multipurpose Project Construction and Administration Bureau at the Yellow River Conservancy Commission (no year a) Project introduction

Information Center of Xiaolangdi Multipurpose Project Construction and Administration Bureau at the Yellow River Conservancy Commission (no year b) Resettlement

Information Center of Xiaolangdi Multipurpose Project Construction and Administration Bureau at the Yellow River Conservancy Commission (no year c) Yimin anzhi [Resettlement]

Information Office of the State Council of the People's Republic of China (2004) China's employment situation and policies, Beijing. http://www.china.org.cn/e-white/20040426/index.htm. Accessed 19 June 2013
International Rivers Network (2005) Nu river dams: approval expected in coming months, 6 Sept 2005. IRN Press Release http://www.internationalrivers.org/resources/nu-river-dams-approval-expected-in-coming-months-3894. Accessed 19 June 2013
Lampton DM (1987) Chinese politics: the bargaining treadmill. Issues Stud 23(3):11–41
Lee N (2007) The development of environmental NGOs in China: a road to civil society? Jamestown Foundation China Brief 6(23)
Lieberthal KG, Lampton DM (eds) (1992) Bureaucracy, politics, and decision making in post-Mao China. University of California Press, Oxford
Lieberthal KG, Oksenberg M (1988) Policy making in China: leaders, structures, and processes. Princeton University Press, NJ
Living River Siam (2003) China Huadian corp to lead Nu river (upper Salween) hydropower plan, press release, June 2003
Makkonen K (2005) Mekong cooperation: the linkages between poverty, environment and transboundary water management in Southwest China's Yunnan province. Paper presented at the conference of the Nordic Association for Chinese Studies, Helsinki, 7–9 June 2005. http://www.nacsorg.com/finland2005/papers/makkonen.pdf. Accessed 19 June 2013
Magee D (2006a) New energy geographies: powershed politics and hydropower decision making in Yunnan, China. PhD thesis, University of Washington, Washington, DC
Magee D (2006b) Powershed politics: yunnan hydropower development under great western development. China Q 185:23–41
McDonald B, Webber M (2002) Involuntary resettlement in China: a model of good practice? Forced Migr Rev 14(July):38–39
McElwee CR II (2008) The environmental impact assessment in China: the first step toward compliant operations. Int Environ Law Committee Newsl 10(4):21–24
Mertha AC, Lowry WR (2006) Unbuilt dams: seminal events and policy change. Comp Politics 39(1):1–20
Mertha AC (2008) China's water warriors: citizen action and policy change. Cornell University Press, Ithaca
Ministry of Environmental Protection (2009) MEP requires to practically enhance planning EIA workm News Release, 10 Sep. 2009. http://english.mep.gov.cn/News_service/news_release/200909/t20090929_161948.htm. Accessed 19 June 2013
Ministry of Environmental Protection (2012) Guanyu jinyibu jiaqiang shuidian jianshe huanjing baohu gongzuo de tongzhi [Notice on Further Strengthening Environmental Protection During Hydropower Construction], Beijing
Ministry of Power and Ministry of Finance (1981) Caizheng bu, dianli gongye bu guanyu cong shuidianzhan fadian chengben zhong tiqu kuqu weihu jijin de tongzhi [Announcement of the Ministry of Finance and the Ministry of Power on the Establishment of a Reservoir Maintenance Fund from the Electricity Costs Generated by Hydroelectric Stations], Beijing
NDRC (National Development and Reform Commission) (1996) Guanyu jianli shuidianzhan he shuiku kuqu houqi fuchi jijin tongzhi [Announcement on the Establishment of a Hydroelectric and Reservoir Post-Relocation Support Fund], Beijing
NDRC (National Development and Reform Commission) (2005) Guanyu jiaqiang shuidian jianshe huangjing baohu gongzuo de tongzhi [Notice on Strengthening Environmental Protection during Hydropower Construction], Beijing
NDRC (National Development and Reform Commission) (2007a) Shuidian gongcheng kexingxing yanjiu baogao bianzhi guicheng: Zhonghua Renming Gongheguo dianli hangye biaozhun, DL/T 5020-2007 [Code For Preparation of Hydroelectric Project Feasibility Study Report: Standards of the People's Republic of China for the Power Industry]. Beijing
NDRC (National Development and Reform Commission) (2007b) Shuidian gongcheng jianshe zhengdi yimin anzhi guihua sheji guifan: Zhonghua Renming Gongheguo dianli hangye biaozhun, DL/T 5064-2007 [Specification of Resettlement Planning and Designing For

Hydroelectric Project: Standards of the People's Republic of China for the Power Industry], Beijing

NDRC (National Development and Reform Commission) (2011) Heliu shuidian guihua baogao ji guihua huanjing yingxiang baogaoshu shencha zanxing banfa [Provisional Measures for the Evaluation of River Hydropower Plans and Environmental Impact Statements], Beijing

Ongley E, Wang X (2004) Transjurisdictional water pollution management in China: the legal and institutional framework. Water Int 29(3):270–281

Osborne M (2006) The paramount power: China and the countries of Southeast Asia. Lowy Institute for International Peace, Sydney (Lowy Institute Paper 11). http://www.lowyinstitute.org/publications/paramount-power. Accessed 19 June 2013

People's Daily (2000) Large hydropower pant to be built in Yunnan, 6 Dec. 2000. http://english.peopledaily.com.cn/english/200012/06/eng20001206_57047.html. Accessed 19 June 2013

Plummer J, Taylor JG (2004) The characteristics of community participation in China. In: Plummer J, Taylor JG (eds) Community participation in China: issues and processes for capacity building. Earthscan, London, pp 36–54

Powell S, Jun M, Pozon I et al (2008) China greening: the emerging role of the public: CLSA Asia-Pacific Markets, WWF Hong Kong, Institute of Public and Environmental Affairs

Qin C, Li J (2005) First public hearing over environment. In: China Daily, 14 April 2005. http://www.chinadaily.com.cn/english/doc/2005-04/14/content_434089.htm. Accessed 19 June 2013

Qin T (2008) Public participation in environmental governance in China: important development in recent years

Qin T (2009) Comments on 'Decision-making framework for large dams in China', written comment on Oliver Hensengerth's presentation at the Sixth Dialogue on Water 15–16 Oct 2009, German Development Institute, Bonn

Shi G, Su Q, Yuan S (2006) Risk and avoidance of risk among migrants of the Xiaolangdi dam project. Chinese Sociol Anthropol 38(3):40–70

Shi G et al (2001) Zhongguo yimin zhengci yu shijian [China resettlement policy and practice] (Yinchuan: Ningxia renmin chubanshe), Yinchuan

Shi J (2008) Nujiang Liuku, Saige dianzhan lieru 'Shi Yi Wu' zhongdian xiangmu [Nu River Liuku and Saige hydroelectric stations listed as major projects in the 'Tenth Five-Year Plan']. In: Zhujiang Sanjiao hezuo xinxi wang [Pan Pearl River cooperation information network], 25 March 2008. http://yunnan.pprd.org.cn/fzgh/wngh/200803/t20080325_41112.htm. Accessed 19 June 2013

Si M (2011) Is hydropower exploitation of the Nu river in China 'a must'? In: The Guardian, 10 Feb 2011. http://www.guardian.co.uk/environment/blog/2011/feb/10/hydropower-exploitation-china. Accessed 19 June 2013

Stanway D (2005) China's hydropower development is an issue for all of Southeastern Asia. Interfax China, 5 July 2005

Stanway D (2007a) China abandons plans for huge dam on Yangtze. In: The Guardian 29 Dec 2007. http://www.guardian.co.uk/world/2007/dec/29/china.waveandtidalpower. Accessed 19 June 2013

Stanway D (2007b) China's renewable energy targets will include large-scale hydro-state planner. AFX News Limited, 4 Sep 2007. http://www.forbes.com/markets/feeds/afx/2007/09/04/afx4077882.html. Accessed 19 June 2013

State Council (1986) Guowuyuan bangongting zhuanfa shuili dianli bu guanyu zhuajin chuli shuiku yimin wenti baogao de tongzhi [Circular of the General Office of the State Council on the Announcement of the Ministry of Power on the Treatment of Problems in Reservoir Resettlement], Beijing

State Council (2006a) Dazhong xing shuili shuidian gongcheng jianshe zhengdi buchang he yimin anzhi tiaoli, Guowuyuan ling (di 471 hao) [Rules of Land Compensation and People Resettlement in Medium and Large Hydraulic and Hydroelectric Projects] (Order of the State Council 471), Beijing

State Council (2006b) Guowuyuan guanyu wanshang dazhong xing shuiku zimin houqi fuchi zhengce de yijian, Guofa (2006): [Opinions of the State Council on Improving the Policies for Follow-Up Support to Residents for Construction of Large and Medium-Sized Reservoirs], Issue No. 26, Serial No. 1205 (20 September 2006), Beijing

State Council (2013) Guowuyuan guanyu yinfa nengyuan fazhan "shi er wu" guihua de tongzhi [State Council Notice on the Publication of the 12th Five Year Plan for Energy Development], Beijing

Taylor JG (2004) The context of community participation in China. In: Plummer J, Taylor JG (eds) Community participation in China: issues and processes for capacity building. Earthscan, London, pp 23–35

Tiqing yifa gongshi Nujiang shuidian huanping baogao [Call for public disclosure of Nujiang hydropower development's EIA report in accordance with the law]. http://www.nsbd.com.cn/NewsDisplay.asp?Id=139335. Accessed 19 June 2013; for an English translation see here: http://journal.probeinternational.org/2005/09/06/call-public-disclosure-nujiang-hydropower-developments-eia-report-accordance-law/. Accessed 19 June 2013

van Wicklin WA III (1999) Sharing project benefits to improve resettlers' livelihoods. In Cernea M (ed.) The economics of involuntary resettlement: question and challenges. World Bank, Washington DC, pp 231–255

Webber M, McDonald B (2004) Involuntary resettlement, production and income: evidence from Xiaolangdi, PRC'. World Dev (32)4:673–690

Wei Y, Fan Y (2006) Zhongguo nengyuan baogao (2006): Zhanlüe yu zhengce yanjiu [China energy report (2006). Strategy and policy research], Beijing: Kexue chubanshe

Wen Wei Po (2006) Shuilibu zhang tan nujiang zhong shengtai, zai guihua [Minister of Water Resources talks about the importance of the Nun River's ecology and revises the plan]; http://paper.wenweipo.com/2006/03/06/CH0603060025.htm. Accessed 19 June 2013

World Bank (1997) China-Xiaolangdi Multipurpose Project Stage II, Washington, DC. http://www.worldbank.org.cn/Chinese/content/867r1186883.shtml. Accessed 19 June 2013

World Bank (2001a) The World Bank position on the report of the World Commission on Dams, Washington, DC

World Bank (2001b) Operational policy 4.37: safety of dams, Washington, DC

World Bank (2007) Project performance assessment report: Xiaolangdi resettlement project, Washington, DC. http://documents.worldbank.org/curated/en/2007/02/9423698/china-xiaolangdi-resettlement-project. Accessed 19 June 2013

Wouters P, Hu D, Zhang J et al (2004) The new development of water law in China. Water Law Rev 7(2):243–308

Yang G, Calhoun C (2007) Media, civil society, and the rise of a green public sphere in China. China Inf 21(2):211–236

Yardley J (2006) China proposes fewer dams in power project to aid environment. In: The New York Times 12 Jan 2006. http://www.nytimes.com/2006/01/12/international/asia/12river.html . Accessed 19 June 2013

Zawahri N, Hensengerth O (2012) Domestic environmental activists and the governance of the Ganges and Mekong Rivers in India and China. Int Environ Agreements Polit Law Econ 31(2):269–298

Zhang C (2007) China reservoir resettlement: policy and practice. Paper presented at the International Experience Sharing Workshop on "Land Acquisition, Resettlement and Rehabilitation, and Benefit Sharing", organized by the Ministry of Rural Development of India, Gurgaon, July 21 to 23, 2007

Zhou P (2003) Liuku Shuidian zhan jiepai [Inauguration of Liuku hydroelectric station). In: Yunnan Daily 19 July 2003. http://www.yn.xinhuanet.com/newscenter/200307/19/content_724446.htm. Accessed 19 June 2013

Zhu T, Lam K (2009) Environmental Impact Assessment in China, Tianjin, Hong Kong. Research Center for Strategic Environmental Assessment at Nankai University, Centre for Strategic Environmental Assessment for China at the Chinese University of Hong Kong

Chapter 4
Towards Responsible Hydropower Development through Contentious Multi-stakeholder Negotiations: The Case of India

Nirmalya Choudhury

Abstract This chapter explores reforms in environmental and resettlement policies in India and the influence of domestic and external actors on the reform process. It also analyses the ways in which environment and resettlement policies have been implemented in a number of hydropower projects. At macro level, the study begins by describing the multilevel processes that govern dam decision-making. It then considers changes regarding environmental clearance and resettlement, and the role which state and non-state actors played in the last three decades. At project level the analysis focuses on how state and non-state actors influence decision-making on the introduction of superior environmental and social standards through changes in policies and laws. The dam projects selected are the Allain Duhangan Project, the Lower Subansiri Project and the Dibang Multipurpose Project. The study argues that, at macro level, it is civil society which has been the major driver of change in the area of resettlement over the last three decades. In the environment arena the changes are the outcome of competing demands from civil society and growth-oriented ministries and departments of the Government of India. At project level, superior social and environmental standards are primarily driven by social movements. International actors like the International Finance Corporation also catalyse the reform process.

4.1 Introduction

The Indian Ministry of Water Resources and such government parastatals as the National Hydro Power Corporation were among those who rejected the World Commission on Dams (WCD) recommendations. The Government of India (GoI)

N. Choudhury (✉)
Council on Energy Environment and Water, New Delhi, India
e-mail: nirmalya.choudhury@ceew.in

W. Scheumann and O. Hensengerth (eds.), *Evolution of Dam Policies*,
DOI: 10.1007/978-3-642-23403-3_4, © Springer-Verlag Berlin Heidelberg 2014

reaffirmed that it would continue its dam-building policy with an overall objective of constructing new storage capacity of around 25 billion cubic metres (BCM) in the next 25 years (GoI Ministry of Water Resources 2001). These reactions may be misleading and hide that India has witnessed ongoing formulation and reformulation of policy pertaining to the resettlement and rehabilitation of displaced people in recent times. In 2004 the Indian Government adopted the first national policy on resettlement, which was reformulated in 2007. In 1994 the first Environmental Impact Assessment (EIA) Notification was introduced to ensure a statutory environmental governance system in project planning. This notification was revamped in 2006 through the enactment of a fresh EIA Notification.

This study seeks to establish whether past policy changes relating to various aspects of the social and environmental sustainability of dams in India have been the result of the gradual influence of global norms or whether they have been driven by internal policy learning processes, or a judicious mix of the two. The aim is to determine whether international norms have induced changes in resettlement and environment planning processes, and to investigate pathways and norm carriers.

Similar to the WCD's recommendations, which ask for inclusive project planning and decision-making, favour a *right and risk* approach and call for the abatement of adverse environmental impacts, the Planning Commission[1] has been discussing inclusive development, while the Ministry of Environment and Forestry (MoEF) and the Ministry of Rural Development have drafted and redrafted policies on environmental impact assessment and resettlement, respectively. These recent developments raise the interesting question whether they are manifestations of the gradual diffusion of global norms among the policy elites in the country, outcomes of a domestic policy learning process through the involvement of domestic governmental and non-governmental actors well insulated from the effect of any international actors and activities or a combination of the two.

It is hypothesised that the extent to which global norms are diffused in the domestic policy arena and the extent to which dam decision-making takes account of socially and environmentally sustainable norms depends primarily on social mobilisation, on a country's embeddedness in the international arena and international institutions and on its dependency on international financial institutions (see introduction of this volume).

4.2 The Strategic Role of Dams for India's Economic and Social Development

India ranks fourth in the world after China, the United States and Russia in terms of the number of its dams. There are more than 4,000 large dams, which play an important role in the country. In India most dams are constructed and maintained

[1] The Planning Commission is an apex planning body.

by the water resource departments of the state governments. Certain dams built on interstate rivers and providing water-related services to more than one state are managed by separate institutions. In recent years there has been a trend towards private-sector operation of hydropower dams.

While many large reservoirs in other parts of the world are specifically intended for water supply, many in India were built primarily for irrigation purposes. But as many cities and towns are running out of water, numerous dams originally built for irrigation are now supplying water for domestic consumption (Shah and Kumar 2008). The National Commission for Integrated Water Resources Development set up by the GoI (1999) has forecast that the total water requirement for domestic purposes in the country as a whole will rise from 43 BCM in 2010 to 62 BCM in 2025 and to 111 BCM in 2050, most of which will come from surface sources (NCIWRD 1999). According to Sengupta (2000), of the 4,291 dams in the country, 96 % will have irrigation as either their sole objective or at least one of their objectives, and only 4.2 % will have power generation as one of their objectives. While it is generally agreed that agricultural production has increased over the years, the actual contribution made by dam-reservoir-based projects to that increase has always been a matter of debate, all the more so as groundwater irrigation has grown in the country in the last three decades (Sengupta 2000).

Large dams also assume strategic importance in India's energy scenario. Large-scale hydropower is now a preferred power generation technology because of its low recurring costs, its lower carbon emissions compared to coal-based thermal power and its ability to supply peak power (TERI 2006). Realising the importance of hydropower for the national energy scenario, the GoI, in collaboration with the Central Electrical Authority, has initiated the preparation of preliminary feasibility studies for 162 new hydroelectric schemes totalling over 50,000 Megawatt (MW).[2]

At national level, the total hydropower potential is 148 Gigawatt (GW), of which only 31 % has already been developed or is under construction. Arunachal Pradesh in the North-East of the country is to have the highest hydropower potential (GoI Central Electricity Authority 2008). Such states as Arunachal Pradesh and Sikkim in the North-East and Himachal Pradesh and Uttaranchal in the North are experiencing rapid development in the hydropower sector. The state governments have identified hydropower as an "avenue for revenue generation".[3] Numerous international agencies, such as the Asian Development Bank (ADB), the World Bank and the International Finance Corporation (IFC), have either conducted feasibility studies (the ADB on India as a whole, the World Bank on the North-East, both in 2007) of the possibility of harnessing hydropower in the country or have funded various governmental and private power development firms and parastatals.

---

[2] http://www.nhpcindia.com/English/Scripts/Hydro_Initiative.aspx. Accessed 6 Aug 2010.

[3] Personal interview with a senior official in the power department in one of these states.

4.3 The Regulatory Framework for Dam-Related Decision-Making

4.3.1 Dam Planning and Decision-Making

Under India's Constitution water is a state responsibility, its overall governance therefore lying within the jurisdiction of the state governments. The Constitution also empowers the Union Parliament to govern interstate rivers and river valleys and arbitrate in disputes over interstate rivers (Iyer 2003). From an operational point of view the federal government plays a part in dam decision-making through its say in financial approval and statutory environmental and forest regulations. The decision-making process of a typical hydropower project can be broken down into (i) the initial planning process, which is a state responsibility; (ii) statutory clearances, which are a federal responsibility and (iii) the post-clearance construction phase, which is again the state governments' responsibility.

(i) *Initial planning process*
The initial planning process begins with the preparation of a preliminary feasibility report designed to identify suitable locations for dam projects and largely based on desk research. In many cases, this report is drawn up by the Central Electricity Authority (CEA), a federal body. On the basis of this report the state government concerned signs a contract with the executing agency on the preparation of a Detailed Project Report. The Detailed Project Report is then sent to the federal government if it is to finance the project with unplanned allocations. Project planning then becomes a federal responsibility.

(ii) *Statutory clearances*
Until the late 1970s, the Detailed Project Report would be sent to the Planning Commission at federal level, which undertook the techno-economic analysis before sanctioning any financial allocations for the project concerned. The Planning Commission then sends the Detailed Project Report to two federal agencies, namely the Central Electricity Authority and the Central Water Commission, for techno-economic analysis. Once the report is cleared and has been returned to the Planning Commission, the latter forwards it for investment clearance. The preliminary analysis is carried out by the Planning Commission's Project Appraisal and Management Division. Subsequently, the project is discussed by the Public Investment Board, an inter-ministerial body chaired by the federal Ministry of Finance. If the Public Investment Board clears the project, it is submitted to the Cabinet Committee of Economic Affairs for approval.

Since the late 1970s, the Planning Commission has required that projects obtain environmental clearance, which was issued by the then Department of Environment. Since 1985, the MoEF has undertaken the process of issuing environmental clearance. The environmental clearance then was an administrative requirement. Since 1994, when the first environmental impact assessment notification was passed, environmental clearance has been a statutory requirement. Administratively, the techno-economic and investment clearances, can take place without a forest and environmental clearance. But to obtain clearance from the Cabinet Committee of Economic Affairs, the project has to have prior environment and forest clearance.

(iii) *Post-clearance construction phase*
Once the project has been cleared, and sometimes before all the clearances have been obtained, the state government signs an implementation agreement with the project developer to enable construction to proceed. During the construction phase the project developer is required to submit periodical (6-monthly) reports to the MoEF on the status of compliance with environmental impact mitigation requirements.

Prior to the amendment of the EIA Notification 1994 in 1997 the entire dam decision-making process was an interdepartmental affair. Since 1997, the environmental clearance process has included public hearings, the only form of public participation in the process. Since the 1990s the public in general have tried to influence decision-making by both participating in the public hearing process and addressing the regulatory authorities.

4.3.2 Environmental Impact Assessment

The "environment" became an official part of public policy and project planning in India in part as a result of India's commitment to the United Nations Conference on Human Environment, and in part through its establishment of the National Committee on Environmental Planning and its subsequent acceptance of this committee's recommendations. At international level, the panel of experts set up by the United Nations Conference on Human Environment argued that the environment should be the boundary condition within which development planning is undertaken. They were particularly critical in their comments on the environmental issues associated with dam (river valley) projects, whose impact would threaten their very purpose. In India the National Committee on Environmental Planning elaborated number of intervention to be taken up at the policy and organisation level. This committee through their recommendations mainstreamed the environment within Indian policy and planning[4] (Sachs 1971; EPW Correspondent 1972a, b; EPW Editorial 1987).

[4] The committee's terms of reference included reviewing policy and programmes with significant environmental implications, advising government, departments and industry on mitigation measures, reviewing existing environmental legislation, regulation and administration, proposing cost-effective solutions to environmental problems, ensuring coordination between the environment and economic policies, promoting research on environmental problems and establishing research facilities wherever possible, increasing public awareness and cooperating with the UN and other international agencies in environmental programmes with global concerns (EPW Correspondent 1972b).

4.3.2.1 Environmental Impact Assessment over the Years

The Environmental Impact Assessment started in India as early as the late 1970s as an administrative requirement for river valley projects. The environment impact assessments were mandated by the Planning Commission. In the early years it was undertaken by the Department of Science and Technology; from 1980 it was entrusted to the Department of Environment. In 1985 the Department of Environment and the Department of Forests were merged to form the MoEF. Since then the MoEF has been entrusted with environmental clearance (Rangacheri et al. 2000).

Until 1994 EIA was carried out as an administrative requirement, but this changed when the first EIA notification was issued in 1994. EIA then had the status of subordinate legislation under the Environmental Protection Act, which had already been passed in 1986. Between 1994 and 2006 twelve amendments were made to the original EIA Notification 1994. Through these amendments more and more projects were exempted from EIA (Nandimath 2009). In 2006 a new EIA Notification was passed, and this was again to be amended in 2009. Table 4.1 shows the changes that have occurred over the last three decades.

In the early years, EIA was largely an interdepartmental process, with no provision for public participation. Moreover the information sought by the MoEF for evaluating the project would not match with the information supplied by the project authority. This then resulted in delays in environmental clearance. As by then the other parts of the project were already completed, the project had already reached the point of no return. Thus even if the regulatory authority was not satisfied with the environmental information provided by the project authority, the project would be given conditional clearance (Maudgal 1991; Mohan 1991).

After 1994, with EIA now subordinate legislation, amendments could be made by bypassing the parliament, which resulted in 12 amendments over a period of 10 years. The subsequent amendments to the 1994 EIA Notification resulted in more and more projects being exempted from EIA (Nandimath 2009). But by then the environmental clearance process had already been identified by part of the government machinery and the industry as an obstacle to the efficient implementation of projects. Civil society also criticised the environmental clearance process as not being conducive to sound environmental decision-making. This led to a couple of events.

First, the World Bank undertook specific activities with a view to strengthening the ministry's environmental clearance process as part of the ongoing Environmental Capacity Building Technical Assistance (TA) project. The aim of this project was to improve the quality of environmental information and strengthen the EIA process (World Bank 2004).

Secondly, a committee chaired by V. Govindrajan (hence the name Govindrajan Committee), the then secretary of industrial policy, was set up to suggest ways of attracting more foreign direct investment to India and to undertake necessary institutional reforms to this end (Nandimath 2009). The committee pointed to the long time taken by environmental clearance, which caused lengthy delays in

Table 4.1 Changes in EIA regulations (author's own compilation)

	1980–1994	1994	2006	2009 (Draft)
Events	EIA becomes part of decision making on water resource projects	First EIA notification	Second EIA notification revoking the first	Draft EIA notification to be finalised
Main actors	Dept. of Science and Technology and later Dept. of Environment	Ministry of Environment and Forestry	Ministry of Environment and Forestry	Ministry of Environment and Forestry
Triggers for change	Required by Planning Commission, Govt. of India	Environmental Protection Act 1986	Govindrajan Committee Report recommending reforms to attract foreign direct investment	Expert committee report of Dept. of Economic Affairs, Ministry of Finance
Influence of international actors	Required by multilateral financial institutions		World Bank's Environmental Capacity Building TA Program	

projects. To ensure speedy environmental clearance as required by the environmental regulations, the Govindarajan Committee recommended that the MoEF should consider setting up a central data centre to serve as a one-stop source of reliable and validated environmental information for the preparation of Environmental Impact Assessments. The Govindrajan Committee's report was accepted for implementation by the GoI (Mukherjee 2002).

On 16 Aug 2004 a draft EIA Notification was posted on the MoEF's website and was followed by four rounds of consultations with federal ministries, state governments and civil society organisations (Nandimath 2009). The draft drew around 500 responses from individuals, research institutes, industry, industrial associations, central government and state government ministries, social activists and non-governmental organisations. It came in for severe criticism from civil society organisations and activists on the ground that the re-engineering process and environmental decision-making had been guided by the desire for industrial growth rather than genuine concern for the environment (Saldanha et al. 2007).

A civil society initiative also undertook political mobilisation by involving various parliamentarians across the political spectrum and members of special parliamentary committees. Finally, on 14 September 2006, the MoEF introduced the 2006 EIA Notification (Saldanha et al. 2007; Menon and Kohli 2007, 2008). There were allegations that the final version of the Notification was discussed exclusively with the industrial associations (Menon and Kohli 2008).

The 2006 EIA Notification struck a fine balance between the competing interests of different stakeholders, but in the end it did not satisfy anybody. Industry and part of the government continued to complain about delays. They perceived EIA as a hurdle to be surmounted before marathon project inception processes could be launched. The activists continued to complain about the EIA's lack of ability to ensure sound environmental decisions.

On 10 Nov 2008 the Department of Economic Affairs of the Ministry of Finance formed an expert group to examine the patterns followed by statutory clearances for industrial and infrastructure projects in India. The expert group included representatives from the financial institutions and the industrial associations, the Industry Advisor of the Planning Commission and the Joint Secretary to the Department of Industrial Policy and Promotion, GoI. The Confederation of Indian Industry, an industrial association, was represented by the Chief Executive Officer of the Indian section of the World Wildlife Fund for Nature (WWF).

The expert group focused on the ten federal government clearances, the first and most important of which concerned the environment. The report pointed out that, more often than not, the environmental and forest clearances resulted in inordinate delays, causing cost overruns and affecting project economics. The expert group recommended that the process of public hearing should be streamlined, that the public hearing process be separated from state pollution control boards and that a strict time limit be imposed for the issue of environmental clearance, after which the project would be *deemed to have been cleared* (GoI Ministry of Finance 2008).

4.3.2.2 The 1994 and 2006 EIA Notifications: Comparing Contents

As the preceding section has shown, in the last one and half decades EIA has come under pressure from various actors with diverse and often contradictory interests. Civil society continued to argue for developmental activity with a minimum environmental impact, as espoused by the UN Convention on Human Environment in the early 1970s. The EIA was also a tool which could be used to oppose large water infrastructure projects on environmental grounds. However, the macroeconomic scenario prevailing in the country had changed, and under the policy of liberalisation the focus was very much on improving the economic parameters and encouraging private capital. What mattered under this regime was that steps were taken to ensure the rapid implementation of projects and to create conditions that led to minimum delays for project developers. In this regard the EIA process was viewed more as a constraint than a tool for promoting developmental goals with a

Table 4.2 Comparison of the 1994 and 2006 EIA Notifications (author's own compilation)

Issues	1994 EIA notification	2006 EIA notification
Coverage and scope	Subsequent amendments of the notification ensured that any new project costing less than US$ 20 million or modernisation project costing less than US$ 10 million would be exempted from EIA	To fasten the clearance process, EIA decision-making was decentralised (certain categories of project were transferred to state level). Project developers were allowed to "secure land" without a clear definition of the meaning and scope of "secure land" [a]
Comprehensive option assessment	Site clearance hints at option assessment on the basis of site but not with respect to the choice of technology	Site clearance and final clearance now replaced with four stages, namely screening, scoping, public consultation and appraisal. No comprehensive option assessment
Decision-making	EIA reports to be assessed by the impact assessment agency of MoEF in consultation with an expert committee drawn from different subject domains and areas of interest, including resettlement experts and representatives of civil society	In reality, expert appraisal committee has been gradually deprived of representatives of civil society and of conservationist and environmental experts; they have been replaced with retired technocrats
Public participation	Any interested person can participate in public hearings and submit oral/ written comments. Later amendments exempted modernization projects from public hearings	Public hearing process detailed, but participation narrowed down to only people affected, and hearing can take place even if very few people attend
Compliance and monitoring	Empowered the expert committee to undertake site visits before, during and after clearance for physical verification of compliance report	Empowers expert appraisal committee to undertake site visits before and during issue of clearance. Silent on post-clearance monitoring
Sanctions	Project liable to cancellation if insufficient or inadequate data submitted more than once	Penalties for providing false or incorrect information, but only if it is proved that the mistake is deliberate

[a] This provision is not clear to the author. Many policy documents in India suffer from such ambiguities, resulting in different interpretations, which can then be contested in court.

minimum environmental impact. So there was a pressure for streamlining the EIA process (Table 4.2).

In the case of river valley and hydropower projects EIA continued to be obligatory, but provision was made for a certain category of projects to be exempted from the EIA requirement (Nandimath 2009; Menon and Kohli 2009). In terms of efficiency, the 2006 EIA Notification was an improvement on the 1994

EIA Notification in that it decentralised the process of granting environmental clearances by transferring some of the projects to the State Level Environment Impact Agency. But, in reality, most of the states could not form a State Level Environment Impact Agency, for a long time, and the projects which had been transferred to the states reverted to the MoEF.

The 2006 EIA Notification broke up the environmental clearance process into screening, scoping, public consultation and appraisal. This was a positive change in the sense that the scoping phase would then define the project developer's terms of reference, ensuring better structuring of the report, comprehensive environmental and social information and a reduction in the mismatch of information required by the MoEF and the information supplied by the project developer.

Ideally, EIA is an instrument that enables the course of action with the least environmental impact to be chosen from among the various sets of possible interventions. Thus, in one sense, EIA per se can be seen as promoting the comprehensive assessment of various options. The 1994 EIA Notification included a provision concerning *site clearance*, meaning that the site chosen for the project had first to be approved by the regulatory authority on the basis of first-level impact assessment and only after the project had obtained *site clearance* could a detailed EIA be undertaken as the basis for *environmental clearance*. Intuitively the *site clearance process* ensured some form of assessing options for project location. However, the 2006 EIA Notification does not require a comprehensive option assessment to be undertaken at any point of the environmental.

There have always been complaints of projects failing to comply with the environmental management plans mandated in environmental clearance certificates (Roy 2008). This despite the provisions made in the EIA Notification 1994 where the expert appraisal committee was empowered to make on-site monitoring of the compliance. The 2006 EIA Notification further dilutes the compliance by not making the post-clearance on-site monitoring mandatory.

The 1994 EIA Notification had already provided for heavy sanctions in case of reporting of false information in the EIA report. In the 2006 EIA Notification the sanctions would not follow automatically, as was in the 1994 Notification, but would be imposed at the discretion of the regulatory authority. The discretion would apply only if it was proved that the project developer had deliberately provided incorrect information.

4.3.3 Regulations for Resettlement and Rehabilitation

With independence in 1947, India entered what is often known as the *Nehruvian Era*, when the main focus was on technology-driven industrialisation, which was expected to trickle down to ensure social equity (Guha 2007). During this era the normative importance of dams was high within Indian polity (Klingensmith 2007). Dam projects were mostly seen as engineering projects and were implemented by engineers who enjoyed high esteem. These projects were often seen as tools for developing the otherwise backward sections of the population, with particular emphasis on tribal people (D'Monte 1986). The resettlement of displaced people and environmental impacts were not considered in the planning and decision-making process, which was dominated by the engineers (Iyer 2007; Mohanty 2005). Land for projects was requisitioned under the Land Acquisition Act 1894, which empowered the government to appropriate private land on the principle of *eminent domain* for *public purpose*. The Act and particularly the concepts of "eminent domain" and "public purpose" have been criticised over the years, but the Act has remained unchanged.

It is generally agreed that dams have displaced many people, especially in tribal areas, but the estimate of the total number displaced varies between 32 and 56 million (Rangacheri et al. 2000; Fernandes 2004; Roy 1999). The World Bank also sees India as one of the problem cases as far as development-induced displacement is concerned. According to some estimates, India accounted for more than 50 % of development-induced displacement in all World Bank projects between 1986 and 1993 (Fox and Brown 1998).

4.3.3.1 Towards a National Resettlement Policy

In the early decades after independence, when dams were equated with development and displacement was seen as an unavoidable cost that the country must pay for it, there was some institutional thinking and learning within the apparatus of the GoI (Table 4.3).

The earliest public policy governing displacement was the T. N. Singh Formula, which called for the employment in the project of one member of every family it displaced. Thus some form of employment-based compensation was already being considered as early as the 1960s. But, with time, the number of displaced families increased and projects became more technology-intensive. With the decline in industry's absorption capacity, the T. N. Singh Formula was abandoned in 1986 (Fernandes and Paranjpye 1997).

Iyer (2007, 3103), a senior government official in the 1980s, observed that:

> Sometime during the 1980s thinking began in the Government of India on the formulation of a policy to govern all future cases of displacement. [...] The subject was discussed many times in the interministerial meetings at the level of secretaries, and at meetings of groups of ministers.

The late 1980s and the 1990s were also a time of considerable controversy in India over the Narmada projects, and particularly the Sardar Sarovar Project, which subsequently became international. In the late 1970s the Narmada Water Disputes Tribunal, a quasi-judicial body formed under the Inter-State Water Disputes Act 1960, had decided that anyone who could prove ownership of

Table 4.3 Early regulations on Resettlement and Rehabilitation in India

	1967	1967	1985	1986
Actor	17-member committee set up by Ministry of Food, Agriculture, Community Development and Cooperation	T. N. Singh	Committee of Department of Welfare in Home Ministry	Standing Committee on Public Enterprises
Issue	Study on basic issues of Land Acquisition Act 1894	Compensation of families displaced by public projects (T. N. Singh Formula)	Rehabilitation of tribal population	Review of T. N. Singh Formula
Implication	Requisition of good agricultural land to be avoided and rehabilitation as moral compulsion of the state	One member of the family would be em-ployed in the project	Rehabilitation policy binding on project developer	T. N. Singh Formula abandoned

agricultural land in the submerged area should receive land in compensation. This decision by the Narmada Water Disputes Tribunal was an important landmark for *land-for-land* compensation in India. But it concerned only land-owners. The people who were dependent on land without possessing the legal recognised titles, or were dependent on common property resources or on providing service to the rural population for their livelihoods were not covered by the *land-for-land* compensation principle.

These drawbacks and the faulty implementation of the resettlement policies in various states in which the Narmada project was located led to the formation of a national and subsequently transnational civil society movement, which eventually forced the withdrawal of the World Bank, which was funding the Sardar Sarovar. The GoI and the state government of Gujarat went ahead with the financing and construction of the dam. The transnational and national campaign failed to stop the dam, and the reforms affecting this project also had little spill-over effects on other projects. However, there were two achievements worthy of note. First, the Narmada controversy resulted in the development of a civil society network, which became increasingly broad, powerful and critical of dam construction in India in the 1980s and 1990s. Second, it sparked a critical debate on how past dam development had impacted on different sections of the population and how, if at all, dams should be built in future.[5]

[5] For more information on Narmada, see Wood (2007), Dwivedi (2006) and Khagram (2004).

Table 4.4 Evolution of a national resettlement and rehabilitation policy (author's own compilition based on Fernandes and Paranjpye 1997; Fernandes 2008)

	1994–95	1995	1998	2004	2007
Title	Draft National Policy on Resettlement and Rehabilitation	Draft National Policy on People Affected by Reservoir Project	Draft National Policy on Resettlement and Rehabilitation	National Policy on Resettlement and Rehabilitation	National Resettlement and Rehabilitation Policy
Leading actor	Ministry of Rural Development	Ministry of Water Resources	Ministry of Rural Areas and Employment	Ministry of Rural Development	Ministry of Rural Development
Other influential actors	Civil society alliance at national level		Civil society alliance at national level		National Development Council, civil society alliance
International actors	Multilateral financial institutions		Multilateral financial institutions		

By the late 1980s civil society in India had organised itself and voiced its demand for a proper resettlement policy. It formed a National Working Group consisting of researchers, academics, social activists, individuals and people affected by the project and drafted a prototype policy known as the *National Policy on Developmental Resettlement of Project- Affected People*. This became the basic policy for the civil society network's subsequent engagement with the government in the early and mid 1990s, when different ministries of the GoI drafted a series of resettlement policies, which never actually materialised into a practical policy (Table 4.4). On each of these occasions the civil society engaged with the respective ministries. The government version in each case was countered with a version drafted by the civil society based on the principles espoused in the *National Policy on Developmental Resettlement of Project-Affected People* (Fernandes 2008, pp. 5–6).

In 1998, the Ministry of Rural Areas and Employment submitted a draft policy entitled National Policy Packages and Guidelines for Resettlement and Rehabilitation. In the same year, a draft amendment to the Land Acquisition Act 1894 known as the Land Acquisition Bill was also prepared. In a bid to reform both the policy and the law governing land appropriation the civil society alliance engaged with the government.

According to Fernandes (2008, p. 6):

> A meeting convened by the Minister for Rural Development in January 1999 ended with an unwritten understanding that a policy would be prepared first in consultation with civil society groups and that a law would then be drafted based on the principles it enunciated. However, the Union Cabinet reportedly rejected the policy in October 1999 and approved

> Land Acquisition Bill 1998 that attempted to reduce the already limited rights of the DP/PAPs (Displaced People/Project Affected People, added by author) under the Land Acquisition Act. The Government fell shortly after it.

By this time the global process of WCD had begun. One of the commissioners on the WCD was a leading civil society activist and the leader of the Narmada movement. Civil society at both domestic and international level was well organised and articulate and was able to contribute its learning and decisions to the WCD.[6] The WCD report published in 2000 was rejected outright by the Ministry of Water Resources in the GoI. The rejection was so strongly worded that, to some extent, it temporarily derailed the mutual learning process, the progressive interaction and the gradual progress achieved in the various drafts of the National Rehabilitation and Resettlement Policies in the 1990s. To quote Iyer (2007, p. 3104):

> Two decades of slow emergence of enlightened thinking were washed out in the flood of rhetoric against what was perceived as an international conspiracy to prevent India from developing.

In 2003, finally, the Ministry of Rural Development put forward the first National Policy on Resettlement and Rehabilitation for Project Affected Families, which was notified in 2004. This policy relaunched civil society activism, and there was considerable interaction between civil society and parts of the government machinery. Another round of intense debate thus followed, and many seminars and conferences were held (Cernea 2006; Iyer 2007). This period of debates, discussions, conferences, workshops and seminars resulted in the National Rehabilitation and Resettlement Policy of 2007.

4.3.3.2 National Rehabilitation and Resettlement Policy 2007: Analysis of Its Contents

The National Rehabilitation and Resettlement Policy 2007 aimed to minimise displacement by specifying various areas to be avoided when locating a project. It also considers comprehensive option assessment, calling on state governments to undertake comprehensive site and technology assessment before deciding on a project. This contrasts sharply with the old draft policies, in which minimising displacement was either not mentioned or was ill-defined. The policy also referred to the need for social impact assessment, which was missing from the earlier draft policies and which civil society had always argued for. The National Rehabilitation and Resettlement Policy 2007 is based on such principles as participation, sustained income generation and improvement of the living standards of the displaced

[6] For international mobilisation, see McCully (2001), Khagram (2004), for domestic social mobilisation, see Fernandes (2008), Iyer (2007).

Table 4.5 National Rehabilitation and Resettlement Policy 2007 (author's own compilation)

Issue	National Rehabilitation and Resettlement Policy 2007
Minimising displacement	Makes pragmatic suggestions, e.g. acquisition of certain categories of land should be avoided, but left to the discretion of the project authority
Comprehensive option assessment	Calls for option assessment based on alternative sites and choice of technology, but subject to discretion of the government concerned. No implementation structure proposed
Pattern of compensation	Better standard of living and sustained income for affected people through shares in companies and land compensation at replacement cost
Public participation	Resettlement action plan in consultation with people and village governance institutions, but no mention of the nature of consultation. Whether the affected people are to be informed or options and acceptance sought from them is not clear
Social impact assessment	Social impact assessment to be carried out, and mitigation plans mentioned, but silence on how the assessment would take place and on its role in project decision-making

population. It provides for the preparation of resettlement action plans in consultation with village institutions. The policy requires the affected population to be offered shares in the project in partial compensation to ensure sustained income generation and ownership of the project for which they have had to sacrifice their assets (Table 4.5).

Although the National Rehabilitation and Resettlement Policy 2007 contained numerous positive provisions, it did not make them binding on the project developer or the government. Most of the above provisions were subject to the discretion of the government or project developer concerned. Thus much of the recommendations are conditioned with words like "may", "to the extent possible" or "if government land is available in the resettlement area". Singh (2006, p. 5308) argues that:

> Perhaps such an approach allows greater flexibility, however past experience conclusively establishes that such flexibility is invariably used against the interests of the displaced people.

While the National Rehabilitation and Resettlement Policy 2007 mentions social impact assessment, it fails to explain how it will be carried out, by whom, at what stage of the project cycle and what impact it will have in project decision-making.

Although the GoI rejected the WCD report, some of its provisions have found their way into National Rehabilitation and Resettlement Policy 2007. As Table 4.5 shows, the policy, at the same time, has struck a fine balance between competing interests.

As a result the implementation of the positive provisions has either been left with some loose ends or has been left to the discretion of the relevant governmental institutions or the project developer.

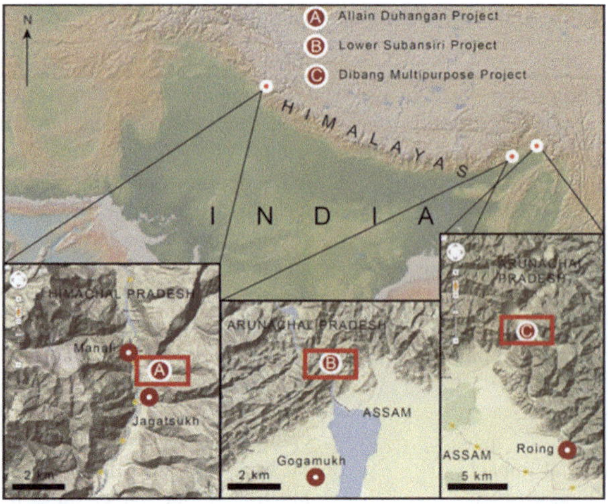

Fig. 4.1 Map of India with dam project sites (author's own compilation using GeoMapApp; GoogleMaps)

4.4 Dam Decision-Making in Practice: The Allain Duhangan, Lower Subansiri and Dibang Multipurpose Dam Projects

This section investigates individual projects in order to arrive at an understanding of the process by which superior social and environmental standards become part of project planning and practice. The case studies were chosen purposely to see whether superior norms, such as one recommended by the World Commission on Dams, have actually had any impact on various actors at project level.

It was therefore decided to focus on those projects where the largest part of the planning, clearance and construction of a hydropower project has taken place since 2000. Three hydropower projects, namely the Allain Duhangan Hydropower Project in Himachal Pradesh, the Lower Subansiri Project on the Assam-Arunachal Pradesh border and the Dibang Multipurpose Project in Arunachal Pradesh were selected for the case studies (Fig. 4.1). The projects differ with regard to (i) the presence or absence of international funding (ii) private or public project development agency and (iii) the current status of the project—planning stage/clearance stage/construction stage (Table 4.6).

4.4.1 The Allain Duhangan Dam Project

The Allain Duhangan project is being developed by Allain Duhangan Hydro Power Limited. Stratkraft Norfund Power Limited of Norway and the International Finance Corporation have equity and debt participation in the project. By 2003 the

Table 4.6 The dam projects investigated at a glance (author's own compilation)

Name of the project	Allain Duhangan	Lower Subansiri	Dibang Multipurpose
Project developer	Private	Public	Public
Current status (as on 2009)	Advanced stage of construction	Construction stage	Clearance stage
Major project purpose	Hydropower	Hydropower	Hydropower and flood control
Capacity (MW)	192	2,000	3,000
International actors	International Finance Corporation	./.	./.
Nature of the project	Run-of-river	Run-of-river	Reservoir
Environmental impact assessment	EIA study repeated in 2003	EIA study carried out in 2001	EIA study carried out in 2007

project had obtained techno-economic clearance from the Central Electricity Authority and forest and environmental clearances from the MoEF. The project authority also possessed letters from the village governance institutions (*panchayat*), giving their consent to the project. The project was then submitted to the IFC for funding in 2003. This was the first time since the Narmada debacle that the World Bank Group (the IFC being its private wing) had been called upon to finance a dam-reservoir project.

When the EIA report on the project appeared on the IFC's website as part of its project disclosure policy, various civil society organisations complained to the IFC that the EIA was not available in the local language. This was a violation of the IFC's own mandate. The IFC translated the report and uploaded it on to its website. On the basis of these documents the non-governmental organisations discussed with the people the facts given in the EIA report and also told them about the implications of the project. This was followed by a series of letters sent to the World Bank and the IFC both by the local populace and by the civil society organisations claiming that the EIA report was faulty and that there had been no involvement of local people in its preparation. Subsequently a meeting between the project developer, IFC staff and representatives of the local community took place. It involved negotiation and persuasion. But both tactics failed, and the local people again wrote to IFC headquarters to describe the course taken by the meeting. The civil society organisations, who were leading the local movement, then forwarded a detailed critique of the EIA report and demanded that a fresh EIA be undertaken. The IFC agreed to this suggestion and subsequently prepared an addendum to the existing EIA report (South Asia Network of Dams Rivers and People, personal communication June 2009; CAO 2004).

By now the civil society organisations were demanding that the public hearing be arranged by an independent panel mutually appointed by the campaigning parties. They also demanded that the decision taken at the public hearing be binding on the project developer. The IFC agreed only to the former demand and set up an independent panel to conduct a fresh public hearing. A team from the Kalpvriksh Environmental Action Group, an environmental NGO, analysed the EIA report and undertook a pre-hearing exercise before the actual public hearing was held. This ensured that local people understood the report and were also able to articulate informed comments to the project developer. In the process the Kalpvriksh team also shared their analysis with the local people and independently sent their response to the World Bank. According to Kalpvriksh, this approach was commendable, since the people attending most public hearings in the country knew little about the project and were largely unaware of their rights and responsibilities at public hearings (Kalpvriksh, personal communication June 2009).

Subsequently, the project developer agreed to compensate the project affected people for the loss of riparian rights by making necessary alternative arrangements to meet local water demand. The project developer and the villagers of Prini, one of the affected villages, signed a memorandum of understanding specifying that the cash compensation for the land requisitioned for the project would be paid at higher than the market rate. Also one member of the landowner's family would be offered employment. The hospital constructed during the project would also act as the primary health centre for the village of Prini. The company would provide pastureland for the people who would be affected as a result of the approach road to the Allain Barrage. At the same meeting the company also agreed that during the construction and subsequent operation of the project, the traditional source of irrigation water in the villages would not be tampered. Moreover a grievance cell would be set up at project level (ADHPL 2004). But besides Prini, there was another village downstream that was affected by the project, namely Jagatsukh.

The Allain Duhangan Project has affected Jagatsukh and Prini in different ways. As the powerhouse and office complex for the project had to be built in Prini, a large area of land was requisitioned there. Land in Jagatsukh was taken to build the approach road to the Duhangan Weir. But the greatest loss for Jagatsukh was access to the River Duhangan, which was diverted into the River Allain by the weir, upstream from the village. According to the villagers in Jagatsukh, much of their livelihood was dependent on the Duhangan and its tributaries. Agriculture in Jagatsukh consisted predominantly of apple orchards (in the uplands) and rice (flat lowlands), the water for which had been obtained from the River Duhangan along traditional irrigation channels. In addition, the source of drinking water in Jagatsukh was a spring, which the people feared would dry up with the construction and tunnelling upstream. They therefore opposed the project, and the Jagatsukh village assembly unanimously passed a resolution opposing the project (Correspondent 2004).

On 15 Sept 2004 local villagers wrote to the IFC's Office of the Compliance Advisor Ombudsman (CAO), criticising the proposed IFC funding of the project. They complained that the EIA documents were flawed, that the public hearing had been based on faulty EIA reports and that people could not, in such circumstances, decide whether the project should go ahead (International Rivers 2004a, b). Despite the protest at project level and although one of the independent public hearing regulators wrote to the IFC objecting to the approval of the project (International Rivers 2004c), the project was submitted to the IFC Board and approved on 12 Oct 2004.

The project was launched, but attempts to influence the decision-making continued. By this time the movement was led by the village of Jagatsukh, which had meanwhile formed a community-based organisation known as Dhomiya Ganga Sangharsh Samiti. The villagers filed a lawsuit with the High Court of Himachal Pradesh, complaining that the project affected their basic right to survival, since it would threaten the availability of water, the absence of which would threaten their lives and livelihoods. The court appointed a high-powered committee to see if the minimum downstream water release from the project was enough for the village of Jagatsukh. The committee appointed by the court calculated the water required by Jagatsukh for drinking and domestic purposes and for consumption by livestock. At its request, the court appointed experts to calculate the water needed for irrigation purposes. The project had meanwhile promised to release 150 l/s downstream water flows, whereas the villagers in Jagatsukh demanded 350 l/s.

The agitation by the villagers of Jagatsukh continued and even went so far as to prevent the project developer's employees from working. This time the project developer filed a lawsuit against the agitators and asked the court to provide sufficient security for the project to continue. The court agreed, and works on the project resumed.

Meanwhile, the IFC's Office of the CAO made periodic visits to the site in an attempt to resolve the differences between the two parties through dialogue. In 2004 and 2005 a series of meetings were held in Prini and Jagatsukh in the presence of the local administration to enable the villagers and the project developer to negotiate. This resulted in a further increase in the cash compensation for the requisitioned land. However, some people in Jagatsukh were still demanding that the project be stopped. According to the 2006 report of the Office of the CAO, the onset of work and the award of contracts to local villagers in the area restored something like peace between the two groups (CAO 2006). During their various visits to the area the representatives of the Office of the CAO ensured that the grievance redressal mechanism had begun to function and also recommended a formal channel of communication between the opposing groups with a view to developing mutual trust (CAO 2008).

In February 2007 a committee was formed by the state government to consider whether the forest clearance had been violated by the project authority. The committee reported that there has been "*continuous and unabated violations of*" various environmental and forest Acts (emphasis quoted from the Civil Writ Petition (2007) noted by Himachal Pradesh High Court). It was alleged in the report that all this had happened despite repeated notices/directions from the Forest Department and State Pollution Control Board. Subsequently the project authority was fined 59.5 million rupees by the Forest Department.

On the basis of this report and the show-cause notice issued by the state government, Dhomya Ganga Sangharsh Samiti filed another lawsuit in the Himachal Pradesh High Court and obtained a temporary injunction on the project. The project developer immediately appealed to the High Court against the injunction. On 26 Sept 2007 the High Court stated in its interim order that, if it was proved that the project developer was no longer violating the rules and was complying with the directions issued by the state government, the injunction could be lifted in view of the importance of the project. The project developer had told the court that the project was scheduled for completion in June 2008 and that a stay order would result in a delay. The court therefore set up another committee of experts, comprising representatives of the district administration and senior officials of the Forest Department, Public Works Department, State Electricity Board, Department of Environment and Pollution Control Board to decide on the injunction. This committee was asked to undertake a site visit within 3 days and to draw up a report within a week. It was empowered by the court to lift the injunction if it saw fit. In its report of 19 Nov 2007 the committee of experts gave the project developer clearance to continue the work on the Allain side of the project and on the Duhangan side until a certain point, beyond which that work was to be restricted to environmental and forest stabilisation, such as the stabilisation of slopes and the development of dumping sites and pastureland. This decision proved that violations of the forest clearance provisions had continued in some areas.

All these developments meant that the project construction faced a delay, the construction cost escalated and the company has had to ask for an increase in the IFC loan, which has been granted.

4.4.2 The Lower Subansiri Dam Project

The Lower Subansiri Project is located on the river Subansiri, one of the major tributaries of the River Brahmaputra. As early as April 1983, the Brahmaputra Board, a river basin agency responsible to the Ministry of Water Resources, undertook a detailed survey and investigation work for a 4,800 MW Subansiri Dam. However, the project could not be implemented because of the project of such a magnitude was supposed to submerge a large area in the state of Arunachal Pradesh. The Brahmaputra Board then carried out a fresh survey and narrowed the choice down to three sites for three projects on the River Subansiri. One of the dam sites was located right on the border between Arunachal Pradesh and Assam. The dam, powerhouse and the storage reservoir were to remain in Arunachal Pradesh, while the office and colony would be sited in Assam, the lower riparian state. In 1998–1999 the Brahmaputra Board began drawing up the detailed project report on the Subansiri Lower Dam, with completion scheduled for 2000. It was not completed on time, and in 2000 the Ministry of Water Resources transferred the project to the National Hydroelectric Power Corporation (NHPC), an undertaking run by the GoI (NHPC 2002). This was one of the first large dam projects anywhere in the north-eastern region of the country, whose power potential had no previously been harnessed, but which had nevertheless always being regarded as the *powerhouse* of India.

By mid-2001 the EIA report on the Lower Subansiri Project was ready, and the public hearing was due to take place in September. In line with the EIA Notification 1994, the summary EIA report was made available to the general public. The Kalpvriksh Environmental Action Group, a national civil society organisation, criticised the EIA report and demanded tough action against the project developer. The criticism was that, though the project was yet to receive necessary forest and environmental clearance, the project developer had proceeded with the construction of the project housing and so violated the Forest Conservation Act 1980 and the Environmental Protection Act 1986. A further cause of the criticism was that the project area was a high conservation area recognised by the Wildlife Institute of India and Birdlife International. Those were not reflected in the EIA report. The EIA summary report was also criticised for undermining the environmental importance of the project area and for data discrepancies.

In September 2001 the public hearing on the project took place. This was one of the first large hydropower projects in the North East, and the people in the area were also unfamiliar with the public hearing process. The hearing was later criticised by Aranyak, a local-level environmental non-governmental organisation (NGO) represented at the hearing, for procedural violations, unsatisfactory answers and unanswered questions (Vagholikar and Ahmed 2003). The public hearing report along with the EIA report was submitted to the expert appraisal committee. The expert appraisal committee recommended environmental clearance for the project in November 2002.

Meanwhile, representatives of Aranyak approached the Supreme Court alleging that, as part of the Tale Valley Wildlife Sanctuary was being flooded, the project required the prior approval of the Supreme Court-appointed Indian Board for Wildlife. Studies were subsequently carried out and presented to the Indian Board for Wildlife in May 2003. The project was cleared subject to stringent conditions. The conditions included that no further projects upstream from the Lower Subansiri Project would be allowed in the future and that the entire catchment area of the Subansiri Dam would be declared a national park. The project developer was also ordered to pay the entire cost of resettling and rehabilitating the people displaced from the national park (Chatradhar 2009).

On 15 June 2003 the first local-level mobilisation took place in the form of a mass meeting organised by the All Mishing Student Union in the Gogamukh project area. The meeting discussed the project and its EIA report in the context of the report of the World Commission on Dams. An appeal was made for a second public hearing, given that people had not been aware of the one held in 2001, since it had been announced in an English-language newspaper that was not widely circulated in the area. Nor had people been properly informed in the EIA report of the downstream impact of the project or of the failure of the project to comply with the requirements laid down by the WCD. Those attending the meeting demanded the abandonment of the Lower Subansiri Project in its current form, given the geological fragility of the Eastern Himalayas, and calling instead for the construction of a smaller, eco-friendly dam (Menon and Kohli 2005, p. 201). Copies of the minutes of this meeting were sent to the regulatory authority and to the project developer. Despite this Lower Subansiri Project received environmental clearance from the MoEF on 16 July 2003, subject to certain conditions.

The conditional environmental clearance is difficult to interpret. While it gives the project environmental clearance and permits work to go ahead, it also calls on the project developer to undertake studies of the impact of the project on aquatic fauna, biodiversity and habitat conservation falling within the project submerged area (Chatradhar 2009). Ideally, these studies should have formed part of the EIA report, on the basis of which the project was to have been appraised. But by giving clearance and at the same time asking for more studies, the MoEF seemed to be indicating that, whatever happened, the project should and would proceed, the studies being undertaken more to legitimise environmental clearance. Thus such studies are not just collections of scientific facts but political tools used for instrumental purposes.

By September 2004 a local community-based institution in the area downstream from the Lower Subansiri Project, the People's Movement for Subansiri Brahmaputra Valley (PMSBV) was already vehemently opposing the project. Along with other local civil society organisations, it arranged a meeting in Arunachal Pradesh, at which both the Assam and Arunachal Pradesh State Pollution Control Boards and the project developer were criticised for their unwillingness to address the concerns of the downstream population (Chatradhar 2009).

While the project developer was going ahead with the construction, the agitation at local level was gaining ground. The PMSBV was also becoming more popular among the local masses and broadening its base by forging relations with other local-level issue-based or mass-based organisations. At national level such civil society organisations as the Kalpvriksh Environmental Group and the South Asia Network of Dams Rivers and People were always present, and there were numerous exchanges of information between these two levels. The PMSBV also linked up with the International Rivers Network (PMSBV, personal communication, April and May 2009).

On International Rivers Day in March 2005 a mass protest against the Lower Subansiri Project was organised by the various local groups on the banks of the

River Subansiri, and a mass petition was submitted to the Life Insurance Corporation of India, which had meanwhile signed a Memorandum of Understanding agreeing to finance the project. In October 2005 a mass rally of 5,000 people was held under the PMSBV banner to protest against the LSP. Memorandums were sent to the Prime Minister of India, the MoEF, the Chief Minister of Assam and the Chief Secretary to the Government of Arunachal Pradesh, and it was demanded that the construction of the Lower Subansiri Project (LSP) should be suspended until a scientific study report had been published on the possible downstream impacts of the project and a public consensus in favour of the project had been obtained in the project areas (Chatradhar 2009). This was the first time that it had been demanded at local level that a project should be halted until a detailed downstream impact study had been conducted. In the following years this issue of a downstream study was to shape the movement protesting against the LSP (Chatradhar 2009).

In late 2006/early 2007, the protest organisation changed its strategy and involved the All Assam Student Union (AASU). The AASU is the most powerful student organisation not only in Assam but in the entire north-eastern region. As information critical of the LSP started flowing from the AASU's platform, more people suddenly began to take an interest. Various government stakeholders and the project developer, perhaps for the first time, started to take the views expressed against the project seriously (Partho Das, personal communication, 20 Apr 2009). While the AASU was not willing directly to oppose the project, it also did not want to ignore the mass appeal against the project. This resulted in a middle course being taken. On 8 Dec 2006 a tripartite meeting of the government of Assam, the project developer and AASU was held. This meeting proved to be a watershed in the events surrounding the Lower Subansiri Project. It was decided that a specialist committee, comprising scientists from three prominent research institutions in the area, would be set up to study the entire downstream impact of the LSP (AASU 2008, pp. 102–103). However, there was a difference of opinion among those attending the meeting over the outcome of the study and its bearing on the project decision-making. In its communications with various authorities the AASU maintained that it had been decided at the meeting that the main work on the dam would not begin until the study was completed and gave the project developer a clean sheet (AASU 2008, pp. 102–103). The project developer, on the other hand, claimed that no such decision, to stop work on the project work until the impact assessment study had been completed, was taken at the meeting (AASU 2008, p. 108).

While AASU involvement immediately brought a change in December 2006, things again drifted, 1.5 years elapsing before the study began. In the meantime the construction of the dam continued. Many of the organisations agitating against the project grew increasingly impatient because they could sense a deliberate strategy behind the delay in the study and the continuation of construction until the project reached the point of no return and became a *fait accompli*. Thus pressure was also mounting on the AASU, which then wrote letters to the Prime Minister, the Parliamentary Standing Committee and the project developer calling for a halt to the work until the study was completed and threatening agitation in the alternative (AASU 2008, pp. 102–103). The study finally began in mid-2008.

The study made some critical comments on the project, particularly on the location of the construction of the dam. But those were contested by the project developer. The state government of Assam initially took an ambivalent position. Over time the state government hardened its position in support of the project (Baruah 2012, p. 46). On the other side the local populace got increasingly agitated with the developments over the project. As a result, since 2011, a road-blockage continue taken place to physically stop any project material reaching the construction site. The state government of Assam came down heavily on the protestors (Baruah 2012, p. 41; Borpujari 2012). As on date the project faces a stalemate.

4.4.3 The Dibang Multipurpose Project

The Dibang Multipurpose Project is a 3,000 MW project which includes the construction of a concrete gravity dam 288 m high and a reservoir with an area of 40.09 km^2. Once completed, the dam will be the highest in India. The project is located in the lower Dibang Valley district in Arunachal Pradesh. The submergence area is around 938 ha, which encompass five villages and around 72 households. Some 14 families have been classified as partially affected: they will lose about 557 ha of land (National Productivity Council 2010).

According to the All Idu Mishmi Student Union (AIMSU), the project developer has been engaged in various survey-related activities since 2002. In April 2007 the Arunachal Pradesh Pollution Control Board announced in the local newspapers that public hearings would be held in Roing and Arzoo. The AIMSU and the Idu Cultural and Literary Society (ICLS), two local organisations in the project areas, requested the State Pollution Control Board to defer the public hearing dates. The public hearings were then postponed until June/July 2007. These two organisations began looking for help outside the state and in other areas where people were more aware of dam-related activities and could give these organisations some guidance. Subsequently, the AIMSU and ICLS came into contact with Kalpvriksh, the Rural Volunteers Centre (RVC), Akajan, and the National Forum of Forest People and Forest Workers, New Delhi. These organisations were already active in the North East in general and more especially in the Lower Subansiri Project area. Their aim was to make people aware of various complexities associated with dam projects. As public awareness increased, the AIMSU and the ICLS informed the Arunachal Pradesh Pollution Control Board that more time was needed to study and understand the EIA report. The public hearings were therefore delayed until the last week of November 2007.

In the meantime the AIMSU and ICLS launched a public awareness campaign in collaboration with the Kalpvriksh Environmental Action Group and the Legal Initiative for Forest and Environment, New Delhi. By the end of this campaign it was generally held that large dams and large hydro projects were not in the interests of the community and their construction would be opposed. However, the local political leaders in the project area wanted to see the project begun quickly and so initiated a countermovement with a view to creating a constituency in favour of the dams in the area. This led to the establishment of another institution, the Dibang Basin Welfare Committee, which organised public meetings to facilitate the forthcoming public hearings, informing the population of the advantages of such mega-projects.

The AIMSU and its partner organisations opposed the project for a number of reasons. First, it was feared that the Idu Mishmi tribe, its members already few in number in the country, would be badly affected by the projects. The agitating organisations argued that the public interest clause should not override the interests of the local people, the Idu Mishmi tribe. The number of people displaced in absolute terms is much smaller than in some of the other mega-projects, such as the Tehri Hydro Electricity Project in Uttaranchal and the Sardar Sarovar Project in Gujarat. But the Idu Mishmi population is already very small, and the project would displace around 8 % of the total number of Idu Mishmi people. The second argument against the project concerns the territoriality of the state's tribal population. If the displaced Idu Mishmi community was resettled in the territory of other tribes, there would be competition for common property resources. This would result in intense intertribal conflicts. Third, the AIMSU has also been arguing against the popular notion that the Dibang Multipurpose Project will usher in development in the hitherto remote and underdeveloped area by providing jobs, awarding contracts and increasing business among the local people. According to the AIMSU, only the "creamy layer" within the community will be awarded, at best, with some insignificant contracts, while the rest of the people will be seen as a source of unskilled labour for the construction site and of domestic help in the houses of the employees of the project developers (AIMSU, personal communication 2009).

The potential downstream impact attributable to the geological features of the project area has also led the movement to assume a spatial pattern in which the AIMSU, which is focusing more on impacts in the submergence area, has collaborated with downstream community-based organisations like Sodiya Mahakuma Suraksha Samiti and Saikhowa Suraksha Samiti. Both these organisations are based in the downstream state of Assam and concentrate more on the downstream impact. The people of Assam were aware that a mega-project was about to be implemented in the Lower Dibang Valley district of Arunachal Pradesh, but they did not have any more information. AIMSU started sharing information on the project with downstream organisations, and gradually the people of the downstream town of Sodiya became aware of the project (Sodiya Mahakuma Suraksha Samiti, personal communication 2009). This process was also aided by the PMSBV, which was then leading the movement concerned about the downstream impact of the Lower Subansiri Project (PMSBV, personal communication 2009).

To understand the role played by the downstream non-governmental organisations, it will be important to consider the broader context of the history of the town of Sodiya. Sodiya was once a prosperous town, but was devastated in 1950 by an earthquake and the ensuing floods. The people therefore remember how destructive floods can be. These concerns were rekindled in 2004 and 2008 by news from other parts of Assam, where flooding was attributed to dams upstream in Bhutan and Arunachal Pradesh. Thus, having accessed concrete information from the AIMSU about the proposed Dibang Multipurpose Project, the downstream organisations initiated active opposition to the project. In this way the Dibang has assumed a unique pattern, with civil society organisations in the upstream state of Arunachal Pradesh and the downstream state of Assam jointly engaged in agitation.

The first public hearing finally took place on 29 Jan 2008. It continued for 12 h. Afterwards the people were sure that the project would be scrapped, since most of those attending the hearing opposed it. The second public hearing was meant to be held on 31 Jan 2008. But the people then discovered that the Prime Minister of India was coming to Arunachal Pradesh to lay the foundation stone for the project on that very day. This attracted considerable criticism from the project area and outside (Veracity 2008; Ghosh 2008) and hardened the position adopted by the agitating organisations. They felt that the public hearing had been no more than a procedural requirement, not a genuine process of public consultation. It was thus decided that there was no need to hold token public hearings, and a second public hearing was opposed. Repeated efforts were subsequently made by the state pollution control board and the project developer to hold the public hearing, but the agitating organisations, with the support of the local population, would not allow it to take place.

The sustained mass movement of local civil society organisations opposing the project and preventing the public hearing from taking place has not only delayed the process of environmental clearance, but with time elicited some favourable responses from the MoEF's Expert Appraisal Committee. The Expert Appraisal Committee instructed the project authority to reapply for the Terms of Reference of the EIA study because the project could not be cleared by 13 Sept 2008[7] and

[7] The Dibang Multipurpose Project had a site clearance based on the EIA Notification 1994. When the Government of India produced the 2006 EIA Notification, it set a period of 2 years (until September 2008) within which the project would be appraised under the 1994 EIA Notification if submitted to the MoEF. As the sustained local community agitation prevented the second obligatory public hearing in the Dibang Valley from being held by September 2008, it was decided at the EAC's 27th meeting in June 2009 that the Dibang Project would have to reapply

also declared that fresh public hearings needed to be held in both the affected districts. The public hearing held in Roing in January 2008, which was widely criticised by the agitating civil society organisation, became ineffective. Apart from this, the new terms of reference for the EIA study issued by the EAC in July 2009 met many of the demands voiced by the agitating organisations.

4.5 Findings and Conclusion

Decision-making on dams in India involves multiple actors at various stages, each with specific functions and responsibilities. Thus it is more akin to a polycentric decision-making setting where the authorities at the respective levels are endowed with certain powers. The role of public participation is formally limited but, because the process of decision-making is long and involves many actors at various levels, influence from the general public and civil society organisations is undeniable. In recent years civil society organisations have influenced decision-making by directly accessing the governmental or government-empowered committees. If violations or complaints are not noted and addressed by the administrative or regulatory system, the people have approached the judiciary and filed lawsuits.

Historically, the role of public participation in the decision-making process has been limited, that process consisting solely of interaction between government departments. This has changed little over the years. What change there has been has occurred in the environmental sphere, where the only organised form of public participation, the public hearing, is to be found. This chapter has shown that the public hearing is the only sounding board for public opinion on dam projects. It thus becomes one of the most contested and controversial arenas. As has been evident over the years from the EIA notifications, the public hearing has been one of the more important areas for reforms.

To perform as a scientific tool which correctly anticipates future environmental impacts and suggests mitigation measures, EIA is based on a number of assumptions. The first is that EIA accurately defines a socio-environmental baseline situation, the second that it correctly identifies impacts and does not underestimate them, the third that it suggests suitable mitigation measures and the last that the mitigation measures are fully implemented. If all these assumptions are to be fulfilled, those involved in the EIA process must focus on *nut-and-bolt*

(Footnote 7 continued)
for the EIA study under the 2006 EIA Notification and also to hold fresh public hearings in the two districts. Thus the public hearing in the Lower Dibang Valley district was considered null and void under the new directions.

issues. Analysis of the EIA Notification and the point of contention in the various cases studied show that there are structural constraints both in the notification and in its actual implementation.

Inadequate baseline information and associated impacts at one level has been the main point of contention in the various projects. There have also been persistent allegations of failure to comply with the environmental management plans. In the notification the monitoring of compliance and the sanctioning of violations are the most weakly addressed. While the 1994 EIA Notification was strong in its insistence on sanctions, there were persistent allegations that the quality of environmental reports was poor (Menon and Kohli 2009). This showed that a strongly worded text is not necessarily translated into practical results. Instead of addressing this issue, the 2006 EIA Notification relaxed the sanctions. Most of the structural constraints in the EIA Notification are not addressed because EIA is often perceived as an administrative hurdle rather than a scientific tool.

The lacunas in the environmental clearance process have often triggered controversies and arguments against dam projects in recent years. The vehement opposition to the projects in the form of criticism of the EIA reports or the EIA process frequently results in ad hoc decision-making at project level. This may strengthen the environmental standards that are then followed in the project concerned, but the structural constraints to proper identification of impacts, mitigation and proper monitoring of the mitigation remain unresolved.

4.5.1 World Commission on Dams and Local Norms

This study also argues that, rather than the WCD norms being gradually diffused in domestic civil society, it was the norms that were circulating in the domestic policy domain which became an international anchor when the WCD was established.

As early as 1989 the National Working Group based its argument for developing a resettlement policy to the benefit of displaced people on the principles of equity, participation and sustainability. More than a decade later, in 2000, the WCD recommendations defined equity, sustainability and participation as the values which would "*run through the entire report*" (WCD 2000, p. 199). The National Working Group's 1989 draft makes a strong case for comprehensive option assessment. It argues that, as projects form part of a realm of "socioeconomic realities" and are "neither formulated nor executed in a social and political vacuum", so "choices should be made from among various alternatives" and decisions on a project should not be a *fait accompli*. The option assessment argument, as opposed to a *fait accompli*, has again been placed within a strong framework capable of sustaining human rights. The National Working Group's draft also argued that *participation* should not depend on the project developer. Rather, it should be the right of the affected population to take part in the decision-making on such projects from "inception to completion, and even thereafter." The draft thus recommended something similar to "gaining public acceptance," as the WCD report was to do later. To make the participation process really effective rather than symbolic, the National Working Group also called for a right of access to information. In the line it follows, in the principles it formulates and in the approach it adopts to human and constitutional rights, the National Working Group's 1989 draft policy is thus similar to the WCD's report. Subsequently, in 1994 and 1998, similar arguments were advanced as an alternative to the drafts proposed by the GoI.

In these circumstances it is therefore not surprising that, although the GoI and the Ministry of Water Resources rejected the report outright; sections of civil society would continue and still continue to espouse it. Indian civil society was also proactive in standing by its views throughout the WCD process. This was catalysed by the fact that one of the leading civil society leader who has been a prominent face in leading an anti-dam movement became one of the WCD Commissioner. On the other hand the government's attitude was not enthusiastic (Iyer 2003).

The era of dam construction until the early 1980s can be described as a period of *innocent ignorance*[8] and the time when social and environmental movements and civil-society-driven consciousness-building began to emerge. These developments have been attributed to transnational norm diffusion and transnational collective action by civil society (Khagram 2004), as the great controversy over the Narmada Sardar Sarovar Project has demonstrated. On the basis of cross-sectional and cross-temporal studies it is argued here that change in India cannot entirely be attributed to international actors. Rather, it is argued that the incorporation of superior standards in national policy documents and steps to ensure that superior practices are followed at project level are outcomes of a long process of negotiation between state actors and such domestic non-state actors as academics, social activists and NGOs. At the same time the multilateral and bilateral development agencies and financial institutions have had a strong *adaptive* influence on the government. Most of the draft resettlement policies prepared in the 1990s referred explicitly to requirements requested by multilateral financial institutions as one of the factors underlying the drafting of a rehabilitation and resettlement policy.

Analysis of the content of government policies drafted by various ministries and agencies over the years reveals a gradual progression in that the resettlement package or resettlement planning per se becomes more inclusive, whereas attitudes remain highly exclusive when it comes to discussing strategic issues, such as *gaining public acceptance* or *comprehensive option assessment*, relating to the selection of a dam from among various options.

[8] This term was coined by R. Ramaswamy Iyer during the Sixth Dialogue on Water held in Bonn in September 2009.

4.5.2 International Finance Institutions as Norm Carriers

The presence of the International Finance Corporation promoted improved standards in the Allain Duhangan Hydropower Project during the clearance process. Public participation in this project prior to investment clearance was better informed than in other, domestically funded projects. It should be realised that, when this project was submitted to the IFC, it had already received all the clearances required by the GoI. It was only because the IFC became involved and had its own information disclosure system that a great deal of information was put in the public domain. This then formed the basis for the lobbying by civil society organisations, which resulted in the appointment of independent regulators for public hearings and of independent organisations to demystify the EIA reports for villagers before public hearings. These procedures are not common in India and can be attributed to the IFC's presence. The reforms, along with a series of negotiations between families affected by the project and the project developer on cash compensation, were driven by NGOs and local community mobilisation. But it would not have been possible, or would have been many times more difficult, if the IFC had not been mandated to upgrade standards which were found to be lacking by the NGOs. In this regard a comparison can be made with the Lower Subansiri Project, where there were no such international actors. As this was one of the first large dam projects in the region, general awareness was initially low in the community and local civil society organisations. It has always been alleged that the people were not adequately informed about the Lower Subansiri Project or about their role, rights and responsibilities during and before public hearings. It was alleged by the civil society organisations represented at the Lower Subansiri public hearing that, when one of the regulators present raised some points in opposition to the project, he was told that, as a regulator, he was required to support it.

In terms of the processes leading to the clearance, the IFC's presence ensured that Allain Duhangan fared much better than Lower Subansiri. But with regard to such issues as whether the local populace had a voice in the project decision-making, and whether the adoption of socially and environmentally sustainable norms was a precondition for project approval, the two projects were similar. While fewer people may have been dissatisfied with Allain Duhangan than with Lower Subansiri, both projects went ahead against a background in which a majority of the people were dissatisfied with the ongoing clearance process and wanted it to be halted, postponed or rejected. The IFC's presence ensured that there was a strong environmental management plan, which the project developer was meant to follow. But in the post-clearance phase the project developer was often pulled up by the state government and fined for contravening the clearance. Thus it was also found that the approved environmental management plan was not strictly monitored even where the IFC was present. In the case of Allain Duhangan many of the reforms following the IFC's clearance of the project were undertaken as a result of lawsuits. Here again, it resembled other projects where the judiciary had been one of the agents in the introduction of progressive reforms. Evidence from Allain Duhangan indicates that, although the presence of international actors may be a *necessary* condition for the application of superior standards, it is not *sufficient* to ensure compliance with them.

4.5.3 Social Movements as Agents of Norm Diffusion

Social mobilisation by civil society organisations has been the main catalytic agent in bringing changes to project decision-making. The pattern of civil society action is quite different in the three projects. If social mobilisation and civil society action are unpacked, it is found that, while NGOs at national and regional level spearhead the mobilisation of civil society, there are also local non-governmental, mass-based and political organisations at project level acting as pressure groups. Some or all of these actors may try to form a coalition to coordinate their activities and so become a pressure group in opposition to the project developer. In the process they will try to influence project decision-making and bring about incremental changes and reforms of projects.

Until the Allain Duhangan Project obtained clearance, the movement was led largely by regional and national organisations supporting the villagers. But at the post-clearance stage the movement was spearheaded solely by the village-based organisation. At that stage the local organisation perhaps continued to maintain some loose coalition with a national organisation through exchanges of information, but there was no coalition in the area of strategic decision-making or action.

In the case of the Lower Subansiri Project, on the other hand, the movement was largely led by national and regional NGOs until the end of 2004. Local mobilisation was still diffuse. At this stage it was the national or regional NGOs which tried to affect decision-making by engaging directly with the regulatory authorities or by filing lawsuits in the Supreme Court. One of the objectives was obviously to stall or at least modify the project, but a by-product of this was that decision-making at regulatory level was far more informed and the clearances issued were subject to strict environmental conservation conditions. The Lower Subansiri Project differed from the Allain Duhangan Project, however, in that the coalition between local NGOs and national or regional NGOs continued in the post-clearance stage and grew in strength. And as social mobilisation at the local level became stronger, it was able to attract sufficient attention from various powerful local actors, such as the politicians in the state and mass-based organisations. When all these actors eventually formed a coalition, they were able to pressure the project developer into rethinking its strategy. This manifested itself in the project developer's commissioning of a comprehensive downstream impact study of the Lower Subansiri Project.

In the case of the Dibang Multipurpose Project two things happened simultaneously. First, by the time this project had become contentious, the people were fairly well aware of the consequences of dam-building. The community in the project area is compact, and the movement was led by strong local organisations from the outset. Thus while Allain Duhangan, seen in hindsight, was unable to prompt a strong mass movement in its area, Lower Subansiri generated strong social mobilisation at local level, but rather late, when the project was already at an advanced stage of construction. In contrast to the other two, the Dibang Multipurpose Project was confronted from a very early stage with a strong local mass movement, which subsequently spread into the areas downstream from the project and generate a unique form of upstream–downstream interstate social mobilisation. At the same time, the various movements were able to forge successful coalitions with NGOs outside the region, including some mass organisations already taking action in the adjoining Lower Subansiri area. Some of what the civil society organisation had learnt from Lower Subansiri was thus transferred to the civil society organisations in the Dibang valley. What proved to be the most effective strategy pursued in the case of the Dibang Multipurpose Project by the local NGOs was to stall the public hearing process. At the same time, the success of this strategy also reveals the weaknesses that exist in the environmental clearance process. There is no doubt that India requires a complete overhaul of the environmental clearance system.

Acknowledgments The author acknowledges the generous support of International Postgraduate Studies in Water Technologies (IPSWaT) for supporting the field research for this study. The author would like to thank Parthojyoti Das, Keshab Chatradhar, Tony Mikrows and Sunder Mahant for their help during the fieldwork. The author would also like to thank the many resource persons in India, who have enormously helped the author in developing a better understanding of environmental governance in India. Author acknowledges the help of Saptarshi Dey in creating a map depicting the study locations.The author would also like to thank Ramaswamy R. Iyer and Peter Mollinga for their comments on an earlier extended version of this chapter.

References

AASU (All Assam Student Union) (2008) Pratwabhan 6th annual Gogamukh conference by All Assam Student Union Dhemaji District, 19–21 Nov 2008

ADHPL (Allain Duhangan Hydroelectric Project Ltd.) (2004) Affidavit of Allain Duhangan Hydroelectric Project, 7 May 2004

Baruah S (2012) Whose river is it anyway? Political economy of hydropower in the Eastern Himalayas. Economic and Political Weekly 47(29):41–52, 21 July 2012

Borpujari P (2012) Damning the dam protestors. http://priyanka-borpujari.blogspot.in/2012/05/damning-dam-protesters.html. Accessed 1 Aug 2012

Cernea MM (2006) Resettlement management. Denying or confronting risks. In: Mathur HM (ed) Managing resettlement in India. Approaches, issues, experiences. Oxford University Press, New Delhi

Chatradhar KC (2009) Lower Subansiri: downstream impact chronology. Unpublished document
Civil Writ Petition (2007) Himachal Pradesh High Court CWP 1421 of 2007, 11 Sept 2007
Compliance Advisor Ombudsman (2004) Assessment report. Complaint regarding Allain Duhangan Hydropower Project Himachal Pradesh. India office of the Compliance Advisor Ombudsman International Finance Corporation Multilateral Investment Guarantee Agency, March 2004
Compliance Advisor Ombudsman (2006) Progress report. Complaint regarding the Allain Duhangan Hydropower Project Himachal Pradesh. India office of the Compliance Advisor Ombudsman International Finance Corporation Multilateral Investment Guarantee Agency, 28 August 2006
Compliance Advisor Ombudsman (2008) Allain Duhangan Hydro Power Project. India Complaint Conclusion Report Office of the Compliance Advisor Ombudsman International Finance Corporation Multilateral Investment Guarantee Agency, March 2008
Correspondent (2004) Clearance to hydel project opposed. Tribune News Service, 18 May 2004
D'Monte D (1986) A dam too far. Economic and Political Weekly 19(47), 24 Nov 1986
Dwivedi R (2006) Conflict and collective action: the Sardar Sarovar Project in India. Routledge, New Delhi
EPW Correspondent (1972a) Environmental planning. What a committee can't do. Economic and Political Weekly 7(18), 29 April 1972
EPW Correspondent (1972b) Development and environment. Economic and Political Weekly 7(25), 17 June 1972
EPW Editorial (1987) Format for environment. Economic and Political Weekly 23(18), 2 May 1987
Fernandes W (2004) Rehabilitation policy for the displaced. Economic and Political Weekly 39(12):1191–1193, 20 March
Fernandes W (2008) The 2006 rehabilitation draft: more of the same?. Mahanirban Calcutta Research Group, Kolkata
Fernandes W, Paranjpye V (eds) (1997) Rehabilitation policy and law in India: a right to livelihood. Indian Social Institute, Pune. Econet, New Delhi
Fox JA, Brown LD (eds) (1998) The struggle for accountability: the world. NGOs and grassroots movements. Massachusetts Institute of Technology, Cambridge
Ghosh P (2008) Public hearing: a mere formality? Livemint.com. Wall Street J. http://www.livemint.com/2008/02/11003825/Public-hearing-a-mere-formali.html. Accessed 26 July 2010
GoI (Government of India) Ministry of Water Resources (2001) Letter 2/WCD/2001/DT (PR) III (2)
GoI (Government of India) Central Electricity Authority (2008) Status of hydro electric potential development CEA, August 2008. http://www.cea.nic.in/hydro/Status%20of%20Hydroelectric%20Potential%20Development.pdf. Accessed 18 Sept 2008
GoI (Government of India) Ministry of Finance (2008) Expert group to examine the schemes of statutory clearances for industrial and infrastructure projects in India, Department of Economic Affairs, Ministry of Finance. http://www.indiaenvironmentportal.org.in/files/Report%20Ministry%20of%20Finance.pdf. Accessed 2 Sept 2009
Guha R (2007) India after Gandhi. The history of the world's largest democracy. Picador Publications, India
International Rivers (2004a) http://www.internationalrivers.org/en/node/893. Accessed 2 Sept 2009
International Rivers (2004b) http://www.internationalrivers.org/en/south-asia/india/affected-people-world-bank-regarding-allain-duhangan-hydro-project. Accessed 2 Sept 2009
International Rivers (2004c) http://internationalrivers.org/en/south-asia/india/independent-observer-world-bank-regarding-allain-duhangan-project. Accessed 31 Dec 2008
Iyer RR (2003) Water perspectives issues concerns. Sage Publications India Pvt Ltd, New Delhi
Iyer RR (2007) Towards a just displacement and rehabilitation policy. Economic and Political Weekly 42(30):3103–3107, 29 July 2007

Khagram S (2004) Dams and development. Transnational struggles for water and power. Oxford University Press, New Delhi
Klingensmith D (2007) One valley and a thousand. Dams nationalism and development. Oxford University Press, New Delhi
Maudgal S (1991) Environmental impacts of water resources development projects. Central Board of Irrigation and Power: environmental impact assessment for water resources projects, Proceedings of regional workshops. Ministry of Environment and Forests, New Delhi
McCully P (2001) The use of a trilateral network: an activists' perspective on the formation of the World Commission on Dams. Am Univ Int Law Rev 16(6):1453–1475
Menon M, Kohli K (2005) Large dams for hydropower in North East India. A dossier. Kalpavriksh and South Asia Network of Dams Rivers and People, New Delhi
Menon M (2007) Kohli K (2007) Environmental decision-making: whose agenda? Economic and Political Weekly 42(26):30
Menon M, Kohli K (2008) Reengineering the legal and policy regimes on environment. Economic and Political Weekly 43(23):7
Menon M, Kohli K (2009) From impact assessment to clearance manufacture. Economic and Political Weekly XLIV 28:20–23, 30 June 2009
Mohan D (1991) Need for environmental impact assessment. Central Board of Irrigation and Power: environmental impact assessment for water resources projects, Proceedings of regional workshops, Ministry of Environment and Forests, pp 16–17, New Delhi
Mohanty B (2005) Displacement and rehabilitation of tribals. Economic and Political Weekly 40(13):1318–1320, 26 March 2005
Mukherjee A (2002) Cabinet secy for implementing panel report on investments. Business Line, 12 Dec 2001
Nandimath OV (2009) Oxford handbook of environmental decision making in India: an EIA model. Oxford University Press, New Delhi
National Productivity Council (2010.) Executive summary of EIA and EMP of Dibang Multipurpose Project. http://envfor.nic.in/divisions/iass/Executive%20Summary_English.pdf. Accessed 2 Sept 2009
NCIWRD (National Commission for Integrated Water Resource Development) (1999) Integrated water resource development: a plan for action. Report of the National Commission for Integrated Water Resource Development (NCIWRD) I
NHPC (National Hydroelectric Power Corporation) Ltd (2002) Executive summary of Subansiri lower hydroelectric project Arunachal Pradesh and Assam (8 × 250 MW)
Rangacheri R et al (2000) Large dams: India's experience. A WCD case study prepared for World Commission of Dams. Cape Town, South Africa
Roy A (1999) The greater common good. India Book Distributor, Bombay
Roy D (2008) Hydropower in Uttarakhand: is 'development' the real objective? Economic and Political Weekly 43(41):19, 11 Oct 2008
Sachs I (1971) Environmental quality management and development planning. Economic and Political Weekly 6(30), 6 Nov 1971
Saldanha LF et al (2007) Green tapism. A review of the Environmental Impact Notification 2006. Environment Support Group, Bangalore
Sengupta N (2000) A brief review. In: Rangachari R et al (2000) Large dams: India's experience. A WCD case study prepared for World Commission of Dams. Cape Town, South Africa
Shah Z, Kumar MD (2008) In the midst of the large dam controversy: objectives, criteria for assessing large water storages in the developing world. Water Res Manage 22(12):1799–1824. doi:10.1007/s11269_008_9254_8
Singh S (2006) Displacement and rehabilitation. A comparison of two policy drafts. Economic and Political Weekly 41(52):5307–5309, 30 December 2006
TERI (The Energy and Resources Institute) (2006) National energy map of India: technology vision 2030. The Energy and Resources Institute and Office of the Principal Scientific Advisor Government of India, New Delhi

Vagholikar N, Ahmed MF (2003) Tracking a hydel project. The story of lower Subansiri. The Ecologist Asia (11)1:25–32, January–March 2003
Veracity (2008) Protest over Dibang project hits the media around the globe: is environmental clearance a mere formality? Veracity, 10 February 2008
WCD (World Commission of Dams) (2000) Dams and development. A new framework for decision making. EarthScan publications Ltd, London
Wood JR (2007) The politics of water resource development in India: the Narmada dams controversy. Sage Publications Pvt Ltd, New Dehli
World Bank (2004) Implementation completion report (IDA 29300) No. 28942. http://www-wds.worldbank.org/external/default/WDSContentServer/WDSP/IB/2005/06/14/000012009_20050614163544/Rendered/PDF/289420rev.pdf. Accessed 2 Sept. 2009

Chapter 5
Sustainable Dam Development in Turkey: Between Europeanization and Authoritarian Governance

Waltina Scheumann, Vera Baumann, Anna-Lena Mueller, Dennis Mutschler, Sylvia Ismail and Thomas Walenta

Abstract The chapter explores reforms of Turkey's environmental, expropriation and resettlement policies as well as the influence of domestic and external actors on dam-related legislation and practices, in particular on the Ataturk and Birecik dams on the Euphrates River, the Ilisu Dam on the Tigris River, the Ermenek Dam on the Goksu River and the Yusufeli Dam on the Coruh River—all of which were built in different periods and under varied political and economic circumstances. The analysis starts with an overview of the strategic role of hydropower in the economic development plans and renewable energy policy of the Turkish government. The chapter singles out two issue areas—environmental impact assessment and expropriation/resettlement—in order to examine legal changes and practices as well as the reasons therein in the context of political developments.

W. Scheumann (✉)
Department of Environmental Policy and Natural Resources Management, German Development Institute—Deutsches Institut für Entwicklungspolitik (DIE), Bonn, Germany
e-mail: Waltina.Scheumann@die-gdi.de

V. Baumann
Division for Economic and Development Cooperation, German Embassy, Tirana, Albania
e-mail: Vera.Baumann@bmz.bund.de

A.-L. Mueller · T. Walenta
KfW German Development Bank, Frankfurt, Germany
e-mail: Anna-Lena.Mueller@kfw.de

T. Walenta
e-mail: Thomas.Walenta@kfw.de

D. Mutschler
Gesellschaft für Internationale Zusammenarbeit (GIZ), Corporate unit AgenZ, Hamburg, Germany
e-mail: dennis.mutschler@giz.de

S. Ismail
German Federal Ministry for Economic Cooperation and Development (BMZ), Bonn, Germany
e-mail: Sylvia.Ismail@bmz.bund.de

W. Scheumann and O. Hensengerth (eds.), *Evolution of Dam Policies*,
DOI: 10.1007/978-3-642-23403-3_5, © Springer-Verlag Berlin Heidelberg 2014

Turkey's aspiration to become a full member of the European Union has influenced its environmental policy; reforms related to expropriation and resettlement have mainly been driven by domestic politics.

5.1 Introduction

At the Fifth World Water Forum in March 2009 in Istanbul, Professor Dogan Altinbilek, a former General Director of Turkish State Hydraulic Works (DSI), took a stand in the ongoing debate on large dams: "*The answer on the worldwide discussion on dams is not* Yes *or* No *but wise planning and implementation.*" His statement expressed a then valid consensus in Turkish society: all political parties represented in parliament, the Turkish public at large and even Turkish non-governmental organisations (NGOs) criticised practices in specific dam cases but did not deny the general need for dams.

Single and multi-purpose dams are expected to provide water and hydro-electricity. But it is particularly hydropower that ranks high on the Turkish government's renewable energy agenda, more so than any other source of renewable energy and technology. As of yet, Turkey has developed only one-third of its vast hydropower potential, which is low compared to other countries from the Organisation for Economic Co-operation and Development (OECD), such as the United States, which has developed 86 % of its potential (DSI 2009, p. 31). Most of the hydroelectric resources that are economical to develop have already been exploited in the OECD countries, and only a few large-scale hydroelectric projects are planned; Turkey and Canada are exceptions[1] (US EIA 2013). The Turkish government is pursuing an ambitious programme to further construct a large number of dams and hydroelectric power plants (HPPs) to satisfy ever increasing energy demands.

This programme is not uncontested. What started as local protests confined to individual dam projects—namely the Ilisu Dam on the Tigris River and the Yusufeli Dam on the Coruh River to mention but a few—has meanwhile evolved into a country-wide civil society movement addressing inadequate expropriation and resettlement practices as well as impacts on the environment, cultural heritage, people's rights and their ways of life. While the benefits of dams are praised, negative environmental and social effects are downplayed by the government; "wise planning and implementation" has yet to be achieved.

This study, like those on Brazil (da Costa, this volume), China (Hensengerth, this volume) and India (Choudhury, this volume), is interested in the dynamics and

[1] In non-OECD countries, hydroelectric power is the predominant source of growth for renewable electricity. Strong growth in hydroelectric generation is expected in China, India, Brazil and in a number of nations in South-east Asia, including Malaysia and Vietnam (US Energy Information Administration 2013).

Fig. 5.1 Dam sites. *Source* Geology and Earth Science 2009

conditions that foster changes by decision-makers regarding dam-related policies. It specifically inquires whether international norms related to environmental protection and expropriation and resettlement have entered the domestic arena and influenced policies.

The study was implemented by the Country Working Group Turkey of the German Development Institute (DIE), Bonn, in 2009, and updated in 2013. After preliminary research in Bonn, the team spent eleven weeks in Turkey, where 60 interviews were conducted with public officials, private companies, researchers, consultants, civil society organisations and NGOs.

The dam projects selected for investigation had to satisfy the following criteria (for details of the methodological approach applied in the country studies, see Scheumann and Hensengerth in this volume): the dam projects need to have been planned and built at different points in time in order to study changes related to environmental and resettlement issues, as mirrored in national regulation and in planning and implementing individual projects; and the funding sources of some projects need to be international in order to study these players' influence *vis-à-vis* others. The dam projects that were eventually selected are shown in Fig. 5.1; technical information is provided in Table 5.1.

This chapter is organised as follows: it first reviews the Turkish government's policies for promoting hydropower development and investigates, in a second step, changes in the institutional decision-making framework. It analyses further whether and how environmental concerns have been integrated into dam planning and project implementation as well as the changes related to the legal and administrative regimes that govern expropriation and resettlement. It then portrays the emergence of a social movement that mirrors an eroding consensus in Turkish

Table 5.1 Technical information on the dam projects investigated

Dam name/river	Construction period	Capacity (MW)	Storage volume (BCM)	Project owner (P)/ Operator (O)
No. 1 Atatürk on Euphrates	1983–1992	2,400	48.50	P: DSI O: DSI/EUAS
No. 2 Birecik on Euphrates	1996–2000	672	1.22	P: BOT 15 years O: Birecik Company
No. 3 Ermenek on Goksu	2002–2010	302	4.582	P: DSI O: DSI/EUAS
No. 4 Ilisu on Tigris	2009–2014	1,200	10.4	P: DSI O: DSI/EUAS
No. 5 Yusufeli on Coruh	2013–2018	540	2.13	P: DSI O: DSI/EUAS

Source Turkish State Hydraulic Works
http://www.dsi.gov.tr/baraj-arama. Accessed 18 September 2013

society regarding the massive dam development programme being pursued. Finally, it concludes with a summary about the extent to which international standards have entered the domestic arena and what the driving factors have been.

5.2 Policies Promoting Hydropower Development

Dam-related decision-making essentially takes place in two policy fields, namely water and energy. The strategies followed in both fields are designed to respond to the country's overall economic and social dynamics, such as high population and urban growth rates; seasonal or even chronic water shortages, which many of Turkey's regions are already facing; and an annual increase in energy demand, which ranges between 6 and 8 % (annual growth rate for electricity is 8.1 %), with a per capita energy consumption that is nearly 20 % below the OECD country average (DSI 2009). To prevent a supply crisis and to decrease reliance on energy imports, the Ninth Development Plan (2007–2013) pushed for diversifying the country's energy sources, in particular towards domestic and renewable energy (SPO 2006, p. 82). The plan further emphasised the role of hydropower by pointing out Turkey's huge, technically viable hydropower potential of 216,000 gigawatt hours (GWh) per year, and its economically exploitable potential of 140,000 GWh per year (Republic of Turkey 2009, pp. 19–20). As of 2009, Turkey's hydropower capacity amounted to 14,553 megawatts (MW), with a further 14,000 MW under construction (IEA 2010, p. 97). Turkey's recent goal is to fully develop its feasible hydropower potential by 2023, when the country celebrates the

Table 5.2 Steps of liberalising the energy sector

Year	Law
1980s	Invention of concession models (BOT, BOO, TOOR)
2001	Electricity Market Law (No. 4628)
2003	Regulation on the Procedure and Principle of Signing a Water-use Right Agreement
2005	Law on the Utilization of Renewable Energy Resources for the Purpose of Generating Electrical Energy (No. 5346)
2007	Energy Efficiency Law (No. 5627)
2010	Law on the Amendments on the Law on the Utilization of Renewable Energy Resources for the Purpose of Generating Electrical Energy (No. 6094)
2011	Renewable Energy Support Mechanism, amendment to Law No. 5346

100th anniversary of the Turkish Republic.[2] This translates into a total of 1,738 hydropower plants, which are expected to produce 140,000 GWh per year (Ministry of Energy and Natural Resources 2010; IEA 2010).

The liberalisation of the energy sector was the way chosen to overcome financial constraints and key to attracting private investors. Liberalisation and deregulation had started already in the early 1980s but gained momentum in the 2000s (Baskan 2011), when the Turkish government responded to European Union (EU) policies and global trends (Table 5.2). The architect of this economic transformation was Turgut Ozal, the then Deputy Prime Minister with responsibility for the economy and the implementation of the International Monetary Fund's structural adjustment programme; he became Prime Minister in 1983 and 1989 (see Table 5.7, Sect. 5.7, which provides information on the political changes, on water resources development and on legislative changes in the two policy fields—environment, resettlement—that we are interested in). From 1981 to 1988, the government under Ozal's leadership facilitated the shift to a free market economy, albeit with various negative impacts on social welfare, such as the deepening of sharp disparities between the lowest income groups and the rich segments of society. When Suleyman Demirel became President in 1993, the government continued to pay particular attention to dam construction, which had already started in the 1950s, when Demirel was the first Director General of DSI. It was his government that decided on the South-eastern Anatolia Project (GAP) and dedicated a huge amount of financial resources towards infrastructure development in the underdeveloped regions of Turkey, among them 22 large dams, 19 hydropower plants and large-scale irrigation infrastructure for irrigating about 1.7 million hectares (Kibaroglu and Baskan 2011; Ozduzen 2008; Table 5.7, in Sect. 5.7).

---

[2] http://www.enerji.gov.tr/index.php?dil=tr&sf=webpages&b=hidrolik&bn=232&hn=&nm=384&id=40699. Accessed 27 August 2013.

The Constitution of 1982 that followed the military coup recognized a “right to live in a safe and sound environment” (Article 56) and established the basic principle that water is a public good under the state’s trusteeship. The authority to explore and manage water resources is vested in the state, and water resources in the domain of private law and private proprietorship are subject to title deed registration. In this way, the state has maintained control over water resources development while at the same time initiating liberalisation in the early 1980s as part of its economic transformation programme.

A first step during the 1980s was the replacement of the classic investment model, whereby planning, financing, construction and operation were carried out only by the public sector using concession models (namely BOT, BOO, TOOR and turnkey schemes) (Table 5.2). The Electricity Market Law (No. 4628) was enacted in 2001 to ensure the development of a financially sound and transparent electricity market operating in a competitive environment. The Electricity Market Law covers generation, transmission, distribution, wholesale and retailing of electricity, and it mandated an independent public institution, namely the Energy Market Regulatory Authority (EMRA), for issuing new licences, determining eligible customers, enforcing third-party access and regulating tariffs (Erdogdu 2007). In 2003, a new law removed access barriers for foreign direct investments to the Turkish market (Foreign Direct Investment Law, 17 June 2003).

The Law on the Utilization of Renewable Energy Resources for the Purpose of Generating Electrical Energy (No. 5346, Renewable Energy Law), which entered into force in 2005 and which was amended in 2011, provides an attractive incentive package. The law entails a fixed, guaranteed price over the first 10 years of operation for companies holding a retail sale licence; retail licence holders must buy 8 % of electricity per year from renewable energy certificate holders; renewable energy generators pay only 1 % of the total initial licensing fee and are exempted during the first eight years of operation (PWC 2012, pp. 10–14; IEA 2010, pp. 103–104).[3]

The Turkish government’s long-standing efforts to exploit its hydro potential have more recently also been nourished by prospects to access the Clean Technology Fund and International Bank for Reconstruction and Development loans, which are available for transforming fast-growing developing countries into low-carbon economies (Islar 2012a, b).

Turkey is party to the United Nations Framework Convention on Climate Change (UNFCCC) and ratified the Kyoto Protocol in February 2009. Turkey has a special status within the Annex I countries, as it did not commit to quantitative carbon emission reductions for the post-2012 period. It is also the only OECD country that does not have national emission targets for 2020 (IEA 2010, p. 19).

[3] Besides, the Energy Efficiency Law (No. 5627) of 2007 aims at increasing the efficient use of energy in order to ease the burden of energy costs on the national economy. It covers principles and procedures applicable to increasing and promoting energy efficiency in industry, power generation, transmission and distribution systems, buildings and transport (World Resources Institute 2009).

Turkey therefore cannot participate in the flexible mechanisms of the protocol, such as the Clean Development Mechanism. But the Turkish private sector engaged voluntarily in carbon trading, with renewable project owners selling their carbon credits in the European carbon market (Dogan 2009). As of January 2011, 109 projects were registered, of which the majority are comprised of hydro and wind projects (PWC 2012, p. 18).

The favourable economic framework has brought about a rush for licences, which is usually led by small national companies, in order to acquire the rights to exploit a river's hydroelectric potential. As a matter of fact, small national companies either link with large Turkish companies or sell the licences. Those companies are, as a rule, more experienced and internationally linked, and since international energy companies have shown a keen interest in taking part in these hydropower projects, joint ventures are often established (Interview BM Group 2009; Interview Dolsar Consulting 2009).

So far, government policies have successfully targeted the exploitation of the country's hydroelectric potential. As of January 2012, hydropower plants were the major recipients of renewable energy investments in Turkey. According to PWC, as of January 2012, there were 832 licences assigned by EMRA with a total capacity of 29,570 MW, of which approximately 15,275 MW is in operation, and 14,295 MW is under construction (PWC 2012, p. 20).

A further push can be expected due to an amendment to the Renewable Energy Law, which the Justice and Development Party (AKP) government had proposed and which passed parliament in December 2010. This amendment removed environmental and social barriers to hydropower development: it now allows the construction of dams in and close to protected areas (Sect. 5.4). It was also decided that cabinet orders can decree "urgent expropriation" of some immovable property based on Article 27 of the Law on Expropriation No. 2942 (Sect. 5.5), thus accelerating hydropower projects at an unprecedented scale.

5.3 Decision-Making Framework: From an Interventionist to a Regulatory State

Until the liberalisation of the energy sector, the Turkish State Hydraulic Works and the Electrical Power Resources Survey and Development Administration (EIE) played an overarching role in water resources development and energy planning, implementation and operation (Kibaroglu and Baskan 2011). Basin-wide hydrologic assessments form the basis of water resources development, and master plans were introduced in the late 1940s, but development only gained momentum in 1953 with the establishment of a central water bureaucracy, namely DSI. DSI assumed a role similar to that of the US Bureau of Reclamation—it was the primary executive state agency for planning and managing Turkey's water resources (including its hydro potential). DSI was established as a legal entity under Law No. 6200 with a distinct mandate and its own budget, and it was put in charge of preparing master plans for river basins and project feasibility studies as well as of project implementation and operation (DSI 2009, p. 7).

The classic dam project (small- or large-scale, single- or multi-purpose) used to be carried out solely by the public sector (Baskan 2011, pp. 83–91). Major water infrastructure projects were state-financed and state-managed. Thereby, project planning, construction and operation were generally the responsibility of DSI and EIE. In addition, a central planning body, that is, the State Planning Organisation (SPO), was established in 1961 to steer economic development by creating a strong industrial base. This also meant that public investments in the water sector were arranged through the national Five Year Development Plan. Financing was approved by SPO and provided by the National Treasury, even though in some cases foreign financial institutions were involved, too, for instance, the World Bank in financing the Seyhan Dam on the Seyhan River in the 1960s, and the Karakaya Dam on the Euphrates River in the 1980s.

For a long time, DSI's investment budget amounted to one-third of Turkey's overall public investment budget. However, while the state investment budget increased from US$6.1 billion in 2005 to US$8.7 billion in 2009, DSI's investment budget increased comparatively slower, that is, from US$1.6 billion in 2005 to US$2.1 billion in 2009 (DSI 2009, pp. 13–14). With around one-fourth of the state's investment budget, DSI remained a powerful player: storage dams for water supply (domestic, industrial, irrigated agriculture) and for flood control are still within the public domain, as are dams and hydropower plants in regions of strategic importance, such as the Euphrates-Tigris river basins.

In the overall administrative setting, in accordance with the Electricity Market Law (No. 4628) of 2001, the major public players—besides the private sector—are DSI and EMRA (Fig. 5.2). DSI and EIE have remained the key institutions for investigating the country's hydropower potential, and DSI became the governmental body to approve private hydropower projects. Following the examination of the technical competence and financial backgrounds of the applicants, companies first have to negotiate with DSI on water-use rights agreements.[4] Only then do they get approval and an overall generation licence from EMRA, on the condition that environmental clearance is given by the Ministry of Environment (Sect. 5.4).

Until recently, in concession and private investment models, environmental impact assessment (EIA) studies were issued only after projects were approved and construction licences granted. Now, feasibility studies, which form the basis for negotiating the water-use rights agreements, include a mandatory chapter on potential environmental impacts.[5] Only then does the EIA procedure start,

[4] According to the Turkish Constitution of 1982, water resources are vested in the state domain, and only the government can transfer/assign user rights. These transfers have inadequately been coined as "selling the rivers". Issues related to the transfer of water-use rights are how existing user rights are dealt with; the duration of the use-contracts and whether the state maintains control over private activities (see Baskan 2011).

[5] DSI/EIE Projelerinde Istenecek (On) Fizibilite Raporunda Yer Alacak Ana Basliklar, Bolum -7. Cevresel Etkiler (Genel).

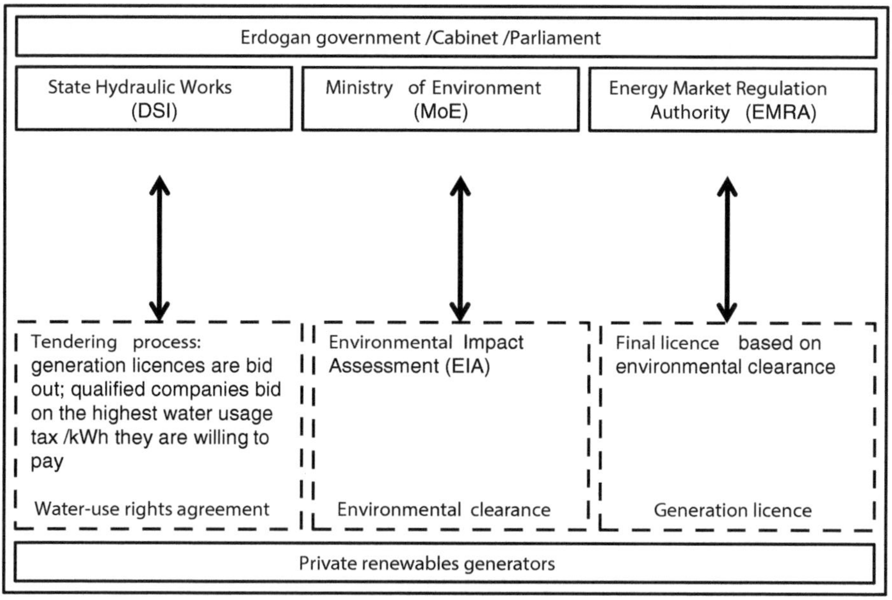

Fig. 5.2 New actors, new roles and new rules. *Source* own compilation

whereby the company prepares the EIA report to be submitted to the respective EIA commissions authorised by the Ministry of Environment. A project developer only receives the final licence from EMRA when the project-related EIA is approved (environmental clearance) by the Ministry of Environment, which is usually the case (Sect. 5.4.1).[6]

Regarding the two issues we examined (environment/expropriation and resettlement), core functions of the state did not change with the liberalisation of the energy sector: setting the legal framework, leading environmental clearance through the EIA procedure, deciding on expropriation, and planning and implementing resettlement.

During the course of sector liberalisation, the role of DSI has been changed, and new actors have been assigned new functions (such as EMRA). But the centralised nature of dam- and hydro-related decision-making has not been altered: central government organisations still decide on water resources development and hydropower exploitation with private investors participating (Kibaroglu and Baskan 2011).

As far as international dam-related norms are concerned, they not only affect individual issue areas such as the environment, expropriation and resettlement, but

[6] A review of the 75 EIAs issued between December 1993 and March 2010 shows that the Ministry of Environment had not given environmental clearance to only one project, that is, the Beskonak Dam on the Koprucay River, because the dam was to be located in the Koprulu Canyon National Park. The project did not take off.

the intent is for them to change the way in which decisions are made. International standards for EIAs and resettlement are geared towards democratisation of the political process and towards a socially inclusive and environmentally friendly development agenda. The centralised political-administrative system in Turkey, where both water resources and energy planning are organised along sectoral lines with strong central bureaucracies, has not opened up yet to civil society and locally elected representatives. As of yet there is no mechanism in the Turkish political-administrative system that allows for the inclusion of locally elected representatives (for instance, of metropolitan cities, provincial towns) and representatives of provinces in decision-making. Local representatives would like to have a voice regarding social and economic development, even more so if development programmes have impacts in their jurisdictions (Interview Yuksel 2009; Onal 2011).

In the next two (Sects. 5.4 and 5.5), we will look into Turkey's dam policy and whether—and how—international environmental and social norms have entered the domestic arena.

5.4 Environment in Dam Decisions

Paving the way towards protecting the environment started in the 1970s following the United Nations Conference on the Human Environment in 1972. In 1973, environmental concerns were addressed for the first time in the country's third Five Year Development Plan. It was not until August 1978 that the first department responsible for the coordination of environment-related activities was established on the initiative of the non-governmental non-profit organisation Environment Foundation of Turkey (EFT). The establishment of the Undersecretariat for the Environment under the Prime Minister had been a key concern of EFT. EFT also played a vital role with regard to the inclusion of Article 56 in the Constitution, which grants the right to a healthy environment and stipulates the prevention of environmental pollution as a duty of the state and its citizens (Table 5.3) (EFT 2007). The EFT moreover drafted and promoted the Environment Law of 1983, which was the first framework legislation for environmental regulations (Innanen 2004, p. 142).

The enactment of the Turkish Environment Law in 1983 and subsequent regulations (Water Pollution Control Directive of 1988, Environmental Impact Assessment Directive of 1993) indicate a growing awareness about protecting the environment. The first issue-specific environmental law with direct consequences for dam construction was the National Park Law of 1983. It prohibited any harmful activities within the borders of protected areas, among them the construction of dams. In 1994, Turkey ratified the UN Convention on Wetlands of International Importance (Ramsar Convention) and thereby committed itself to preserving and protecting its wetlands, which also implies restrictions for dam construction. Today, Turkey has designated 12 areas as Ramsar sites, among them the Goksu River delta, which will be affected by a series of dams to be built (one is the

Table 5.3 Legislation related to environment

Year	Event	
1973	3rd Five Year Development Plan (1973–1977)	Environmental concerns
1978	Undersecretary of Environment established under Prime Minister	
1979	4th Five Year Development Plan (1978–1983)	Environmental issues gain significant weight
1982	Turkish Constitution, Article 56	Right to live in a safe and sound environment
1983	Environment Law; National Parks Law	
1990	6th Five Year Development Plan (1990–1994)	Makes reference to "sustainable development"
1991	Ministry of Environment established	
1992	Environment Commission of the Parliament established	
1993	EIA Directive	
1994	Party to Ramsar Convention	
1997	Party to Convention on Biodiversity; first revision of EIA Directive	
2001	EU Accession Partnership	
2002, 2003, 2008	Second, third and fourth revisions of EIA Directive	Alignment with EU EIA Directive
2009, 2011	Revisions of EIA Directive	
2010	Amendment to Renewable Energy Law	Allows construction of renewable energy plants in or close to protected areas
	Draft Nature and Biodiversity Conservation Act submitted to parliament	Withdrawn by AKP faction due to Taksim Gezi Park protests
2013	Suspension of Provisional Art. 3 from EIA Directive, amended to Environment Law	Gained status of law

Source own compilation

Ermenek Dam on a tributary of the Goksu River, for which an environmental management plan has been concluded between the Sixth Regional Directorate of DSI Adana and the local Authority for the Protection of Special Areas; Table 5.4).

Table 5.4 EIA requirements of dam projects investigated

Dam project	EIA	Environmental issues
Ataturk Dam 1983–1992	Prior to EIA Directive	GAP RDA issued environmental studies on sedimentation and water quality
Birecik Dam 1994–2001	Exempted from EIA requirements	Protection of bald ibis by NGO in cooperation with Regional Directorate DSI Minimum downstream flow (500 m^3/s) is covered in Turkish-Syrian Protocol (1987)
Ermenek Dam 2002–2013	EIA approved on 3 Dec. 1999	EIA approved, conditioned a minimum in-stream flow for river stretch between dam site and outlet of pressure tunnel Management plan of Regional Directorate DSI in cooperation with local Authority for the Protection of Special Areas
Ilisu Dam 2007–2009	Exempted from EIA requirements but conducted on ECAs' request	EIA report revised in 2002, again updated in 2006 Environmental Monitoring and Management Plan
Yusufeli Dam 2013–2018	Exempted from EIA requirements but conducted on ECAs' request	An alternative option (three dams) was proposed by a local association to reduce negative social impacts; DSI decided on two dams (Artvin and Yusufeli)

Source own compilation

However, environmental concerns have only partly been integrated into Turkey's water and renewable energy policies—among them hydropower—when it comes to striking the balance between the utilisation and protection of its rivers and ecosystems (Kibaroglu and Baskan 2011, pp. 20–24). It might not be a coincidence that Turkey has not yet aligned with the EU Water Framework Directive (Sumer and Muluk 2011), which attempts to balance the tradeoffs between utilisation and protection in setting up a comprehensive framework. Insiders and water experts assumed that the government would strive to develop Turkey's water resources and would dedicate significant efforts in finishing its major dam and hydropower projects before 2015 (Interview Yildiz and Ozbay 2009; Interview Tuncok 2009)—the year that was then regarded as the earliest possible date for Turkey's EU accession.

Due to its horizontal cross-sector nature, an environmental impact assessment is regarded as a practical tool to serve environmental objectives by integrating environmental criteria into the planning and decision-making process of infrastructure projects, including dams (see IAIA 2013).[7] It is viewed as an essential planning instrument that requires the systematic assessment of environmental, social and cultural impacts deriving from infrastructure projects as well as the

[7] Online at: http://www.iaia.org/about/. Accessed 20 August 2013.

identification of mitigation measures. Moreover, it is the only institutionalised instrument in Turkey that allows for direct participation of the public in dam project decision-making.

We therefore decided to take a closer look at dam projects issued during the pre- and the post-EIA era; the cut-off date was February 1993, when the EIA Directive was adopted. The Ataturk and Birecik dams were planned and built in the pre-EIA era; Ermenek, Yusufeli and Ilisu fall into the post-1993 era (Table 5.4).

As shown in Table 5.4, EIAs were not always carried out in the post-1993 era due to legally authorised exceptions (EIA Directive, Provisional Art. 3), in this way aligning with the EU EIA Directive and at the same time completing its major dam and hydropower projects before 2015. Instead, EIAs were only issued at the request of international financiers and export credit agencies (ECAs), and as consequences of court verdicts.

5.4.1 The Turkish EIA Directive: A Legal Turning Point

Environmental protection did not start with Turkey's orientation towards the EU. Nevertheless, Turkey's aspirations to become a full member of the EU had a major influence.

The EIA Directive was prepared on the basis of Article 10 of the Environment Law from 1983. Since its adoption in February 1993, the Turkish EIA Directive has been continuously revised (23 June 1997, 6 June 2002, 16 December 2003, 17 July 2008) in order to align it with the EU EIA Directive of 1985 (85/337/EEC) and its amendments. However, further revisions were made in 2009, in 2011 and the Directive was recently changed again in 2013.

The process gained momentum in 1999, when Turkey became an accession candidate, and even more in 2001, when the EU Council concluded the Accession Partnership with Turkey. The National Environment Action Plan 1998 (SPO 1999) and the National Program for the Adoption of the Acquis 2001 (SPO 2001) are both committed to adopting EU environmental standards and regulations at a pace feasible for integration with the EU in the long term. This was a national priority (Innanen 2004, p. 143; Izci 2005).

One of the short-term priorities of the Environment Chapter of the Progress Report Turkey (European Commission 2007, 2009), to which accession negotiations refer to, was to transpose the EU EIA Directive into Turkish law. Consequently, the European Commission's annual progress reports have reported continuously on—and indicated the progress made in—this field. While the European Commission's Report of 2000 highlighted huge differences between Turkey and the EU in terms of environmental standards (Izci 2005), the latest OECD Turkey Environmental Performance Review applauded that "*the regulatory framework for environmental impact assessment of projects has been strengthened and steps launched for the integration of strategic environmental assessment of policies*" (OECD 2008, p. 22). According to the European Commission's Progress Report of 2011, the EU EIA Directive "*is now largely transposed.*" However, "*there are … concerns around a large number of planned hydro-power projects for which neither an EIA nor a strategic environmental assessment has yet been carried out. Transposition of the SEA Directive is still at an early stage*" (European Commission 2011, p. 100). The Progress Report further mentions "*…growing concern about the possible adverse effects on potentially protected species of flora and fauna as a result of the building of new large water and energy infrastructure*" and that "*…environmental protection requirements are still not taken into account in the implementation of infrastructure projects*" (European Commission 2011, p. 102). It notes no progress on nature protection, which is particularly important, since the assessment of impacts has to be judged on the basis of objectives/thresholds as defined in national law; further, "*the amended by-law on the protection of wetlands has weakened the protection status, and falls short of the requirements of the Convention on Wetlands of International Importance*" (European Commission 2011, p. 101).

The Strategic Environmental Assessment is, and remains, particularly relevant. This becomes clear when looking at the Coruh River, where a cascade of several dams (reservoir and run-of-river types) is under construction, and more are planned to be built on its tributaries (Klaphake and Scheumann 2011). At present, environmental clearance is given only for individual projects and fails to consider the cumulative impacts of the many dam and hydroelectric power plant projects that are auctioned and licensed along the river system. This has already led to a verdict of a local court, which argued that the Ministry of Environment would not fulfil its obligations, as defined in Law No. 4856 Regarding the Establishment and Duties of the Ministry of Environment and Forestry (Konak 2011–12, pp. 207–220). Cumulative and synergistic environmental impacts would need to be assessed by the Ministry of Environment because these impacts are beyond the scope of individual dam and HPP projects. PWC, arguing from an investor's perspective, states that "*the assessment of environmental impacts … would have been much healthier had a river basin planning study for each river basin been carried out in advance of licensing.* […] *The same issue complicates feasibility studies, obliging investors to revise their original feasibility studies after licensing and creating delays. The absence of river basin planning also prevents power plants along the same river to be connected to the system in the most efficient way*" (PWC 2012, p. 21).

In December 2012 the Turkish Ministry of Environment and Urbanization and the Turkish Ministry of Forestry and Water Affairs commissioned sample guidelines for the Cumulative Environmental Impact Assessment (CEIA) for hydropower projects in Turkey, in cooperation with the Energy Sector Management Assistance Program (ESMAP) of the World Bank (World Bank/ESMAP/TC Cevre ve Sehircilik Bakanligi 2012). The status of the CEIA guidelines is not yet clear.

Table 5.5 EIA criteria for dams (2003 and 2013)

Type	Annex I (2013)	Annex II (2013)	Annex I (2003)	Annex II (2003)
Dam with reservoir	Reservoir volume equal to/ higher than 10 MCM	Reservoir volume equal to/ higher than 5 MCM	Reservoir volume equal to/ higher than 100 MCM or 15 km^2 reservoir surface area	Reservoir volume equal to/ higher than 10 MCM
HPP and Run-of-river	25 MW and higher	1–25 MW	50 MW and higher	10 MW and higher

Source own compilation

5.4.2 Post-1993 Development: Exemptions of EIA Requirements

Dams have been subject to EIA requirements ever since the very first EIA Directive in 1993. Article 6(3) of the EIA Directive of 2008 defines the status of EIAs in project decision-making and reads as follows: "*No incentive, approval, permission, construction and usage license can be given, no investment can be initiated, nor any tender awarded for projects subject to this By-Law unless 'Environmental Impact Assessment is Positive' decision or 'No Environmental Impact Assessment is Required' decision is made*" (Republic of Turkey 2008).

Although the 1993 provisions for dams were rather soft, stricter criteria were first developed in the revisions of 2008. If one compares the Turkish EIA Directive of December 2003 with the revisions made up until and during 2013, Annex I now makes EIAs mandatory for dam projects with a reservoir volume of 10 million cubic metres (MCM) and more, and for HPPs and run-of-river-type projects of 25 MW and more (Table 5.5).

Though this is in line with the EU EIA Directive,[8] Provisional Article 3 of the Turkish EIA Directive has given ample room for bypassing the EIA requirements. Provisional Article 3 had survived all revisions of the EIA Directive until April 2013, when it finally became an amendment of the Environment Law by parliamentary decision. Not only did the EU Commission (Progress Report of 2011) raise concerns about this addendum to the EIA Directive because it authorised exceptions, in 2008 the chamber of environmental engineers (TMMOB–CMO) also issued a lawsuit demanding the suspension of Provisional Article 3, which was rejected by the General Assembly of Administrative Chambers of the Council

[8] Annex I (15.): "Dams and other installations designed for the holding back or permanent storage of water, where a new or additional amount of water held back or stored exceeds 10 million cubic metres." Annex II (g): "Dams and other installations designed to hold water or store it on a long-term basis (projects not included in Annex I)"; http://ec.europa.eu/environment/eia/full-legal-text/9711.htm. Accessed 24 March 2010.

of State *(Danistay* in Turkish).[9] The TMMOB–CMO, together with one of the most influential environmental NGOs in Turkey—the Turkish Foundation for Erosion Control, Afforestation and Nature Protection (TEMA)—appealed, which then was acceded to in December 2010 by the *Danistay*. In the years that followed, Provisional Article 3 was subject to numerous lawsuits issued by the TMMOB–CMO. Although the Ministry of Environment changed Provisional Article 3 in accordance with the *Danistay's* ruling of 2011, a Circular from the Office of the Prime Minister decreed that the EIA requirements would not apply if a project was part of investment programmes approved prior to the enactment of the EIA Directive of 1993.[10] Again TMMOB sued in May 2012, this time resulting in the *Danistay's* verdict that an EIA is obligatory.

The plaintiffs were eventually successful and Provisional Article 3 was suspended from the EIA Directive. However, the AKP government proposed amending the Environment Law of 1983, which was finally decided by the parliament in April 2013 (Resmi Gazette No. 28661, 29 May 2013). Now, with this decision, Provisional Article 3 obtained the status of a law, meaning that plaintiffs can no longer approach the *Danistay*.

As it stands now, Provisional Article 3 of the Environment Law exempts projects from EIA requirements if they were part of investment programmes concluded prior to 23 June 1997; if planning and tendering was concluded on 5 April 2013; or if projects had already started operation.

Prior to this legislative change, civil society and individuals sued the respective ministry for applying the EIA Directive's Provisional Article 3 to specific dam projects. Court verdicts stopped, for instance, the construction of the Konaktepe Dam in the Munzur Valley, which was declared a National Park in 1971. In July 2005, the court ruled that Provisional Article 3 did not apply to this project because its application design (one of the mentioned eventualities) had not been prepared before 7 February 1993, the cut-off date specified for exemption. In the absence of an EIA report, the court decided that the project lacked legality, all the more as the dam was planned inside the boundaries of a National Park. This decision was overruled by the *Danistay*, which then suggested that upon request of international financiers, an EIA—in line with World Bank standards—should be prepared.[11] The licence for the Konaktepe Dam and HPP was finally awarded to Konaktepe Electricity Generation A.S. in 2010 (Interview Tigrek 2010).

Opponents to the Yusufeli Dam on the Coruh River filed a lawsuit demanding not only the application of the EIA Directive of 1993 for the DSI-favoured two-dam option (Artvin Dam and Yusufeli Dam) (Table 5.4), but proposed instead an alternative option to be investigated (Box 5.1). This is remarkable in view of international standards: the World Commission on Dams has proposed a

---

[9] *Danistay* is the highest administrative court, whose rulings are final in the sense that there is no other institution to appeal to.

[10] This was also as a reaction to the ongoing Ilisu Dam disputes.

[11] According to the Istanbul Water Tribunal (2009), an EIA was not issued.

"comprehensive options assessment" as one strategic priority to explore alternatives that would deliver benefits to societies and avoid social costs (WCD 2001, pp. 221–224). An alternative option has also been proposed for the Ilisu Dam by Yalcin and Tigrek (2011), but it was not officially considered.

Box 5.1 Protests against the Yusufeli Dam on the Coruh River

The mayors of Yusufeli and Kilickaya, together with the Local Culture Association, filed a lawsuit against the Ministry of Energy and Natural Resources demanding the application of the EIA Directive of 1993 for the DSI-favoured two-dam option (Artvin Dam and Yusufeli Dam). The Local Culture Association proposed instead three dams, which would reduce the number of people that need to be resettled (414 instead of 1,954), prevent the town of Yusufeli from being inundated and would reduce the amount of land being inundated by 12 %. DSI argued that the two-dam option would account for significantly lower investment costs (US$2,230 million as compared to US$2,506 million) and would produce 380 GWh more energy each year. Over the course of 50 years, the benefit of the two-dam option over the three-dam alternative is expected to be US$1,140 million. DSI decided in favour of the two-dam alternative, supported by an assessment made by the Japan International Cooperation Agency.

The plaintiffs were successful in the Provincial Court, which unanimously ruled that the construction of the Yusufeli Dam was unlawful without an EIA and that the exemption rule of the EIA Directive did not apply. Without an EIA, the realisation of the two-dam option—based only on the feasibility study of 1986—would not fulfil the common welfare criteria. However, when it came to an appeal, the plaintiffs lost.

Source Interview Saglam and Sahin (2009), review of legal documents

Recent legislation has weakened the status of protected areas. In December 2010, an amendment (Law No. 6094) to the Renewable Energy Law (Law No. 5346) passed parliament and allows for the construction of dams in—and close to—protected areas.[12] Moreover, a draft Nature and Biodiversity Conservation Act was submitted to parliament in the same year but withdrawn by the AKP faction at

[12] Article 5: The first sentence of paragraph three of Article 8 of Law No 5346 is amended as follows, and the following paragraphs are added to follow the article: "*Permission shall be granted for the establishment of electrical energy production facilities based on Renewable Energy Resources in national parks, nature parks, nature monumental and nature preservation sites, preservation forests, wildlife promotion sites, and special environmental preservation site provided that an affirmative opinion of the Ministry, or of the regional conservatory board in the case of natural conservation areas, is obtained*" (original translation) (http://www.invest.gov.tr/en.../RENEWABLE.ENERGY.pdf. *Accessed* 26 *August* 2013).

the very last moment because of the Taksim Gezi Park protests (Hurriyet, 5 June 2013).[13] The draft was intended to shift authority from local protection boards to a National Biological Diversity Board, which can grant permissions for dams/HPPs to be built in protected areas. It was feared that the national board will overturn the local boards' decisions, revoking the protection status of all nature preserves (see Table 5.3, which shows major legislative and institutional changes related to the environment).

5.4.3 Broad Public Rights Vis-à-Vis a Weak Status of the EIA

The revisions of the Turkish EIA Directive made since 1993 have incrementally improved the means for the public to participate in the EIA procedure.[14] As it stands now, the Turkish EIA Directive provides for public participation in the scoping phase and for public meetings (Art. 10) and written submissions by the public (Art. 12) in the EIA report. The term "public", as used in the Turkish EIA Directive, allows in principle for everybody to participate, and is thus broader than the requirement of the EU EIA Directive, which restricts the right of participation to the "public concerned". NGOs and researchers can be invited to the EIA commissions, which usually consist of the project owner, representatives of the General and Provincial Directorates of the Ministry of Environment as well as other administrative bodies.

The EIA reports are supposed to be made available to the public, including non-technical summaries, by displaying them at the provincial governors' offices or the Provincial Directorates of the Ministry of Environment. During participation meetings, the public may ask questions, and the EIA commission and the project developer are obliged to answer them. All comments made by the public must be recorded and taken into account during the subsequent stages of the EIA process (Turgut 2003, p. 167; Coskun 2005). Written submissions from the public, that is, comments, questions and recommendations, can be sent to the Ministry of Environment or its Provincial Directoates, and the EIA commissions and the Ministry of Environment need to take in account the reservations and recommendations made and to inform the public of its overall decision.

---

[13] The protest started on 28 May 2013 and initially contested the urban development plan for Istanbul's Taksim Gezi Park but later addressed a wide range of concerns such as freedom of the press, freedom of assembly and what was perceived as the government's encroachment on Turkey's secularism.

[14] Although Turkey has not yet signed the Aarhus Convention, the National Programme for the Adoption of the acquis envisioned improving access to environmental information in order to align with this convention (SPO 2001, 403ff.). As amended, Art. 30 of Environment Law No. 2872 meets the criteria of Art. 9 of the Aarhus Convention on the right to access environmental information; the Right to Information Act, which came into force in April 2004, "*seems to give more room for environmental NGOs* [...]" (Izci 2005, p. 96). The revised EIA Directive of 2013 further rules that information brochures, surveys and public meetings have to be issued.

In the cases analysed, participation meetings were organised as part of the EIA process. However, the process lacks as of yet proactive promotion and transparency: a non-technical summary was not always included in the EIA reports; and a search of the Ministry of Environment's website, which displays EIA decisions, revealed that public disclosure of information is rather poor. This refers to the final EIA reports; the EIA commissions' recommendations; the Ministry's approval document and whether and how the public's concerns have been addressed (as requested in Art. 14(2)); conditional requirements (mitigation measures) that ought to be implemented by project developers; reports on compliance-monitoring to be undertaken by the EIA Department of the Ministry of Environment; and reports by the project developer to be submitted to the environmental bureaucracy.

According to the EIA Department of the Ministry of Environment, people frequently write comments and submit questions. But the EIA Department views "*public participation in the final EIA a checking procedure. No important issue comes up anymore*" (Interview MoEF EIA Department 2009). This is why the World Wide Fund for Nature (WWF) Turkey and the Industrial Development Bank of Turkey (TSKB)[15] raised fundamental critiques. Both have pointed to the rather late stage of the project cycle at which the public is able to raise its voice (WWF-Turkey 2003; Interview Acinci TSKB 2009). WWF-Turkey has been criticising in particular the fact that the EIA process starts only after planning and project design is almost completed, and thus leaves little room for project modifications (Interview Eryilmaz WWF-Turkey 2009).

Despite its weak status, representatives of the public used the EIA instrument to its benefit, for instance in the case of the Ermenek Dam: the mayor of Ermenek, who was invited by the Ministry of Environment to the EIA commission, put four items on the agenda: (i) the need to build a treatment plant for water that will otherwise flow untreated into the reservoir; (ii) additional compensation payments because 30 % of farmland would be inundated; (iii) the relocation of a historical 700-year-old bridge; and (iv) farmers should be permitted to use water from the reservoir for irrigation. The local meeting, which was announced in a local newspaper as well as by spokespersons and loudspeakers in the town, was attended by the surprisingly high number of 350 people, who all supported dam construction and their mayor's proposals. As a result, a wastewater treatment plant was built (financed by DSI), and payment rates for farmland were eight times higher than required by law (paid by the private company) (Interview Mayor of Ermenek 2009). This result was achieved because of the mayor's negotiation abilities and because the number of people affected was rather small (~ 150), when compared with the number of people affected by the Ilisu Dam (>50,000). In addition, the private company involved also made commitments to corporate social responsibility practices (Interview BM Group 2009).

---

[15] TSKB holds the leadership in financing renewable energy projects—it has financed 65 hydroelectricity projects with a total capacity of 2,150 MW (PWC 2012, p. 17).

5.4.4 International Actors Insist on EIAs and Demand Higher Environmental Standards

Since the late 1980s, it is no longer multilateral development banks (for instance the World Bank[16]) that play an important financial role in the Turkish dam business but rather private companies, banks and export credit agencies (ECAs). Those involved in the Turkish dam business have demanded the application of higher environmental standards—referring to the OECD's Common Approaches for Officially Supported Export Credits and Environmental and Social Due Diligence (Common Approaches) and the Equator Principles—, improved baseline studies as well as the strict implementation of mitigation means.

Also, the 83 lawsuits that were filed against dams and HPPs resulted in the cancellation of dam projects in the first instance on the ground where the submitted EIA reports were inadequate, misleading or where the projects were to be built in protected areas (see Konak (Konak 2011–12) on verdicts of the Provincial Court of Rize).

In two of the dam cases included in our sample (see Yusufeli Dam in Box 5.2 and Ilisu Dam in Box 5.3), international financing was at stake because no EIA was issued by Turkish authorities (Yusufeli), and because the EIA report was of poor quality (Ilisu).

Box 5.2 EIAs for the Yusufeli Dam on the Coruh River at the request of foreign financiers

Turkish authorities had not required an EIA for the Yusufeli Dam—a large dam in every respect (Table 5.1)—arguing that the EIA Directive is not applicable because the reconnaissance study (1969), the master plan (1982), the feasibility study (1986) and final project design (1990) were all dated prior to the EIA Directive.

In the 1990s, DSI granted a construction licence to a private consortium led by Spie Batignolles TP, involving private companies from France, the United Kingdom, Belgium and Spain. In 2008, the Yusufeli Dam became part of the government's investment plan. In February 2009 the Turkish government called for tenders for the final design; the bid was won by Dolsar, a Turkish engineering and consultancy company.

An EIA was carried out later by Dolsar at the request of the consortium and the ECAs involved. When the international consortium withdrew because of unsettled EIA issues, financing was provided by the national treasury and construction started in February 2013.

Source Interview Dolsar Consulting (2009)

[16] The Karakaya Dam on the Euphrates River (1986) was the last dam co-financed by the World Bank. The Bank considered financing the Kayraktepe Dam on the main stem of the Goksu River but declined.

Box 5.3 EIAs for the Ilisu Dam on the Tigris River at the request of foreign financiers

Studies for the Ilisu Dam on the Tigris River started as early as 1954 but only in 1997 did the dam become part of the public investment plan. The Ilisu Engineering Group was assigned the task of performing an EIA study, which was completed in 2001.

This EIA was revised in 2002 at the request of the British Export Credits Guarantee Department because the previous EIA study did not correctly assess the environmental impacts due to poor baseline data, on which suitable mitigation measures were to be based and subsequently recommended. The second consortium (Austria, Germany and Switzerland) asked for an update of the revised study, which was finally made available in English and Turkish in 2006.

Reports of an independent Committee of Experts released in December 2007, in May and November 2008 and in February 2009 concluded that the environmental terms of reference had not fully been met by the Turkish implementing agencies. The terms covered, among other things, the construction of upstream water treatment plants capable of ensuring that the water quality in the reservoir is maintained; maintenance of adequate downstream flows at all times; and development of a detailed plan to preserve as much of the archaeological heritage of the Hasankeyf district as possible. Controversies between the international consortium and the Turkish government could not be settled, and the governments of Austria, Germany and Switzerland finally cancelled their export credit guarantees on 7 July 2009.[17]

Source Sub-Committee on Environment February (2007), May (2008), November (2008), February (2009); External Monitoring Group Report June (2009); Encon (2006)

Despite the enactment of the EIA Directive in 1993, major deficiencies related to integrating environmental issues into dam planning by means of EIAs are as follows:

- The EIA requirements, as ruled in Annex I (dam projects), are not applied according to the criteria as established in the Turkish EIA Directive; the possible exemptions are too broad.
- The scope of EIAs does not extend to downstream areas, some of which are designated Ramsar Sites and protected under national law.
- Approval of dam projects located in and affecting protected areas is not linked to strict environmental measures.

---

[17] http://www.oekb.at/de/unternehmen/Presse/pressetexte/Seiten/070709-ausstieg-ilisu.aspx. Accessed 8 April 2010. Agreed minutes between OEKO, Euler Hermes, Exportrisikogarantie and GD Devlet Su Isleri, 6 October 2006.

- The public's access to information is limited, and public participation is not proactively promoted.
- In-stream flow requirements are not based on specific ecological conditions; existing uses of rivers' waters are not considered prior to project construction and operation.
- The 1971 Law on Aquatic Flora and Fauna (Su Urunleri Kanunu, Art. 22) is not applied, which made the construction of fish passes and fish ladders mandatory.
- Monitoring of mitigation measures by the respective government department are not made on a regular basis.
- The Provincial Directorates need to be strengthened so that they are able to fulfil their mandate, namely to assess whether an EIA is necessary for Annex II dam projects (and all other projects/activities, as listed in the EIA Directive Annex II) (Interview MoEF EIA Department 2009; OECD 2008, p. 100; Interview MoEF Balikesir 2009; Interview MoEF Sanliurfa 2009).

5.5 Expropriation and Resettlement

An estimated 350,000 people were involuntarily resettled as of 2005 due to dam projects (Akyurek 2005),[18] and approximately 250,000 more will be resettled due to dam projects that are currently in the planning stages (DSI 2009, p.79). Even more so than with environmental issues, Turkey's expropriation and resettlement practices have been subject to harsh criticisms, and legislation and practices have been far below international standards.

Expropriation and resettlement fall under two legal and administrative regimes (Table 5.6), which are analysed in Sects. 5.5.1 and 5.5.2, respectively. Whereas expropriation used to be in the hands of DSI, resettlement was under the umbrella of the Ministry of Public Works and Settlement (Akyurek 2005, p. 46) until 2006, when DSI was mandated to coordinate both expropriation and government-assisted resettlement.

Unlike the analysis of changes in the environmental (EIA) field, there is no cut-off date for expropriation and resettlement procedures, but rather gradual reforms have taken place that are driven by domestic processes and are influenced by international actors.

5.5.1 Changes Related to Expropriation

In 1956, the first law on expropriation (No. 6830) passed the Turkish parliament and introduced *public benefit* as a necessary requirement. The Constitution of

[18] Data are not available after 2005.

Table 5.6 Legislation related to expropriation and resettlement

Year	Legislation	Content
1934	Resettlement Law 2510	Eligible for resettlement are immigrants, refugees, nomads (not those affected by infrastructure projects)
1956	Expropriation Law 6830	"Public benefit" as a condition for expropriation
1970	Resettlement Law 1306	Eligibility criteria for resettlement; two types of resettlement
1982	Constitution (Article 46)	Expropriation for "public benefit"; compensation payments in advance
1983	Expropriation Law 2942	Value Assessment Commissions include local representatives; new methods for determining compensation payments
	Amendment (4650) to Expropriation Law 2942 in 2001 and 2004	Strengthened individual rights: formal agreement with property owner is pre-condition for expropriation and transfer of land title
2006	Resettlement Law 5543	Project owner responsible for resettlement; Local Resettlement Commissions presided over by local authority
	Amendments to Resettlement Law 5543 in 2007, 2008, 2011, 2012, 2013	
	Amendments to Expropriation Law 2942 in 2010, 2012 and 2013	"Urgent expropriation" can be applied in renewable/hydropower projects by cabinet decree

Source own compilation

Turkey of 1982 conditions expropriation of private property on the prevalence of a *public benefit* decision and the payment of compensation in advance.

Because this law had made no distinction between land expropriation, for example a primary school and expropriations for large-scale infrastructure projects, modifications became necessary. In 1983, Expropriation Law No. 2942 was adopted to allow for the smooth implementation of the then planned South-eastern Anatolia Project (GAP), which comprises a large number of large dams and HPPs and affects a large number of people.[19] The new Law No. 2942 ruled that representatives of municipalities affected nominate three out of five members of the Value Assessment Commission, set up to assess the value of the assets

[19] The new ruling falls considerably behind practices that were implemented in the mid-1970s, when the first dam on the Euphrates River, the Keban Dam, was planned and built. During the 1970s, policies were aimed at recovering—and even improving—the living standards of the people that had to be resettled (Scheumann and Irmak 1991).

expropriated based on factors such as soil classes, location and size (Nisanci 2005). It was also decreed that compensation payments were to be disbursed in cash and in advance or, under legally defined circumstances, in instalments of equal amounts within 5 years (DSI 2009b).

During the expropriation and resettlement processes in the Ataturk Dam case, it was shown that value assessments and compensation payments were the most critical issues (Box 5.4). A study revealed that the number of people and families affected, as well as the land that would be fully or partly inundated or affected by other activities such as construction work, was not correctly assessed; costs were underestimated as a consequence (Scheumann and Irmak 1991, pp. 36–38). The valuations made between 1986 and 1988 varied greatly because the commissions involved referred to different laws; land prices had changed; different data were given for the same plots of land concerning productivity; and because different soil classes and different uses were referred (instead of rain-fed agriculture, for example, irrigation).

Finally, high inflation was an issue. More than a year passed between the work of the commission and the issuing of notifications about the compensation values. Since the year in which notification is given to the farmers is so important, higher payments should have been given, because the property value would have increased in the meantime. Moreover, long periods of time passed between the announcement by the commission concerning compensation and the actual payments, leading to depreciations due to inflation rates of around 75 %.

Numerous people went to court to obtain higher payments, but not all people affected could pay the resulting court costs (especially as the lawyer gets 10 % of the compensation payments and the plaintiff must pay the appraisers appointed by the courts).

In most cases, the courts decided in favour of the plaintiffs, who were successful to the degree that expropriation costs and expropriation budgets increased ex-post. Furthermore, the assets of communities (towns, villages) such as schools, water supply, community houses, health-care centres and roads were not registered and assessed.

People affected also filed lawsuits at the European Court of Human Rights, referring to the European Convention on Human Rights, which Turkey has ratified (it is a signatory and a member of the Council of Europe). Plaintiffs referred to Article 1 of the Convention, which protects property rights (Council of Europe 2004, p. 2). Between January 1992 and March 2003, one quarter of the cases brought to the European Court of Human Rights concerned expropriation (354 out of 1,357 cases), 60 % of which concerned practices of DSI (Yomralioglu et al. 2007). In most cases, complaints were related to delayed compensation payments, which is a particularly crucial issue against the background of persisting high inflation (Nisanci et al. 2007, p.6).

It was the GAP Regional Development Administration (GAP RDA) that was confronted with the negative impacts of one of the core infrastructure projects of GAP: the Ataturk Dam (expropriation was then the mandate of DSI). GAP RDA therefore issued a survey in 1994 that was conducted by the Sociology

Association. This survey should serve as a basis for developing a new approach. As GAP RDA President Olcay Unver said: "*To ensure sustainable development, these communities should be subject to resettlement in the most appropriate manner and new areas of employment should be created. In addition, there is a need for such measures to be taken as to ensure that expropriation compensations paid are channelled into productive areas*" (Sociology Association 2004, iii).

Box 5.4 Expropriation and resettlement in the Ataturk Dam case

Expropriation based on Law 2942 (1983)

In total, 25,700 parcels stretching over 43,300 hectares were expropriated. The value of the expropriated assets was assessed by a commission assigned by DSI. For those opting for cash compensation (self-resettlement), payments were released—according to the law—in cash instalments over a period of five years without adjustments for high inflation. However, compensation payments were below market prices because the values assessed were on the basis of the real estate tax statements of those affected. Many of the people affected filed lawsuits in order to increase compensations, although they had to bear costs at that time. In the end, the court decided in favour of the plaintiffs in about 90 % of the cases. As a result, expropriation costs for DSI increased by 30 % and amounted to one-tenth of total construction costs.

A planner from the GAP regional office explained: "*We had a very bad experience with the Ataturk Dam. People did not get their money on time. Some spent their money on gambling; others went and blew it all. They did not know how to spend it well. Some left their money in the bank, but because of high inflation rates* [more than 100 %] *their money quickly devalued. Others started businesses but lacked experience and lost everything*" (Morvaridi 2004, p. 29).

Newspaper articles report that in 2007 and 2008, there were still 53 families that had not received adequate compensation (they wanted land, not cash) (Zaman, 4 May 2007); one district authority voluntarily paid compensation to ten families that had not received payments from the government for 21 years (Zaman, 30 December 2008).

Resettlement based on Law No. 1306 (1970)

More than 100 hamlets and villages, as well as the land and homes of more than 55,000 people, were fully or partially inundated. Official information on resettlement arrived shortly before the resettlement started in 1988. According to data from 2002, 1,131 families preferred to be resettled by the government. Out of these resettled families, 169 chose rural and 315 chose urban resettlement, whereas 416 received land to construct new houses.

Akca et al. report that "*the majority of the resettled individuals had to accept less land and less family income* [...] *because the price of land in the newly developed resettlement areas was greater than expected due to land speculation*" (2013, p. 104). The Sociology Association's survey (1994) found that 67 % of the resettled people wanted to return to their original habitations; 49 % of the families that had organised their resettlement on their own were said to suffer from economic hardship. In 2002, 231 families were still living in temporary accommodations.

The inhabitants of Old Samsat had the choice between two new resettlement areas, but decided on a third location supported by their parliamentarian representatives. The mayor of New Samsat mentioned that only a few houses had been built at the time when resettlement started; electricity was provided after some weeks but potable water only became available the following two years; roads were constructed and financed by the municipality and not the government; and irrigation schemes near New Samsat were only installed in 2006. Villagers complained about inferior land/soil fertility, as compared to their former lands next to the Euphrates riverbanks.

In 2003, the Turkish newspaper *Radikal* reported that resettlement of 230 families had not been resolved (out of 987 who were eligible for resettlement) (20 September 2003).

Source Sociology Association (1994); Interview Aslan, Mayor of New Samsat (2009); Interview Head of Agricultural Office Samsat (2009); Akca et al. (2013); Altinbilek and Tortajada (2012); Morvaridi (2004)

Asset owners also used informal ways to lobby for change. A high number of people stated their dissatisfaction via letters to DSI, the Ministry of Interior and SPO (Interview DSI 6th Regional Directorate Adana 2009; Interview Ozdalga 2009). People affected also approached their parliamentarians and expected their representatives to lobby for change in parliament (Interview Caglar 2009).[20] DSI Regional and Provincial Directorates have been frequently contacted by lobby groups and mayors to change expropriation provisions and procedures. DSI officials emphasised the sensitivity that DSI had shown for the concerns of civil society, and the Provincial Directorates would constantly report on inadequate practices to the General Directorate (Interview DSI 6th Regional Directorate Adana 2009).

It was not before the former Vice-President of the World Bank Kemal Dervis became the Turkish Minister for Economic Affairs and the Treasury (he held this

[20] Turkish citizens frequently contact their respective parliamentarians. A high number of citizens travel to the capital Ankara to get an appointment with their parliamentarian—the numbers of visitors in parliament total 3,000 per day; http://www.haberturk.com/gundem/haber/665165-tbmmye-2-milyon-184-bin-ziyaret. Accessed 18 September 2013.

position from 2001 to 2002) that a fundamental shift took place. He initiated an amendment to the Expropriation Law in 2001, which was part of a package explicitly aiming at aligning with EU requirements, and at meeting the World Bank's standards. The amendment provided for a timely disbursement of compensation payments, and even more important, it strengthened the rights of the people affected (Kudat 2001; Yomralioglu et al. 2007).

Expropriation procedures only start upon approval of a "public interest" by the respective ministry (i.e. Ministry of Natural Resources and Energy), as ruled by the amendment (2001, No. 4650) to the Expropriation Law, and it is only public agencies who are allowed to acquire/expropriate land either temporarily or permanently. The project owner (either public or private) must initiate cadastral works to identify properties and ownership, which are to be displayed on a geographical expropriation map. Affected asset owners are informed of the expropriation via an officially registered letter indicating a company's/DSI's interest to purchase the land through negotiation and that the company/DSI would pay the current market price of the land or, if applicable, the usufruct right for the land, in cash. Subsequently, all data and documents related to the value and legal status of all properties are collected by the Value Assessment Commission. A valuation report is prepared for each parcel. The commission may consult agricultural or real estate experts in order to determine the value of land and buildings. If an agreement is achieved, expropriation is registered at the land office and the compensation payments are paid. Compensation must be paid to the owner before the investor can enter the land.

In case of disagreements, that is, in cases where the expropriation is not concluded through negotiation and purchase, it is now the project owner/developer (not the individual affected) who has to approach the court with the request to set the expropriation value and register the immovable property in the name of the project owner. The local court would then reassess and set an asset's expropriation value (Akyurek 2005, p. 53). In line with the court judgement the expropriation fee is then deposited in an escrow account at a commercial rate of interest in the name of the owner(s) who is(are) given access to the proceeds once the disputes are solved.

An agreement has to be found in no more than three court procedures over two months' time. Only after legal procedures are completed and payment of compensation is made, is the title deed transferred and registered. People affected still have the right to file lawsuits in a higher court to obtain higher compensation payments (Box 5.5). In that case, the plaintiffs have to bear the costs though. This regulation resulted in a higher number of agreements being concluded between the expropriating agency and the people affected (83 % since 2001), and in a reduced number of court cases (Cay and Evren 2007, p. 4). Above all, the new law and its amendments have increased transparency of the valuation process, and reduced political influence in the composition of the commission (Kudat 2001).

Box 5.5 Expropriation in the Ermenek Dam case

Expropriation based on the amended (2001) Expropriation Law

DSI informed the people affected about two weeks before the expropriation process started. This appears to be rather short notice, although no specific time is required by law. The compensation payments were based on market prices obtained from the Provincial Directorate of the Ministry of Agriculture and multiplied with coefficients for land use and soil quality, respectively. Social surveys were carried out and efforts were undertaken to gain complete information on properties. Based on records of the cadastral office, 1,500 parcels were expropriated, affecting approximately 300–400 people. Thirty to forty percent of the expropriated parcels were subject to lawsuits filed by DSI. Asset owners did not agree on the compensation payments first proposed by DSI. After renegotiating, DSI increased the compensation payments significantly, so that most of the people accepted. Those who still did not agree went to court again, which then decided—based on legal regulations—on a lower compensation payment than negotiated with DSI. In the end, the overall costs of the expropriation amounted to about 10 % of total infrastructure costs.
Source Interview Mayor of Ermenek (2009); Interview BM Group (2009); Interview Gurel, DSI Ermenek (2009)

Although the Expropriation Law strengthened civil and property rights of citizens in 2001, the parliament decided in December 2010 that cabinet orders can decree "urgent expropriation" of immovable property based on Article 27 of the Law on Expropriation No. 2942,[21] which weakened the affected people's right to file lawsuits.[22] It can only be speculated whether the decisions made in December 2010 were a reaction to the tedious negotiations over expropriations, which hindered the ongoing construction of the Ilisu Dam.

The Expropriation Law was again amended in May 2013 and in July 2013. The first amendment ruled that if properties were confiscated between 9 October 1956 and 4 November 1983, without issuing an official process, the owner could sue for

[21] http://www.erdem-erdem.av.tr/erdem-erdem.php?katid=12110&id=15139&main_kat=15132. Accessed 24 August 2013.

[22] See for instance: The Resolution of the Council of Ministers dated 09.11.2012 on the Urgent Expropriation of Immovable Property by the Energy Market Regulatory Authority on behalf of the Treasury for the establishment of Bedirduzu-2 Hydroelectric Electricity Plant in Erzincan, Cayirli district, published in the Official Gazette dated 06 Dec. 2012 and numbered 28489. For more examples see: http://www.erdem-erdem.av.tr/erdem-erdem.php?katid=12110&id=15139&main_kat=15132. Accessed 20 September 2013.

his/her rights. The second amendment decreed that compensations would not be granted for buildings/facilities that were built after expropriation had been publicly announced.

5.5.2 Changes Related to Resettlement

The first Turkish resettlement law of 1934 (No. 2510) focused on the resettlement of immigrants, refugees and nomads (Table 5.6). Hence, when Turkey started construction of its first large dam, the Keban Dam on the Euphrates River, in the late 1960s, it lacked appropriate legislation for dam-induced resettlement. The number of settlements to be re-located amounted to 175, and these were home to about 30,000 people.

The Resettlement Law No. 5543 of 1970 also applied to large-scale dam-induced resettlement. It stipulated that people whose assets were to be expropriated could choose either (i) to be resettled by the government (government-assisted resettlement), or (ii) to receive cash compensation and settle at locations of his/her choice (self-resettlement). Governmental resettlement could be either urban resettlement or rural resettlement. The government was supposed to finance the municipal infrastructure, and people affected had to contribute towards costs by depositing their compensation payments into a Special Resettlement Fund; additional credits are given by the government if costs of new houses exceed compensation payments (which is an indication that compensation payments based on market prices do not cover replacement costs). Eligible for resettlement assistance are people whose assets are affected entirely or partly, and people without title deeds who have been settled in the area for at least 3 years before the project was planned.

During the 1980s and 1990s, it was the poor resettlement record of the Ataturk Dam (Box 5.4) in particular that led the parliament to establish a commission to get evidence on practices and impacts (Interview Ozdalga 2009). The topic was also on the agenda of talks held between Turkish authorities and the World Bank. The World Bank pushed for a decentralisation of the resettlement process and demanded the preparation and implementation of Resettlement Action Plans (RAP),[23] and asked for the inclusion of this instrument in national law.

The first RAP was prepared at the request of the World Bank and developed in cooperation with DSI for the Tahtali Dam (1986–1999), built during the course of the Izmir water supply project (Altinbilek et al. 1999, p. 298f.). A RAP was also prepared for the Birecik Dam and issued by the GAP RDA (Box 5.6). In due

[23] Resettlement Action Plans aim at minimising socio-economic effects of projects on the affected people through a fast reestablishment of their individual productivity and their communities' social structures. Resettlement Action Plans comprise measures that aim at enhancing coordination of governmental institutions, to increase participation of people affected and to ensure that expropriation and resettlement plans match the pace of the construction works.

course, foreign ECAs also began to request RAPs as a requirement for their support of the Ilisu Dam and Yusufeli Dam (Cernea 2006, p. 16f, Box 5.7).

But Resettlement Action Plans have not yet become a legal requirement and are not anchored in Turkish law. As Cernea mentions: "*The RAP* [Ilisu Dam] *is not a basis that legal recourse could use to enforce provisions it includes* [...]. *The affected population cannot take legal action if the promises contained in the RAP are not respected*" (Cernea 2006, p. 17; Cernea et al. 2008, Box 5.7).

Box 5.6 Expropriation and resettlement in the Birecik Dam case
(same legislation as Ataturk Dam)

Expropriation based on Law No. 2942 (1983)
When impoundment started in 1999, nine villages were fully inundated and three villages as well as the Halfeti district centre were partly set under water. About 30,000 people were affected. After an independent commission had evaluated property values, and after these calculated compensations were disbursed, most people went to court. They were often successful in claiming higher payments. However, people who had no land titles were not compensated, and many families are still waiting for the government to provide them with new houses.

According to the mayor of Halfeti, compensation payments were significantly higher than in other GAP regions. Compensation payments for the 850 households that were expropriated were fully disbursed, but delays had occurred.

Resettlement based on Law No. 1306 (1970)

It was the first time that GAP RDA developed a RAP that aimed at delivering timely and proper information to the people affected and provided for increased participation. Social surveys were carried out in the 13 most affected villages, comprising interviews with 1,307 families. An information and advisory office was opened in Halfeti, and people were informed about dam construction and their rights and options in public meetings.

Whereas 49 families decided in favour of rural resettlement, 151 families preferred urban resettlement. The new locations were determined by the General Directorate of Agricultural Reform and the General Directorate of Rural Affairs. The village of Halfeti had the opportunity to choose between three alternatives.

The people were supposed to also benefit from the GAP Sustainable Development Umbrella Program—advisory services were offered, and a multi-purpose community centre, a primary school, a hospital and a shopping centre were opened. Education programmes and consultancy services were provided in cooperation with the GAP Entrepreneur Support Centres to support the affected people in finding new sources of income. The Food and Agriculture Organization of the United Nations financed training for new agricultural practices in Meteler village (Birecik), turkey farming in the village of Asagicardak (Nizip) and organised workshops to discuss the feasibility of investments planned.

However, the mayor of Birecik (the provincial capital, which is subject to in-migration) doubted the efficiency of the programmes because they would not have reached the villages. Villagers who had been relocated complained that their new houses were overcrowded and unfinished. The people who had opted for rural resettlement complained about having been compensated with less-fertile land.

In 2003, the Turkish newspaper *Zaman* found that 90 inhabitants of Halfeti were evicted from their new homes because they did not pay rents. The inhabitants claimed they could not do so because they had not received their compensations for the last four years (Zaman, 17 June 2003).

Source Interview GAP RDA (2004a, b; 2009a, b); Interview Gencek, Mayor of Halfeti (2009); Interview Yuksel, Mayor of Birecik (2009)

Until the resettlement law was revised in 2006, it was DSI that carried out expropriation whereas the Ministry of Public Works and Settlement was responsible for the resettlement process. This strict administrative separation also resulted in poor coordination between both agencies. Therefore, DSI had long ago requested to also become entitled to plan and implement resettlement (Altinbilek et al. 1999, p. 299). The revised Law No. 5543 now states that funds necessary for financing resettlement are allocated to the project owner, usually DSI, which meant that expropriation and resettlement would be in one set of hands. The law also stipulated the setting up of Local Resettlement Commissions, which were mandated to identify the people eligible for resettlement. These commissions are supposed to include personnel from the local branches of the relevant ministries and institutions and should be presided over by a high-level administrative authority (usually the Governor of the province or the county).

Meetings are to be held in the municipalities affected to announce resettlement procedures as well as eligibility criteria and the way to apply for government-assisted resettlement. This information is then to be published in a Declaration of Resettlement. Within 60 days after this announcement, individuals who choose to be resettled by the government or project owner have to apply in writing and include information about their social situations. On this basis, a Local Resettlement Commission decides whether a family is entitled to rural or urban resettlement. The project owner then chooses the new resettlement area together with the provincial government. Transfer costs to new settlements would be covered, and assets of municipalities (infrastructure, mosques, etc.) would be financed by the government (for details see Hazar and Ozbek n.d.).

Box 5.7 Expropriation and resettlement in the Yusufeli Dam case (same legislation as Ermenek Dam)

Expropriation based on Law No. 4650 (2001)

The total area inundated was 3,241 hectares, including 735 hectares of agricultural land. Altogether 12,164 people and 2,499 assets were affected. According to DSI, the majority of the people affected preferred cash compensation, whereas Eberlein et al. (2010) reported from interviews that most people preferred land-to-land compensation. People only decided in favour of cash compensation if the designated relocation area was unsuitable for agriculture. The amount of compensation was criticised for being insufficient for purchasing new homes offered by the authorities as well as for the construction of public infrastructure.

Resettlement based on Law No. 5543 (2006)

The construction of the Yusufeli Dam required the relocation of Yusufeli town and three villages, and affected a further 17 villages. In July 2006, a Resettlement Action Plan was developed that aimed at achieving the pre-resettlement living standard.

The wish of the Yusufeli inhabitants to be resettled as a whole community was not feasible due to the hilly and rocky surroundings of Yusufeli. The Turkish government decided on the location for the new Yusufeli District Centre in July 2005, but inhabitants claimed that the location proposed was unsuitable, as it did not allow for income-generating activities.

Source Interview Saglam, Mayor of Yusufeli (2009); Eberlein et al. (2010)

Major weaknesses related to expropriation and resettlement planning and implementation are as follows:

- Current legislation related to compensation payments considers the market value of depreciated assets, whereas international standards favour replacement values (Cernea 2006, p. 19).[24] The compensation payments for agricultural land are calculated using the "net income approach"; the "cost approach" method is used for the valuation of structures and infrastructure; and "peer assessments" value housing plots (Hazar and Ozbek n.d.).
- Complex issues arise if individual rights of heirs are unregistered; if inheritance issues are unresolved; if individual shares of jointly owned land cannot be identified but consent must be obtained.

---

[24] Market value is the value an asset owner would get if he/she sold his/her asset in a competitive market; replacement value refers to the amount that one would have to pay to build new assets.

- Rights of tenant farmers are not protected since the law is based on compensating owned land/assets (Kudat 2007).
- Delays of expropriation processes still occur because capacities of local offices of public agencies to be consulted are inadequate (Kudat 2007; World Bank n.a.).
- Turkey's policies favour cash compensation and self-resettlement over government-assisted programmes. Agencies concerned do not actively encourage people to opt for government-assisted resettlement. One constraint for government-assisted resettlement is the shortage of suitable land for realising the "land-for-land" packages as well as for new resettlement areas, which are preferably chosen where the state owns land.
- Current legislation does not stipulate the preparation and disclosure of resettlement action plans. However, if international financiers are involved resettlement action plans are common practice.
- Budget allocations are neither sufficient for expropriation nor for resettlement. DSI's expropriation budget in 2008, for instance, added up to only about one-sixth of the sum that DSI had applied for (DSI 2009, p. 79). In addition, initial cost-estimates for expropriation have frequently turned out to be too low because costs were assessed based on information collected for the feasibility studies (sometimes dating back 20 years), while the final costs often exceeded the estimated costs by two or three due to an increased number of eligible people and increased costs.

5.6 From Individual Actions to the Emergence of a Social Movement

Until the late 1990s and the beginning of the 2000s, organised protests against large dams were unheard of, although people were affected by unfair value assessments and compensation payments as well as inadequate resettlement practices. Nevertheless, they individually filed lawsuits at local courts, and some even approached the European Court of Human Rights, which was instrumental in changing expropriation legislation and practices.

In Turkey, civil society, in its relation to the state, has seen dynamic and fundamental shifts, with the number of NGOs and civil society organisations significantly increasing over the past decades (Adem 2005, p. 72). More importantly, their role and relation to the political elite has been changing. This can best be demonstrated by the nature of the NGOs. According to Adem (2005), the founders of the Turkish Association for the Conservation of Nature and Natural Resources, established back in the 1950s, were high-level bureaucrats. Many of the founding members of the Environment Foundation of Turkey, which was created as a non-profit organisation in 1978, were lawyers and enjoyed good relations with the state apparatus. They were thus able to push for institutionalising environmental issues, among them the establishment of the Undersecretariat for the

Environment under the Prime Minister—a key concern of EFT, along with creating the legal basis for environmental protection by drafting and promoting the Environment Law of 1983.

During the 1990s, a new generation of environmental organisations emerged that demanded the protection of civil rights and freedom as well as raised environmental issues. Major events were the Habitat Forum in 1996—an international meeting in Istanbul when hundreds of Turkish civil society organisations met—and the earthquake in 1999, for which civil society organisations mobilised thousands of volunteers and donations (TUSEV 2005). When finally the Turkish government accepted the Copenhagen Criteria in 2001, it created political space by granting freedom for/of associations. As the Third Sector Foundation of Turkey (TUSEV) observed, civil society organisations "*are shifting from being loose, informal groups to more structured and organized institutions*" (TUSEV 2005, pp. 14–15).

Today, several groups are active and demand a voice in decision-making on water resources development and management, among them experts, academics and intellectuals; civil society organisations and non-governmental organisations; and the Union of Chambers of Turkish Engineers and Architects (a public umbrella organisation with 23 chambers and about 300,000 members). They all are sceptical about the AKP government's massive river development programme.

The debate over dams has also become part of party politics and an instrumental position for the political opposition (Kaygusuz and Arsel 2005; Kadirbeyoglu 2005). Opponents of the Ilisu and Yusufeli dams were, among others, elected mayors of villages, towns as well as of metropolitan cities. They were not members of the governing AKP party but of opposition parties: the mayor of the town of Yusufeli was a member of the Nationalist Movement Party; the mayors in the south-eastern provinces and districts affected by the Ilisu Dam—namely in the provinces of Diyarbakir, Batman, Siirt, Sirnak, Bismil and Dargecit—were members of the Kurdish Peace and Democracy Party. The mayors allied themselves with local NGOs and sued respective ministries in court in absence of any other effective institutionalised means of taking part in decisions that would affect their constituencies.

What had historically started as local protests confined to individual dam projects has meanwhile turned into a social movement that concerns inadequate expropriation and resettlement practices and the negative impacts on the environment, cultural heritage, people's rights and their ways of life. The enactment of the Turkish EIA Directive in 1993 served as an instrument for people affected and the broader public to influence specific dam projects—in addition to using the legal means of the juridical system and the European Court of Human Rights. But participation in the EIA process gained high levels of interest only if it concerned expropriation.[25] This

[25] In the case of ongoing resistance against a coal power plant near the city of Gerze, it was even reported that the strategy of opponents was to prevent a company from completing the EIA in order to stop the project.

has changed with the Ilisu and Yusufeli dams and the many dams and HPPs to be built in the Black Sea Region (Islar 2012a, b; Konak 2011–12).

However, it would be misleading to interpret these protests only as being driven or inspired by the AKP government's dam-building programme. This movement has been nourished by diverse sources that are part of broader critiques of the state's role in a neo-liberal environment:

- Decentralisation: The mayors of the metropolitan city of Diyarbakir and the provincial capital of Birecik raised the issue that they would have no voice in deciding on the purpose of dams, their locations and whether they wanted them, but that they would have to bear the social and economic costs of in-migration without receiving financial support for infrastructure (Interview Yuksel, Mayor of Birecik 2009; Baydemir 2011).
- Commercialization: The Turkish Water Assembly, founded in January 2010, unites local people and their representatives with NGOs, lawyers and scientists who advocate a "new water policy" considering the environment, rural development and the interests of local people (Turkish Water Assembly 2011).[26]
- Liberalisation: The TMMOB–CMO, which has more than 300,000 members, condemns the liberalisation of the energy sector and complains about the selling out of national resources; it argues in favour of water resources being a public good and being held under the state's stewardship (Onal 2011).

As diverse as this movement is, civil society organisations, environmental NGOs and elected local representatives successfully allied with social movements across Europe and internationally (see Atzl in this volume), but they have not found influential allies within the government. On the contrary, the attitude of Erdogan's AKP government has hardened. His key allegations are: that the anti-dam protests are externally controlled; that they are linked to Kurdish separatism and nationalism; that they impair the nation and its hydro-based modernisation.

[26] Talking in the name of representatives of more than 50 rivers, Lawyer Yakup Okumusoglu, a member of the Turkish Water Assembly said: "Nature of Turkey is facing an ecocide that has been never experienced in its history. The main cause of this unacceptable situation is that the rivers and streams which give life to our land have become subjects of epidemic HEPP and dam projects. More than 1,700 HEPP and dam projects were planned leaving rivers with no or very little natural water flow for the living of local people, maintaining of the cultural diversity and nature.... Rivers and river basins of Turkey have become subject for purely commercial development, with no considerations on the rural issues and the environment.... However, the Ministry of Environment insists consistently with proceeding this HEPP and dam policy with no change. Therefore, even court decisions involving the suspension or cancelation of HEPP and dam projects are not implemented. Such an attitude and policy is unacceptable for any national and international standards." (original translation, online at: *english.turkiyesumeclisi.net/Access 25 November 2013*).

Table 5.7 Political environment of dam-related reforms

Political environment	Water resources/ hydropower	Environment	Expropriation	Resettlement
1945–1959 Central economy, one-party system	State-led development by central bureaucracies		(1956) in public interest	Law (1934) only applies to refugees and nomads
1960–1969 Military coup Civil govt. promotes import substitution	Cont.			
1970–1979 Military coup Govt. promotes protectionist model	Cont.	Undersecretary of Environment (1978)		Law (1970) applies to people affected by large infrastructure
1980–1993 Military coup Integration into global economy, liberalisation of energy sector	Constitution (1982): water as a public good Promotion of GAP	Environment Law/National Parks (1983); Ministry (1991); EIA Dir. (1993)	Constitution (1982): "public benefit" Expropriation Law (1983)	
1993–2000 President Demirel	Dam program Decentralisation of large-scale irrigation schemes	Party to Ramsar and Biodiversity conventions		
1999–2009 Accession candidate; Copenhagen criteria (2001); UNFCCC + Kyoto Protocol (2009)	Promotes renewable energy policy	Revisions of EIA Directive to align with EU acquis	Strengthening of civil rights (2001) Adopting EU + World Bank standards	Expropriation + resettlement by project owner (2006)
2010–	Incentives for hydro; amendment to Renewable Energy Law	Renewable energy projects in protected areas; suspension of Art.3 in EIA Directive, but amended to Environment Law	Cabinet can decree "urgent expropriation" for renewable projects	

5.7 Findings and Conclusion

When the Turkish government subscribed to low carbon to drive its energy agenda in the 2000s, it prioritised the use of river water for generating hydropower over other uses, and gave the energy sector advantageous terms for using rivers. During the course of the government's economic programme (i.e. the full exploitation of the country's hydropower potential), DSI rejected outright the World Commission on Dams' report (WCD 2000) because it feared a veto on dam development and suspected that it was driven by external, non-legitimised actors (WCD 2001).[27] However, DSI and the Turkish government have always referred to the World Bank's standards, which are considered to be more conducive for the Bank's clients' interests.

So far, our analysis has identified moderate moves that Turkish governments have made until now towards international standards.

Legal reforms in Turkey related to environmental planning and practices were obviously influenced by aspirations to become a full member of the European Union (Table 5.7 shows the political environments of dam-related reforms); a clear indication of that was the re-drafting of the Turkish EIA Directive in 2008. But Turkey had, at the same time, very selectively and cautiously aligned with the EU's environment and water acquis. Provisional Article 3 of the Turkish EIA Directive allowed for broad exemptions of the EIA requirements. In 2010, the possibilities for exempting dam projects from EIA requirements even increased after the article gained the status of law when it was amended to the Environment Law by parliamentary decision. The 2008 version opened the EIA procedure to civil society and NGOs, but its influence on the design of dam projects and overall decision-making remained limited. However, with growing political space for civil society organisations and an increased recognition of civil rights following the recognition of the Copenhagen Criteria in 2001, the EIA procedure has been used intensively and has been instrumental in advocating for international environmental and social standards.

Reforms related to expropriation and resettlement were the result of multiple factors—not the least being public debates on the poor resettlement record concerning the Ataturk Dam and the numerous court cases, including those at the European Court of Human Rights, which exposed Turkey's policies internationally. A major shift was made in 2001, when the Turkish parliament strengthened civil and property rights, and thus accelerated alignment with the Copenhagen Criteria and World Bank standards, as suggested by Kemal Dervis. The new ruling strengthened the rights of people affected, and abandoned a practice that would be better characterised as the confiscation of property.

A forerunner and innovative player was GAP RDA, but also Regional Directorates of DSI, which wanted to change course after the Ataturk Dam experiences

[27] Members of the World Commission on Dams were not representatives of states (see Scheumann and Hensengerth as well as Nymann et al. in this volume).

and invented a Resettlement Action Plan for the Birecik Dam in the late 1990s. One can only speculate about the reasons: GAP RDA was dedicated towards overall regional development—it was the Regional Development Administration that made GAP a project not only for hydraulic infrastructure (dams, hydropower plants) but also for creating opportunities for the economic and social development of the whole region. However, Resettlement Action Plans have not yet become a legal requirement and are not anchored in Turkish law.

Turkey saw an increase in civil society activities when international financiers and export credit agencies of OECD countries became involved in dam financing (Ilisu and Yusufeli dams). The presence of international actors provided additional leverage to civil society, whose action is geared towards ensuring that international norms are actually complied with *in practice*. NGOs and civil society groups have referred to international norms and allied with transnational and European partners able to exert pressure on international consortia via their home countries and private banks and companies (see Atzl in this volume). The private sector (Interview BM Group 2009; Interview Dolsar Consulting 2009) has also been engaged in applying international standards, driven by its integration into the global economy and carbon markets.

However, a change in government policies came in 2010, when the AKP government removed environmental and social barriers in order to accelerate hydropower development on an unprecedented scale.

Statements and activities of civil society organisations, lawyers, professional organisations, elected mayors and the academic elite show that the pro-dam consensus is eroding. The Turkish government has yet to make the change in direction that is required if "wise planning and implementation" is to be achieved. Towards this end, overall framework conditions are key, because the implementation of international dam-related norms not only takes place in individual issue areas—such as the environment, expropriation and resettlement—but also in the ways in which decisions are made. This includes, and relates to, the degree of decentralisation, the democratisation of the political process and whether the government follows a socially inclusive and environmentally-friendly development agenda. The political-administrative system in Turkey—where both water resources and energy planning is organised along sectoral lines with strong central bureaucracies—has not yet opened up to civil society and locally elected representatives, but only to the private sector.

Acknowledgments We are most thankful to Professor Dr Aysegul Kibaroglu from Okan University Istanbul, Dr Sahnaz Tigrek from Middle East Technical University Ankara, Dr Vakur Sumer from Selçuk University Konya, Professor Dr Muzaffer Yucel and Professor Dr Sevilay Topcu from Cukurova University Adana, Professor Gokhan Orhan from University Balikesir and many other academic colleagues throughout Turkey.

We gratefully acknowledge the support we received: from water policy and water law experts; from the public authorities concerned with dam planning and environmental, expropriation and resettlement planning and project implementation; from politicians and members of the Turkish parliament; from private sector representatives involved in consultancy services and dam construction; and from non-governmental organisations engaged in the dam debate.

References

Adem C (2005) Non-state actors and environmentalism. In: Adaman F, M. Arsel (eds.), Environmentalism in Turkey between democracy and development? Ashgate Publishing Company, Aldershot, pp 71–86

Akca E, Fujikura, R Sabbag C (2013) Atatürk dam resettlement process: increased disparity resulting from insufficient financial compensation. Int J Water Res Dev 29(1):101–108

Akyurek G (2005) Impact of Atatürk Dam on social and environmental aspects of the Southeastern Anatolia Project, thesis submitted to the Graduate School of Natural and Applied Sciences of the Middle East Technical University, Ankara

Altinbilek HD, Bayram M, Hazar T (1999) The new approach to development project-induced resettlement in Turkey. Water Res Dev 15(3):291–300

Altinbilek D, Tortajada C (2012) The Atatürk Dam in the context of the southeastern Anatolia (GAP) project. In: Tortajada C et al (eds) Impacts of large dams: a global assessment. Springer, Heidelberg, pp 171–199

Baskan A (2011) Liberalization of Turkey's hydroelectricity sector. In: Kibaroglu A et al (eds) Turkey's water policy. National frameworks and international cooperation. Springer, Heidelberg, pp 83–91

Baydemir, O. (2011) Opening speech of the second water infrastructure symposium, 16–18 Sept, Diyarbakir, Mayor of the City of Diyarbakir

Cay T, Evren N (2007) Kamulastirma uygulamalarinda uzlasma basarimi, Ankara

Cernea M, Guoqing S, Hazar T, Kir Y (2008) Institutions and capacity building for resettlement in Ilisu. Report on the second field visit of the Committee of Experts-Resettlement, 10–19 March 2008

Cernea MM (2006) Comments on the resettlement action plan for the Ilisu Dam and HEPP project, prepared for the Berne Declaration, Switzerland and the Ilisu Dam campaign Europe. http://www.evb.ch/cm_data/studie_des_ehemaligen_umsiedlungs-weltbankexperten_Prof._Dr._cernea_23.02.2006.pdf. Accessed 10 Nov 2009

Committee of Experts; Sub-committees on Environment Feb 2007, May 2008, Nov 2008, Feb 2009, External Monitoring Group Report, June 2009. http://www.agaportal.de/.../coe_environment_20080213.p.... Accessed 27 Nov 2013

Coskun A (2005) An evaluation of the environmental impact assessment system in Turkey. Int J Environ Sustain Dev 4(1):47–66

Council of Europe (2004) Protocol to the convention for the protection of human rights and fundamental freedoms, as amended by Protocol No. 11. Council of Europe (European Treaty Series No. 9), Strasbourg

Dogan YP (2009) Turkey to become key carbon market to kick CO_2 habit, In: Today's Zaman (March 4)

DSI (Devlet Su Isleri) (2009) Water and DSI, 5th World Water Forum Istanbul, Ankara

Eberlein C et al (2010) The Ilisu Dam in Turkey and the role of export credit agencies and NGO networks. Water Altern 3(2):291–312

EFT (Environment Foundation of Turkey) (2007) Environment foundation of Turkey. http://www.cevre.org.tr/. Accessed 20 May 2009

Encon (2006) Yusufeli Dam and HEPP: environmental impact assessment, Ankara

Erdogdu E (2007) Regulatory reform in Turkish energy industry: an analysis. Energy Policy 35(2):984–993

European Commission (2007): Turkey 2007 progress report, Commission staff working document, Brussels

European Commission (2009): Turkey 2009 progress report, Commission staff working document, Brussels

European Commission (2011) Turkey 20011 progress report, Commission staff working document, Brussels

GAP RDA (South-eastern Anatolia Project Regional Development Administration) (2004a) Birecik Dam and hydraulic power plant. http://www.gap.gov.tr/English/Dergi/D581997/birecik.html. Accessed 22 May 2009

GAP RDA (South-eastern Anatolia Project Regional Development Administration) (2004b) The planning and practical application experience of the resettlement, employment and economic investments of the people affected by the Birecik dam: the approach of the GAP administration. http://www.gap.gov.tr/English/Oproje/birecik.html. Accessed 22 May 2009

GAP RDA (South-eastern Anatolia Project Regional Development Administration) (2009a) GAP—UNDP sustainable development programme. http://www.gap.gov.tr/gap_eng.php?sayfa=English/gapundp.html. Accessed 22 May 2009

GAP RDA (South-eastern Anatolia Project Regional Development Administration) (2009b) Resettlement, employment and economic investments of people affected by Birecik Dam, a project for planning and implementation. http://www.gap.gov.tr/English/Sosprj/birecik.html. Accessed 12 April 2009

Hazar T, Ozbek AK (n.d.) Expropriation, resettlement and rehabilitation (with some cases) in Turkey, only via Google. Accessed 24 Aug 2013

IAIA (2013) Website of International Association for Impact Assessment. http://www.iaia.org/. Accessed 20 Aug 2013

Innanen SER (2004) Environmental impact assessment in Turkey: capacity-building for the European Union accession. Impact Assess Proj Apprais 22(2):141–151

International Energy Agency (2010) Energy policies of IEA countries, Turkey 2009 Review, Paris

Islar M (2012a) Struggles for recognition: Privatization of water use rights of Turkish rivers. Local Environ 17(3):317–329

Islar M (2012b) Privatized hydropower development in Turkey: a case of water grabbing? Water Altern 5(2):376–391

Izci R (2005) The impact of the European Union on environmental policy. In: Adaman F, Arsel M (eds) Environmentalism in Turkey between democracy and development? Ashgate Publishing Company, Aldershot, pp 87–100

Kadirbeyoglu Z (2005) Assessing the efficacy of transnational advocacy networks. In: Adaman F, Arsel M (eds) Environmentalism in Turkey between democracy and development? Ashgate Publishing Company, Aldershot, pp 101–116

Kaygusuz, K./M. Arsel (2005): Energy politics and policy, in: F. Adaman/M. Arsel (eds) Environmentalism in Turkey between democracy and development? Ashgate Publishing Company, pp 149–165

Kibaroglu A, Baskan A (2011) Turkey's water policy framework. National frameworks and international cooperation. In: Kibaroglu A et al (eds) Turkey's water policy. Springer, Heidelberg, pp 3–25

Klaphake A, Scheumann W (2011) Coruh river basin: hydropower development and transboundary cooperation. In: Kibaroglu A et al (eds) Turkey's water policy. National frameworks and international cooperation. Springer, Heidelberg, pp 251–261

Konak N (2011–12) The emergence of environmental concerns and the judicial system: River-based hydro schemes in Turkey. Int J Environ Cult Econ Soc Sustain 7(6):207–220

Kudat A (2001) Key elements of the new Turkish expropriation law and implications for the BTC project. https://www.google.de/search?q=Kudat%2C+A.+%282001%29%3A+Key+elements+of+the+new+Turkish+expropriation+law+and+implications+for+the+BTC+project&ie=utf-8&oe=utf-8&aq=t&rls=org.mozilla:de:official&client=firefox-a&channel=np&source=hp&gfe_rd=cr&ei=SrXfUvLaF8KX_Ab2voDYBA. Accessed at 22 January 2014

Ministry of Energy and Natural Resources (2010) Strategic plan 2010–2014 Ankara, TC Turkey. http://www.enerji.gov.tr/…/ETKB_2010_2014_Stratejik. Accessed 25 November 2013

Morvaridi B (2004) Resettlement, rights to development and the Ilisu Dam, Turkey. Dev Change 35(4):719–741

Nisanci R (2005): The production of pixel-based urban land value maps with nominal valuation method using GIS, Ph.D. thesis, Karadeniz Technical University, Trabzon

Nisanci R, Uzun B, Yomralioglu T (2007) Land valuation issues of expropriation applications in Turkey, seminar on Compulsory Purchase and Compensation in Expropriation and Takings, FIG Commission 9, Helsinki

OECD (Organisation for Economic Co-operation and Development) (2008): Environmental performance reviews—Turkey. OECD Publishing, Paris

Onal T (2011) Opening speech of the second water infrastructure symposium, Diyarbakir, 16–18 Sept 2011

Ozduzen N (2008) General information of GAP and the Ataturk Dam. http://www.adiyamanli.org/gap/info_on_gap.htm. Accessed 12 April 2009

PÖYRY (2010) Hydropower in Turkey: potential and market assessment, commissioned by INTPOW, Norway

PWC (2012) Turkey's renewable energy sector from a global perspective. http://www.pwc.com.tr/en/publications/industrial/energy/pdf/renewable-report-2012.jhtml. Accessed 18 Sept 2013

Republic of Turkey (2008) By-law on environmental impact assessment (EIA), Turkish Official Gazette, Number: 26939 (revisions in 2009, 2011, 2013)

Republic of Turkey (2009) Turkey water report, Ankara

Scheumann W, Irmak A (1991) Das Umsiedlungskonzept der türkischen Regierung am Beispiel des Atatürks Staudamms. TU Int 12/13:36–38

Sociology Association (1994) Survey on the problems of employment and resettlement in areas which will be affected by dammed lakes in GAP region, executive summary, Ankara, June

SPO (State Planning Organisation) (1999) National environmental action plan of Turkey, Ankara

SPO (State Planning Organisation) (2001) National programme for the Adoption of the Acquis, Ankara

SPO (State Planning Organisation) (2006) Ninth development plan 2007–2013, Ankara

Sumer V, Muluk C (2011) Challenges for Turkey to implement the EU water framework directive. In: Kibaroglu A et al (eds) Turkey's water policy. National frameworks and international cooperation. Springer, Heidelberg, pp 43–67

TUSEV—(Turkiye Ucuncu Sektor Vakfi) (2005) Civil society in Turkey: An era of transition, CIVICUS Civil Society Index Country Report for Turkey. https://www.academia.edu/742021/Civil_Society_in_Turkey_CSI_Turkey_Country_Report. Accessed 25 Nov 2013

Turgut N (2003) EIA with reference to the EU Directive. Environ Policy Law 33(3–4):163–169

Turkish Water Assembly (2011) http://english.turkiyesumeclisi.net/.../2010/.../turkish-water. Accessed 25 Nov 2013

US Energy Information Administration (2013) International energy outlook 2013. http://www.eia.gov/forecasts/ieo/. Accessed 27 Nov 2013

WCD (World Commission on Dams) (2000) Dams and development: A new framework for decision-making. Earthscan Publications Ltd., London

WCD (World Commission on Dams) (2001): ICOLD members—Turkey: general Directorate of State Hydraulic Works (DSI): response to the final report of the World Commission on Dams. http://www.dams.org/report/reaction/icold_turkey.htm. Accessed 15 May 2009

WWF Turkey (World Wide Fund for Nature Turkey) (2003) Ermenek Dam and Hydropower project: evaluation of ecological issues, research and evaluation report, Istanbul

World Bank/ESMAP/T.C. Cevre ve Sehirlik Bakanligi (2012) Sample guidelines: cumulative impact assessment for hydropower projects in Turkey, Ankara

World Resources Institute (2009) Policies in Turkey. http://projects.wri.org/sd-pams-database/turkey. Accessed 15 May 2009

World Bank (n.a.) Second renewable energy and energy efficiency project. Resettlement policy framework, RP 762. http://www-wds.worldbank.org/external/default/WDSContentServer/WDSP/IB/2009/02/05/000334955_20090205051942/Rendered/PDF/RP762RP0P112570work0draft0Jan262009.pdf. Accessed 4 Dec 2013

Yalcin E, Tigrek S (2011) Investigation of alternative solutions to Ilisu Dam and HEPP, Civil Engineering Department, Middle East Technical University, Ankara, Turkey (file with author)

Yomralioglu B, Uzun B, Nisanci R (2007) Land valuation issues of expropriation applications in Turkey, Seminar on compulsory purchase and compensation in land acquisition and takings, FIG Commission 9, Helsinki, Finland, 6–8 Sept

Newspaper

Hurriyet 5 June 2013
Zaman 30 December 2008; 4 May 2007
Radikal 20 September 2003

Interviews

Aslan AK (2009) Mayor of New Samsat, 13 March, New Samsat
Head of Agricultural Office New Samsat (2009) 13 March, New Samsat
BM Group (2009) CEOs, 24 February and 6 March, Ankara
Caglar A (2009) Professor at Hacettepe University, 26 Feb, Ankara
Dolsar Consulting (2009) CEOs I. Aker, S. Aker, A. Bacak, 28 April, Ankara
DSI General Directorate Ankara (2009) A. Icten, head of expropriation department, 29 April, Ankara
DSI 6th Regional Directorate Adana (2009) Bulent Selek and Turan Yilmaz, 3, 9 and 10 March, Adana
DSI Ermenek (2009) M. Gurel, 6 March, Ermenek dam construction site
ENCON (2009) 30 April, Ankara
GAP RDA (2009) Mehmet Acikgoz, Regional Director, 12 March, Sanliurfa
Gencek A (2009) Mayor of Halfeti, 16 March, New Halfeti
Kalaycioglu S (2009) Professor at Middle East Technical University, 26 Feb, Ankara
MoEF (Ministry of Environment and Forestry) Ankara (2009a) Mustafa Satilmis, EIA Department, 20 Feb and 28 April, Ankara
MoEF (Ministry of Environment and Forestry) Ankara (2009b) 28 April 2009, Ankara
MoEF (Ministry of Environment and Forestry) Adana (2009): 4 March, Adana
MoEF (Ministry of Environment and Forestry) Sanliurfa (2009) 12 March, Sanliurfa
MoEF (Ministry of Environment and Forestry) Balikesir (2009) 25 March, Balikesir
Ozdalga H (2009) Member of Parliament and chairmen of the Committee of Environment, 23 Feb, Ankara
Saglam Y, Sahin O (2009) Mayor of Yusufeli Town and head of administration, 19 March
Sozkesen U, Tezcam A (2009) Mayor of Ermenek and Kaymakan, 6 March
Tigrek S (2010) Professor at Middle East Technical University, Ankara
TSKB (2009) R. Acinci, 23 March, Istanbul
Tuncok K (2009) DHI Water and Danish Hydraulic Institute, 24 Feb, Ankara
Yildiz D, Ozbay O (2009) Private sector consultancy, 18 Feb, Ankara
Yucel M (2009) Professor at Cukurova University, 4 March, Adana
Yuksel AK (2009) Mayor of Birecik, 16 March, Birecik
WWF Turkey (2009) C. Eryilmaz, 22 Feb, Ankara

Chapter 6
NGOs as Strategic Actors in the Promotion of Sustainable Dam Development

Sara Eichert

Abstract Considering the heterogeneity of actors and strategies employed in transnational advocacy networks, as well as the modern governance debate where some activists express their concern about the power shift away from government towards the private sector, this chapter explores the various strategies non-governmental organisations (NGOs) use to persuade private actors to promote sustainable dam development and considers why some NGOs choose to adopt a more or less collaborative attitude towards the dam industry. It analyses the strategies of two international NGOs in a recent industry-driven, multi-stakeholder dialogue, the Hydropower Sustainability Assessment Forum (HSAF), and their positions on its outcome, the 2010 International Hydropower Association's Hydropower Sustainability Assessment Protocol. The analysis suggests that ideological differences might account for the choice of more or less collaborative approaches.

6.1 Introduction

Non-governmental organisations (NGOs) and so-called transnational advocacy networks (TANs) (Keck and Sikkink 1998) have attracted the attention of scholars in various academic fields. Authors have highlighted the capacity of these non-governmental actors to use their "moral authority" to influence norms and practices, to create new norms and to disseminate them (Florini 2000; Finnemore and Sikkink 1998; Nadelman 1990; Sikkink 2002). One prominent and often cited example is their successful participation in the independent, multi-stakeholder World Commission on Dams (WCD). As several case studies have further demonstrated, NGOs' alliance with domestic civil society opposition has succeeded in

S. Eichert (✉)
German Development Institute (DIE), Bonn, Germany
e-mail: seichert@ucm.es

W. Scheumann and O. Hensengerth (eds.), *Evolution of Dam Policies*,
DOI: 10.1007/978-3-642-23403-3_6, © Springer-Verlag Berlin Heidelberg 2014

halting and altering large hydroelectric dams around the world (Khagram 2000, Khagram et al. 2002). Recent case studies, on the other hand, have also revealed limitations and asymmetries of the strategies pursued by transnational anti-dam networks towards particular dams, since the chosen NGO strategies may influence the overall success of a network campaign.[1] In fact, analysis of these rather vaguely defined networks has failed to explain individual NGO strategies and tactics and has taken little or no account of the existence of more or less collaborative interactions between NGOs and public and private actors.

Furthermore, the lack of commitment to WCD guidelines by the dam industry, funders, and national governments of countries where the majority of dams have been and are being developed (e.g. China, India, Russia, and Turkey), has left the future of the development of hydropower sustainability standards open. Yet the social and environmental standards set by international institutions and financial sectors, such as the International Finance Corporation's (IFC) performance standards, the World Bank's safeguard policies, and the Equator Principles should be highlighted as milestones in the pressure that civil society has continued to exert in the last 30 years.

In this context, NGOs today pursue various strategies to influence stakeholder groups who take part in the elaboration of sustainability standards for water and energy infrastructure development. Notwithstanding the broadly accepted perception that NGOs have fundamentally transformed the established relationship between business and government, they also have a direct impact on public policy and corporate strategies (Doh and Teegen 2003, p. xv), and increasingly interact with various stakeholder groups, including the private sector (Ottaway 2001). Some NGOs today increasingly seek strategic partnerships and dialogues with companies and government, whereas others generally decline to collaborate, especially when industry is involved (Åhlström and Sjöström 2005).

This chapter explores the various strategies NGOs use to persuade private actors to promote sustainable dam development and considers why some NGOs choose to adopt a more or less collaborative attitude towards the dam industry. For this purpose, I will compare the strategies of two international NGOs, the World Wide Fund for Nature (WWF) International and International Rivers (IR), which are currently engaged in the promotion of sustainable dam and hydropower development. I will analyse their strategies in a recent industry-driven, multi-stakeholder dialogue, the Hydropower Sustainability Assessment Forum (HSAF), and their positions on its outcome, the 2010 International Hydropower Association's Hydropower Sustainability Assessment Protocol.

[1] See Atzl (2012) in this book for a discussion on chosen pathways and techniques in the anti-Ilisu Dam campaign.

6.2 Theoretical Framework and Methodology

6.2.1 Literature Review

The literature on transnational advocacy networks[2] (Keck and Sikkink 1998) and on the assessment of non-governmental organisations in world politics has concentrated mainly on norm dissemination (e.g. Florini 2000; Finnemore and Sikkink 1998; Khagram et al. 2002; Nadelman 1990; Sikkink 2002) and the advantages of NGO participation for global governance architecture and its limits in this respect (e.g. Mathews 1997; O'Brien 2005; Simmons 1998; Weiss and Gordenker 1996) or explored the tensions within networks of NGOs and other civic organisations (e.g. Jordan and Van Tuijl 2000; Nelson 1997; Atzl 2012). Despite the wide acceptance of Keck and Sikkink's (1998) analytical framework of the boomerang pattern in the exploration of the pathways and techniques used by TANs to influence the status quo of norms and practices, some authors soon took a critical view of the structural differences inherent in these networks (Nelson 1997). They further questioned the supposed homogeneity of TANs by pointing out that the "exchange of information or services may be dense, but irregular and unbalanced. The quality of the ties between different members of a network is hardly ever constant and mutually recognized rules in a campaign are far from common" (Jordan and Van Tuijl 2000, p. 2062). Hence, advocacy campaigns by these rather loosely defined TANs, though focusing on a common issue, can take quite different and also competitive approaches (Jordan and Van Tuijl 2000, p. 2062). Tensions within TANs may result from internal power and resource asymmetries and from divergent standpoints on advocacy tactics and strategies envisioned by the constituent members of the network (see Rohrscheider and Dalton 2002 and Hogenboom 2003, cited in Dobner 2010, p. 278). Atzl (2012), for instance, analyses the different pathways and techniques chosen to influence decision-makers involved in the construction of the Ilisu Dam on the River Tigris in Turkey. His case study departs from the assumption of internal power, resource asymmetries and possible divergence of tactics and strategies employed by the constituents of TANs and concentrates primarily on analysing the actual capacity of each actor to influence and control the anti-Ilisu campaign's internal agenda-setting. His case study reveals a certain "asymmetry" of the objectives and pathways chosen by local, national and international anti-Ilisu campaigns. The strategy pursued by some international and European anti-dam networks of campaigning against financial support from European export credit agencies for the Ilisu Dam has resulted in the withdrawal of credits and can be categorized as a definite success for norm dissemination at international level. At local level, however, such major objectives as halting the construction of the dam have not

[2] Loosely defined as "actors working internationally on an issue, who are bound together by shared values, a common discourse, and dense exchanges of information and services" (Keck and Sikkink 1998, p. 2).

been achieved, since "the campaign's strategy has led into a situation where it has lost its most important pathway of influence" (at European and international level) for bringing about changes in Turkish legislation, for example (Atzl 2012). Considering these advances in the analysis of the role of NGOs and other members of advocacy networks and in norm dissemination focused on sustainable hydropower and dam development, there are few publications today that analyse the various motives and strategies of these NGOs in terms of whether they are likely to engage with or confront the hydropower industry.

On the other hand, the initiative discussed in this chapter is not the first of its kind to assess and try to improve the corporate social responsibility of an industry and/or specific companies for their actions and practices. In the past two decades there have been various multi-stakeholder, collaborative initiatives, such as the *Forest Stewardship Council* created in 1993, the *Ethical Trading* Initiative launched in 1998 and the 2002 *Extractive Industries' Transparency Initiative*, to mention but a few. These collaborative initiatives too are consistent with the academic debate and the emergence of new modes of governance described, for instance, by the German sociologist Mayntz (2001). Modern governance is defined as a new and more cooperative form of governance in which decision-making tends to be less hierarchical and more efficient. It is based on the assumption that specific problems or issues cannot be resolved by one actor alone, but need to involve public, private and civil actors—and, more specifically, those actors who cause the problem, who are being affected by it and who have the power and resources to solve it (Mayntz 2001, p. 5). In that sense, multi-stakeholder initiatives have emerged together with new direct and indirect governance mechanisms, such as self- and co-regulation (of companies in accordance with the principles of corporate social responsibility, for example) and public–private partnerships. Consequently, NGO participation in multi-stakeholder dialogues today is perceived as almost normal—they provide such resources as knowledge, power and norms to other stakeholders in formal negotiations (Brühl 2009, p. 33), legitimacy and reputation (see Austin 2000; Glasbergen and Groenenberg 2001) and access to NGO networks.

Nonetheless, one of the main concerns expressed about governance, and one that is referred to in the subsequent discussion of the motivation and strategies of the entities we have selected for analysis, is the power shift away from government towards the private sector, which can be seen as one of the primary problems associated with governance. In fact, governance can and has been understood as "governing without government" (see Rhodes 1996). "Collaborative" governance efforts have often been criticized for their failure to include important stakeholders and their consequent lack of legitimacy, transparency and social accountability. The often voluntary nature of these initiatives similarly raises concerns about the weakening of existing standards and the failure to develop legally enforceable regulatory mechanisms.

Furthermore, to distinguish and explain the selection of NGO strategies and to examine the possible explanations for their divergence, it is convenient to review the literature that classifies different types of NGOs. They may be classified according to their activities, their ideologies (i.e. the beliefs and attitudes that explain their behaviour), the type of benefit they are seeking, their field of action or simply the geographical scale of their activities. This chapter analyses the two entities selected for analysis—WWF and IR—on the basis of their activities and ideology with a view to understanding strategies of environmental NGOs active in the field of sustainable hydropower.

"Operational NGOs", are defined by the World Bank (2010) as NGOs "whose primary purpose is to fund, design, or implement development-related programs or projects". They are service-oriented and "traditionally apolitical" (Parker 2003, p. 88). On the other hand, "advocacy NGOs" defend and promote specific issues or causes with the aim of influencing the policies and practices of public, private and civil institutions. They are not "behemoths weighted down by long-term investments in offices and staff in various parts of the world". They "exist with few staff, and through electronic linkages (…) maintain vast and loose networks" and prefer "to take on a monitoring (and thus potentially antagonistic) role rather than a collaborative role" in relation to business (Parker 2003, p. 86). Hybrid NGOs (e.g. CARE and Save the Children International) are more likely to endorse a collaborative approach than advocacy NGOs (Parker 2003, p. 86). They are located between the classical apolitical operational NGO and the "more public political activist" advocacy NGO (Parker 2003, p. 87). They also "constitute a strategic entry point for MNEs (multinational enterprises) because they have already constructed the foundations for relationships that recognize interdependence and where mutual respect can be found in different priorities and thresholds" (Parker 2003, p. 89).

Complementing this classification, Zald and McCarthy (1980) introduce an interesting categorization of NGOs based on their ideology: reformative or radical. According to their classification, hybrid NGOs can be identified as "reformative", whereas "more radical", advocacy NGOs would tend to adopt a "potentially antagonistic role", as referred to in Parker (2003) above. The difference between radical and reformative NGOs lies, however, in the assumption that, while the latter see companies as both part of the problem and part of the solution, the former do not (Den Hond and De Bakker 2007, p. 903). Winston (2002) anticipates the expansion of NGO-business interactions because of the power shift "away from government towards corporations" (Winston 2002, p. 72) and also follows the ideology-based differentiation of NGOs. He distinguishes between NGOs that act as *engagers*, who "try to draw corporations into dialogue in order to persuade them by means of ethical and prudential arguments to adopt voluntary codes of conduct" and those which act as *confronters*, who "believe that corporations will act only when their financial interests are threatened, and therefore take a more adversarial stance toward them" (Winston 2002, p. 71). Moreover, the NGO split is reflected in the dilemma over whether to engage in corporate social responsibility (CSR) or to advocate enforceable state regulation as a compulsory means in contrast to the voluntary CSR approach (Winston 2002, p. 71). From this classification Van Huijstee (2009, p. 40) identifies two divergent activist attitudes towards the private sector: collaborative and confrontational NGO strategies.

As a result, some authors have recently begun to analyse inter-sectoral dialogues (Van Huijstee and Glasbergen 2007), cooperative strategies outside environmental NGOs' networks, i.e. NGO-company dialogues, and inside such networks, i.e. cooperation among network members and within departments of one NGO (Doh et al. 2003) and new partnerships emerging between the private sector and environmental NGOs (Glasbergen and Groenenberg 2001, Glasbergen 2007). These partnerships ultimately resulted in some environmental NGOs abandoning their "antagonistic relation with the business community" (Glasbergen and Groenenberg 2001, Glasbergen 2007).[3] The following theoretical framework will seek to shed light on the reasons for an NGO's choice of a particular strategy.

6.2.2 Theoretical Framework

A promising analytical approach developed by Den Hond and De Bakker (2007) assumes that divergent NGO strategies and the protest tactics they employ are due to ideological differences between these activist groups—that is, the beliefs and attitudes cited by NGOs to justify their behaviour when confronted with a given problem (Den Hond and De Bakker 2007, p. 903)—rather than their interest or identity (see Rowley and Moldoveanu 2003). They assume that the general objective of both radical and reformative NGOs is to influence social or institutional change by de- and/or reinstitutionalizing existing corporate or sectoral norms and practices, deinstutionalization meaning the "abandonment" of institutions, reinstitutionalization their "replacement" (Rowley and Moldoveanu 2003, p. 904). For instance, reformative NGOs intent on deinstitutionalization, i.e. the non-application of existing norms and practices to the construction of hydropower dams, may use *consequential arguments*, that is, grant moral legitimacy to the sector because of what it *does*. Radical NGOs are more likely to use *structural arguments*, that is, to grant moral legitimacy because of what the sector *is*.

More specifically, a reformative NGO will therefore discredit the unsustainable performance of company X in the development of dam Y and then call for the improvement of the company's performance. It will then suggest, for example, a partnership or dialogue that could subsequently have positive reputational effects on the company (granting it *pragmatic legitimacy*). However, radical NGOs are more likely to go a step further and not only criticize company X's performance, but possibly its product and the whole sector.

This means that reformative NGOs are more likely to employ a *symbolic gain* or *material gain* strategy to replace existing norms (reinstitutionalization), while radical NGOs, pursuing confrontational strategies, usually with the aim of deinstitutionalizing, or ousting existing norms and practices, will pursue *symbolic* or

[3] See also Martínez-Alier (2002) on the change of discourse in various environmental movements.

material damage strategies. Symbolic strategies are resource-efficient, depend less on the mobilization of large numbers of people and are therefore more commonly used than material strategies. Examples of the logic of symbolic gain are positive publicity and cooperation in the form of partnerships or dialogues. NGOs using symbolic damage tactics may seek to cause negative publicity by writing letters or emails, submitting petitions, and organizing marches and rallies (Rowley and Moldoveanu 2003). NGOs that employ symbolic damage tactics tend to document abuses and employ "moral shaming" (Winston 2002). In practice, however, NGOs will often pursue a mixture of confrontational and collaborative tactics (Brunnengräber 2001; Winston 2002).

Van Huijstee and Glasbergen (2010) recently adopted the aforementioned analytical approaches in a comparative in-depth case study that analysed the strategies used by WWF Netherlands (symbolic gain) and Friends of the Earth Netherlands (symbolic damage) to influence a Dutch bank's responsibility for sustainable finance. They eventually added the proposition, first, that both reformative and radical NGOs will try to reinstitutionalize corporate or industrial practices, second, that symbolic damage and gain strategies employed by NGOs in parallel interact and third, that the two strategies will reinforce each other by first deinstitutionalizing norms and practices by means of symbolic damage tactics and then inducing a company to institutionalize alternative practices or norms in the reinstitutionalization process. In summary, both types of strategy may be necessary to effect social or institutional change. Hence, as Den Hond and De Bakker (2007, p. 904) also point out, for institutional change to occur it is important that existing institutions, i.e. norms and practices before they are replaced, are deinstitutionalized.

6.2.3 Methodology

This chapter follows a primarily qualitative approach aimed at describing and understanding the choice of strategy used by NGOs to promote sustainable hydropower standards. For this purpose, I will compare two cases: the strategies pursued by two international NGOs to promote sustainable hydropower. Besides analysing the general strategies they adopt to achieve hydropower sustainability, I will discuss their tactics and their attitudes towards business in a recent industry-led, multi-stakeholder dialogue, the Hydropower Sustainability Assessment Forum (HSAF), and their motivation to engage in or oppose the process.

Cases were selected on the basis of the general objective of this chapter of comparing the strategies adopted by individual and different international NGOs in their quest for sustainable hydropower in the private sector. The forum was not in fact a private actor and included both NGOs and government agencies However, it is regarded as a multi-stakeholder initiative primarily driven and owned by the hydropower industry, and specifically by the International Hydropower Association

(IHA).[4] Moreover, the use of the 2010 Hydropower Sustainability Assessment Protocol (referred to as the "Protocol", "2010 Protocol" or "IHA Protocol" in the following) and of drafts and earlier versions is partly controlled by the IHA, as in the cases of the release of assessment results.[5] As our theoretical framework suggests, NGOs generally pursue one of two strategies, a symbolic gain or a symbolic damage strategy, vis-à-vis the business community, their goal being both to deinstitutionalize the status quo and to reinstitutionalize new norms/practices. We have therefore chosen the WWF and IR as our examples. Furthermore, previous case studies by Van Huijstee and Glasbergen (2010) and Parker (2003) have categorized the WWF as usually adopting a collaborative approach to the industry. However, no such classification can be found in the literature on IR, although the above classification of NGOs indicates that it is primarily an advocacy NGO. Both the WWF and IR are non-profit and non-governmental organisations that focus on national and international environmental and social issues, but have a particular interest in the sustainable planning, construction and operation of hydropower and dam infrastructure. They engage with a variety of stakeholders at local, national, regional, international and transnational level.

IR was founded in Berkeley, California, in 1985. This non-profit organisation operates in over 60 countries throughout the world and has staff working on sites in Brazil, India and Thailand, for example. Its head office is located in the United States. Its key focus is on the sustainable development of river basins worldwide, but its approach is oriented towards human rights and social issues far more than the WWF's. Its overall organisational goal is to "protect(s) rivers and defend(s) the rights of communities that depend on them", promoting sustainable alternatives to the otherwise destructive water and energy planning of river basins (IR no year).

The *WWF*, which was founded in 1961 and is registered under Swiss law, can be categorized as an international non-profit, non-governmental environmental NGO (officially a foundation) with 30 independent national offices and some five associate offices worldwide. Its key concerns are environmentalism and nature conservancy. Its organisational structure is network-based. The WWF also heads and coordinates "the *WWF Network* of offices around the world, through developing policies and priorities, fostering global partnerships, coordinating international campaigns, and providing supportive measures in order to help make the global operation run as smoothly as it can".[6]

---

[4] The IHA describes itself as a non-profit, unincorporated (i.e. voluntary) and mutual association of its members, who are both individuals and organizations active in the hydropower sector or a related field. Its head office is located in London, United Kingdom, see http://www.hydropower.org/about-iha/; http://www.hydropower.org/about-iha/current-members.html for a full list of its current corporate members and sponsors. Accessed 04 October 2012.

[5] See http://www.hydropower.org/sustainable_hydropower/hsaf_Hydropower_Sustainability_Assessment_Protocol.html for the purpose and use of the Protocol. Accessed 04 October 2012.

[6] See http://wwf.panda.org/who_we_are/organization/. Accessed 23 August 2012.

The data I collected came primarily from two sources: documentation (NGO strategy chapters, press releases, website material, HSAF documentation, e-mail exchanges) and semi-structured interviews in support of my documentary analysis to ensure the validity of my comparative case study. Interviews were conducted with some of the stakeholders involved in the HSAF process and with IR and WWF policy experts and coordinators. Forum members were generally asked about their expectations and their motivation (or lack of motivation) to participate, their view of the forum, their perception of forum dynamics, and their perception of NGOs and the forum outcome. Both IR and the WWF were asked to give a more general assessment of organisational strategies with regard to hydropower sustainability. Interviews were conducted during August 2012 with the following organisations: IR, WWF, German Agency for International Cooperation (GIZ), and Oxfam International, and with a senior staff member of an international organisation (which did not want to be identified).

6.3 NGO Strategies for Sustainable Hydropower

This section first discusses the general strategies adopted by the NGOs selected to promote sustainable hydropower, and then analyse their symbolic gain and symbolic damage strategies in the HSAF and the drawing up of the Protocol. Finally, I compare the two strategies, using the theoretical and conceptual frameworks.

6.3.1 The WWF's Global Water Security Initiative

WWF International launched its Global Dams Initiative shortly after the WCD process. Post-WCD publications by WWF, for instance, criticized the failure of six dams to comply with WCD priorities and shed light on the weak influence that the United Nations Dams and Development Project, which was established by UNEP to disseminate the WCD's recommendations, had had on policy (WWF 2005). The initiative has recently been renamed the WWF Global Water Security Initiative, as which it has become one of the three pillars of WWF's Freshwater approach.[7] Yet it still has the same mission: to ensure that "water and energy needs are met in a sustainable manner without the destruction of the very ecosystems that provide the basis of a clean and stable water supply" (WWF 2007, p. 3). Coordinated by WWF, the Global Water Security Initiative represents the WWF's position on both hydropower and water infrastructure and entails activities aimed at providing knowledge on and building capacities for water infrastructure development within the organisation itself.

[7] The other two pillars are Water Stewardship and Habitat Protection.

It is very important to mention that the WWF national offices work independently (with the support of WWF International) on country-specific strategies, since political or cultural constraints (e.g. non-democratic regimes) may have a major impact on the choice of the WWF's strategies at operational level. The WWF Sustainable Hydropower Specialist and coordinating member of the WWF Water Security Initiative, Dr Jian-hua Meng, gave a clear example of this during his interview: whereas the WWF national offices in some countries rely on tactics that target individual dam projects, the WWF makes the strategic choice in others, such as China, of collaborating with state-led water planning agencies and the water ministry in the establishment of basin-wide development plans, which is more likely eventually to succeed in ensuring the sustainable development of a river basin than is opposition to a single dam project.

On its website, WWF International strongly emphasizes that it "actively engages with banks and hydropower companies to ensure broad-based improvement of practices" (WWF no year). Numerous examples of WWF collaboration with both public and private actors can be found: the aforementioned WWF commitment (since 1980) to work together with the Chinese government for the sustainable development of the Yangtze river basin; cooperation with dam financiers, e.g. the *Investing in Nature* partnership with HSBC Holdings; environmental impact assessments of the Ta Trach Dam in Vietnam on behalf of the export credit agency Japan Bank for International Cooperation; and collaborating with public agencies to improve existing dams in South Africa (WWF 2005, p. 13; WWF 2007). In addition to its active engagement in these collaborative activities in the form of partnerships or dialogues largely based on knowledge exchange, the WWF sees itself as a facilitator of or mediator in dialogues and negotiations,[8] an example being consensus-based solutions in such countries as Turkey, where the debate between dam proponents and opponents is highly polarized. In such cases, the WWF tries to facilitate dialogue and mediate between the two fronts (interview with Dr Jian-hua Meng). An example of the WWF's international engagement is its recent coordination of a side-event panel entitled *Scalable Solutions—Managing water for economies, communities and nature* at the Rio + 20 UN Conference on Sustainable Development held in Brazil in 2012.[9] The overriding objective of this event was to demonstrate the positive effects of an integrated collaborative approach to managing water security problems and the trade–offs with energy and food in six WWF initiatives around the globe, including one relating to China's Water Planning and the Sustainable Hydropower Initiative in the Mekong Basin.

---

[8] See Islam et al. (2013, p. 130) for a recent discussion on whether mutually advantageous results are possible in water negotiations, given that the various stakeholders involved defend competing interests. The authors highlight the limits to and the importance of shifting from "competitive, zero-sum thinking to cooperative efforts to create values".

[9] Jointly with the Asian Development Bank, the Agence Française de Développement, the French Global Environment Facility and the Inter-American Development Bank.

In trying to determine how necessary dams are for development, the WWF concludes that hydropower dams are "both blessing [sic] and a curse" (WWF 2007, p. 3). Publications point to both the advantages and the disadvantages of dam infrastructure. Moreover, "there is room for essential and well-planned dam development for water and energy supply (…) to consider all options, both large-scale and small-scale, as well as demand-side and supply-side" (WWF 2007, p. 43). Dr Jian-hua Meng pointed out that the organisation does not oppose (large) hydropower dams, since they are perceived as an essential part of energy mixes worldwide. Yet the promotion of hydropower sustainability should be based on a "comprehensive assessment of all options, alternatives and impacts. This should include careful site choice, a cumulative assessment of impacts of dams in river basins, as well as extensive mitigation measures for those dams that are given the go-ahead" (WWF 2007, p. 3). He added that the WWF is no "anti-dam organisation, we are no anti-development organisation either. We are looking for good solutions", i.e. solutions that support "right dams at right places but keep out the wrong ones".

6.3.2 International Rivers' Attitude Towards Large Dams

Alongside the WWF, in its central position within the international environmental movement, IR has also been identified historically as an essential actor in the process of linking domestic anti-dam activism around the world (Khagram 2000). Its efforts in this regard date back to the mid-1980s and have since resulted in its occupying a strong coordinating leadership position in the global anti-dam movement/network (Nelson 1997, p. 432). At an earlier stage, it supported and organized conferences of dam-affected people worldwide, where together they articulated and declared new norms and practices for dam-building and sustainable development. Together with other anti-dam groups, it called for the establishment of an independent World Bank review commission on dams, which eventually led to the formation of the World Commission on Dams (McCully 2001). IR soon expressed its view on dams:

> While our analysis of the problem remained the same, we now understood that the destruction of rivers was as much a social and human rights issue as environmental. We started to see the importance of dams as centrepieces of inappropriate development ideology (…) (Williams 1998, quoted in Khagram 2000, pp. 92–93).

Accordingly, IR essentially derived its "legitimacy from the respect of the individual groups we worked with all over the world. We saw ourselves as one part of a larger citizen-based movement working on these issues" (Williams 1998, quoted in Khagram 2000, p. 93). Consequently, IR today acts on behalf of its partners, including the people affected, against "governments, banks, companies and international agencies (IR no year). At local level, where dams are built, IR pursues a capacity-building strategy aimed at raising awareness of existing standards, participatory rights and sustainable alternatives. Indeed, much of the work IR engages in is advocacy-based and aimed at raising awareness; it also calls and campaigns for the "most destructive dams not to be built" and works with the media to attract wide public attention, as Zachary Hurwitz, Policy Program Coordinator, mentioned when interviewed.

At international level, much of IR' work is concentrated on supporting and strengthening existing international environmental, social and human rights standards, such as the World Bank's Safeguards and the International Finance Corporation's (IFC) Performance Standards. It also generally supports strong policy commitments by national governments, dam financiers and builders. However, unlike the WWF, IR hardly interacts with or partners the dam industry and has been targeting it both in country-specific campaigns and in its current *Stop Dam Greenwashing* campaign. This campaign disapproves of the dam-building industry's efforts to justify dam construction in a general discourse on greenhouse gas reduction, climate change mitigation, and "clean energy" or a clean development mechanism, which amounts to "a public relations offensive" to "repackage hydropower as a green, renewable technology" (Imhof and Lanza 2010). Their advocacy work manifests itself, for instance, in frequent publications, newsletter updates, letters to financiers, builders and governments, and assessment reports on dam and development issues.

Although IR is usually known to endorse the World Bank's Safeguards and the IFC's Performance Standards, a recent publication, *Infrastructure for whom?* (IR 2012), openly criticizes the infrastructure strategies of the Group of 20 and the World Bank for disregarding the social and environmental costs of infrastructure projects. After the World Bank's withholding of support for infrastructure projects during much of the 1990s, recent reports indicate renewed multilateral development-bank support for "large-scale, centralized and private investment" infrastructure projects, such as the proposed Grand Inga power trading scheme on the River Congo. Hydropower projects are justified by their contribution to economic growth, a country's need for energy, and positive climate change mitigation effects (IR 2012). IR, which mostly represents dam-affected communities' interests and relates its demands to the existing standards referred to above, sees this approach as misguided because it does not consider the true "energy and water needs of the poor", it does not, as a rule, benefit them, and it does not specifically strengthen social environmental and human rights safeguards.

During the interview, Hurwitz explained that IR does not, of course, have a black-and-white position on hydropower, or on large, multi-purpose dams. Like the WWF, it supports the WCD's strategic priorities and policy principles, promotes "decentralized, small-scale renewable energy" as an alternative to proposed large-scale dam projects, and cumulative environmental impact assessments rather than mere project-based assessments, and stresses the importance of stakeholder participation, for instance. However, Hurwitz believes that the economic interests of the dam industry and governments clearly outweigh the interests of the worst-affected people. In fact, another example of its attitude towards the dam industry and funders is its criticism of the predominately economic approach of the World

Water Council (WWC),[10] seeing it as a "lobby group heavily weighted with engineering and construction companies, dam-building state agencies, and water supply corporations" (IR 2003, p. 1) that does not represent those affected by dam infrastructure. Hence, IR sees its mission as being to ensure the accountability of public and private funders of dam projects, to monitor compliance with standards and to promote "stricter standards which can redirect resources from destructive dams into benign alternatives" (IR no year).

In summary, IR and WWF are addressing the same issue, identifiable as an interest in promoting the sustainability of dam and hydropower projects, neither being, strictly speaking, opposed to (large) dams and hydropower infrastructure per se. However, there are clear nuances. IR does not criticize dam-building as such, although "better options" may be available. It is critical of the very priorities set in water and energy infrastructure development that justify hydropower and dam projects, such as the *need* of the rural poor for water and energy, whereas large-scale dams meet not those needs but, say, urban energy demand. For hydropower to be a sustainable option, it should be based on the infrastructure *needs* of the people, not solely on economic incentives. For IR, social, environmental, technical and economic aspects do not have the same priority when it assesses dam projects. Yet, although both the WWF and IR recognize the role private investment can play, the latter is more resolute in emphasizing that private-sector participation should take the form of "appropriate measures", meaning that standards should be strengthened first (IR no year; IR 2012; WWF 2003). In addition, IR places far greater emphasis on broad-based public participation when pointing out, for instance, that the G20 lacks the legitimacy "to set norms and take decisions in dialogues" that do not "include most of the poor countries and population groups" (IR 2012, p. 19). Altogether, this implies a clear difference in the vision of how development and industry are perceived and in the source of credibility the two organisations rely on, the latter depending essentially on the people most affected by water and energy infrastructure projects.

6.4 NGO Strategies Towards the Hydropower Sustainability Assessment Forum Process

The *Hydropower Sustainability Assessment Forum* was "an industry-driven and -owned initiative" to review the 2004 Hydropower Sustainability Guidelines and 2006 Hydropower Sustainability Assessment Protocol (Locher et al. 2010, p. 44).

---

[10] The WWC, founded in 1996, is an international multi-stakeholder network of more than 300 member organizations, ranging from government organizations, corporations, NGOs, academic institutions and think tanks to (international) financial institutions actively engaged in the water sector. It is known for organizing the World Water Forum, an international forum that has taken place every 3 years since 1997, on the last occasion in Marseilles in 2012. For more information on the WWC, see http://www.worldwatercouncil.org/. Accessed 23 August 2012.

The Hydropower Sustainability Assessment Protocol adopted by the International Hydropower Association (IHA) in 2006 was in fact the sixth protocol drawn up by the IHA or external researchers and consultants in a continuous process of evaluation and trial assessments. The forum was officially launched in March 2008 and sought to "establish a broadly endorsed assessment tool to measure and guide sustainability performance in the hydropower sector (...) rather than establish a new standard" (Locher et al. 2011, p. 1), which was eventually to became the 2010 Hydropower Sustainability Assessment Protocol. In other words, the Protocol was not intended to replace the WCD's recommendations, but rather to build on the WCD's store of information on hydropower sustainability (Locher et al. 2011, pp. 52–53).

The Protocol is divided into a background document and four assessment tools (Early Stage, Preparation, Implementation and Operation), representing the four distinct stages of a hydropower project cycle. These sections are further subdivided into environmental, social, economic/financial and technical sustainability aspects, each then being assessed on a scale of 1–5 (1 = absence or poor practice; 3 = basic good practice, and 5 = proven best practice) by reference to six assessment criteria (Assessment, Management, Stakeholder Engagement, Stakeholder Support, Conformance/Compliance, and Outcomes). The Protocol thus adopts a "gradational approach"[11] (see IHA 2010a, pp. 9–19). As we will later discuss, the scoring method of the 2010 IHA Protocol and the wording of each score was initially criticized (see Sect. 6.4.2).

Following the publication of the draft HSAP in August 2009, the first (January–February 2009) and second (September–November 2009) public consultation phases and a trial assessment of the draft, the final Protocol was adopted in November 2010. The IHA Protocol process still continues, focusing currently on the actual assessment and implementation of the new Protocol and the establishment of the Hydropower Sustainability Assessment Council. A future Governance Committee and chambers representing the various stakeholder groups involved in the Protocol process will make up the Council. However, an Interim Governance Committee[12]—guiding the development of both the Charter and Terms and Conditions for the application of the IHA Protocol—is currently in charge of populating the chambers.

The 14 members of the forum (excluding its coordinator and chairperson) represented developing countries, developed countries, NGOs, the financial sector and the hydropower sector (Locher et al. 2010; see Table 6.1). As for the general criticism voiced, for instance, by IR that the HSAF was a "self-selected circle" and represented neither dam-affected people nor southern NGOs, it was said during the interviews that the forum coordinator and members considered including other social NGOs, especially one active in the field of involuntary

[11] The Early State Section does not contain any scoring statements, however, since it merely provides guidance (IHA 2010a).

[12] For further details, see http://www.hydrosustainability.org/Governance.aspx (accessed on 13 March 2013).

Table 6.1 Forum members (Locher et al. 2010)

Sector	Forum members
Developing countries	China Institute of Water Resources and Hydropower Research, PR China China Hydropower Engineering Consulting Group Corporation, PR China Ministry of Energy and Water Development, Zambia
Developed countries	Department of Energy, Norad, Norway National Energy Authority, Iceland Federal Ministry for Economic Cooperation and Development (BMZ)/GIZ (observer)
Environmental NGOs	WWF International The Nature Conservancy
Social NGOs	Oxfam International Transparency International
Financial sector	Environmental and Social Risk Management, Citigroup Global Markets Inc., member of the Equator Principles Financial Institutions Senior Water Resources Specialist, World Bank (observer)
Hydropower sector	International Hydropower Association Hydro Tasmania

resettlement. However, according to the document containing the IHA's responses to the questions raised in the first consultation phase, the forum did not "claim to be fully representative of all stakeholder groups" (IHA 2009, p. 3). Instead, it was decided to invite external speakers to the nine formal meetings and to convene two public consultation phases (see IHA 2009; IHA 2010b for further details of the consultation phases).

As one HSAF participant, Kirsten Nyman, said, the IHA's selection criteria might have included cooperation with such "constructive" NGOs as the WWF and Oxfam. Although we cannot explain the details of the IHA's selection of NGOs, the findings of other studies of NGO-company dialogues show it to be unlikely that NGO attitudes alone determined the selection of NGOs by corporations or sectors of industry, although they generally prefer to work with NGOs perceived as "constructive". NGO expertise and the possible reputational impact of an NGO on a company or sector are equally important in the selection process (see Van Huijstee and Glasbergen 2007, p. 303).

6.4.1 Symbolic Gain: WWF Collaboration in the HSAF

Although, as pointed out above, the forum was officially launched in March 2008 and remained in existence until the adoption of the final Protocol in November 2010, it was set up as a result of earlier discussions between the WWF, The Nature Conservancy (TNC) and the International Hydropower Association (IHA) in 2007.

In other words, the WWF was involved in the process as a crucial environmental NGO representative and a vital protocol partner from the very beginning. After the general rejection of the WCD guidelines by some of the major dam-building countries and in the context of growing sustainability debates and continuing protests against large dams, the International Hydropower Association (IHA) eventually revised its own internal criteria for hydropower planning and construction. This revision resulted in the development of the Hydropower Sustainability Guidelines (IHA 2004) and the Hydropower Sustainability Protocol (IHA 2006). The WWF interviewee stressed that the IHA's production of these two documents was "a first step and not a bad one" (interview with Dr Jian-hua Meng).

When the IHA presented its guidelines, the WWF and TNC approached the IHA's executive director with the idea of not only reviewing its 2004 guidelines but also working on a protocol concerning the actual "measurement" of the hydropower sector's sustainability performance. Eventually, relations with the IHA were established, and it was decided to involve more stakeholders, especially from outside the hydropower sector. On the WWF's recommendation Oxfam International, for instance, was approached jointly by the IHA and WWF. On the other hand, the WWF interviewee, speaking from his own experience of the current post-HSAF process of establishing the Protocol's Governance Council, said that the Council was having some difficulty finding NGOs that both had the capacity to participate in an intense, almost three-year HSAF process and were actually interested in participating. The WWF still had the capacity in terms of resources and expertise and was interested in taking part in the HSAF process.

Before the analysis of the WWF's strategy and motivation with regard to the forum process and its Protocol begins, it is also interesting to note that social and environmental NGOs generally tried to support each other's position by matching environmental and social policy positions throughout the forum process. Furthermore, interviewees generally observed that all four NGOs tried to participate constructively in the negotiations, showing both their limitations and their willingness to compromise. Accordingly, both the WWF and TNC often played the role of mediator between different stakeholder groups. All the interviewees emphasized that they did not regard the WWF as being particularly different from the other NGOs, having a comparable collaborative attitude within the forum process, although they recognized nuances in levels of agreement on social and environmental demands put forward by the NGOs. Even the GIZ representative interviewed acknowledged that she had expected both environmental and social issues to require backing in the discussion with the hydropower sector—expecting each of the two parties (NGOs and the industry) to be strongly opposed to the other's position. Yet environmental issues were far easier than the whole range of social issues, since many environmental positions had been generally accepted by the industry and governments, whereas Oxfam International's strong policy commitment to the strategic priority of free, prior and informed consent proved to be more challenging for the industry and others.

As for the WWF's motivation and strategy, WWF International's Water Security Initiative expert, Dr Jian-hua Meng, who is involved in the current creation of the Governance Council for which the Protocol provides, explained that one reason for the WWF's desire to participate in the forum was generally to ensure the broad-based, industry-wide application of the principles it advocates in its Water Security Initiative. The WWF representative also said that choosing a certain partner for dialogue—in this case the IHA, which, according to its organisational values, seeks leadership in the sector—was part of the WWF's strategy. It also meant choosing an entity regarded as capable of leading industry-wide change, and Jian-hua Meng of the WWF specifically applauded the companies that ultimately endorsed the 2010 IHA Protocol. By cooperating with the IHA, the WWF also hoped to be considered more credible by both the industry and the regulators who have traditionally opposed the WCD principles and guidelines. It therefore demonstrated its intentions of participating and advancing in hydropower sustainability discussions. The WWF representative also referred to the few results that demanding enforceable state regulation and international standards has so far achieved. From his point of view, the assumption that the industry's interests influence public decision-making and vice versa clearly makes it less likely that stringent standards will be developed.

Furthermore, the WWF's experience has demonstrated that, if it collaborates with one private actor likely to have a positive reputational impact, for instance, as it has done in the case of the sustainability partners who subscribe to the Protocol by giving it credibility, more may follow and expect the same treatment. Rather than opposing one dam at a time, the WWF seeks industry-wide change, because its goal is "market transformation": "We want bad projects to disappear, to raise the expectations of financiers, governments and even consumers" of the dam industry. The WWF representative said it was like comparing dams to cars. Governments want good dams just as consumers want good cars. This line of argument suggests that, as contracting authorities, governments will choose hydropower project developers with a good record of best practice over those who set poor sustainability targets, "bad dams" possibly resulting in negative effects on governments' reputations among their citizens. This indicates that the WWF is adopting a multi-level approach and, without neglecting the importance of state regulation and cooperation with governments, it has also chosen to work on an industry-wide scale to effect change as outlined above.

Thus, like most interviewees, the WWF representative stressed the essentially consensus- and compromise-based nature of the 2010 IHA Protocol. He added that the WWF was pleased with the Hydropower Sustainability Assessment Protocol and the forum process as such, and that its expectations had been met. He also said in the interview that the WWF was convinced its decision to participate in the forum process had been the right choice, that the protocol was a "good one", and that the forum could be seen as a clear success.[13] As for the general criticism of the

[13] He also said, however, that the WWF was not pleased that basin-scale hydropower planning was not adequately covered by the protocol, but rather mainstreamed, because a stand-alone page—that is basin-scale hydropower planning treated as a separate topic—might have triggered China's resignation from the forum.

wording used to describe good and best practice scores, the WWF security expert referred to “nuances”. While other interviewees clearly expressed their concerns, Kirsten Nyman, for instance, indicating that the wording of the scores “can be misleading” when actually applied, it should be emphasized that this method is useful in comparisons of projects with current practice in the sector, but each user still has to define his own benchmark in order to make a final quality statement. The WWF's commitment to the process is also evident from its continuing commitment to the post-forum institutional embedding efforts. While Oxfam International, for example, declined to join the post-HSAF process,[14] which is designed to set up stakeholder-based chambers for the further review and monitoring of the actual application of the Protocol, the WWF remains a member of the HSAF Council and will be represented in the environmental chambers. Although it is too early to make any assumptions about greater understanding and trust as a result of the HSAF process, it was said that relations among forum members depended on the individuals attending the forum “rather than being entrenched in the various organisations” (interview 5). The further institutional embedding of these relationships throughout the post-forum process may, however, further improve relations among participating organisations.

6.4.2 Symbolic Damage: International Rivers' Response to the HSAF Process and Protocol

During the 30 months of the HSAF, IR monitored its proceedings closely. However, it did not begin to make public statements on the subject until late 2008. A few months after the forum's official launch, IR published a report in December 2008 making a critical comparison of the IHA's 2004 guidelines and the 2006 protocol with the WCD's strategic priorities and policy principles and the Performance Standards of the World Bank Group's International Finance Corporation (IR 2008).

After the HSAF's key components document was published in January 2009 and the first public consultation phase opened, IR again published a report a few months later, this time criticizing the draft protocol as a major setback for the WCD's “rights-and-risk approach (…) which relied on negotiated, legally enforceable agreements with dam-affected communities” (Bosshard 2009a). IR' Policy Director, Peter Bosshard, criticized its scoring method; the exclusion of dam-affected people from the dialogue; the failure to comply with existing international conventions, social and environmental standards and national legislation; and the wording of various parts of the Protocol. He also argued that, by measuring only process and performance criteria “with equal weight”—a reference to the protocol's scoring method—the document was essentially “flawed”

[14] Although its representative recognized important advances in the new protocol, Oxfam decided to monitor the process from an external vantage point and to help communities to understand the Protocol and its application (interview with Michael Simon).

because it did not "require or propose any compliance mechanisms" and essentially ignored the actual dam project outcome (Bosshard 2009a). In short, "low performance in crucial aspects (for example social impacts, human rights, governance) can be offset by a higher performance in other, less relevant aspects (for example project management, environmental impacts or economics)".[15] In addition, many aspects had not received the attention they deserved in the protocol, examples being existing dams, human rights and risk assessment for dam-affected people, displacement, compensation, a basin-wide approach and cumulative environmental impact assessments (Bosshard 2009a).

Although the dam industry may actually be concerned about the sustainability of their products, Bosshard (2009b) perceives the forum proposal as a way "to replace clear minimum standards with consulting reports and management plans. It asks affected people to trade in hard-won rights for a lot of consultants' speak, that is, trade in IFC's Performance Standards and World Bank's safeguard policies as important benchmarks for a voluntary industry-driven HSAP." In early 2009, IR (2009) also sent a letter to the IHA, endorsed by 35 civil society organisations, requesting broader participation. The letter expressed concern that dam-affected people and civil society had not participated in the selection of forum members, formal meetings and external inputs or in the first public consultation phase. The IHA responded by promising to include comments in the subsequent forum process and also to invite those who had endorsed the letter to the second public consultation phase in September and November 2009.

Most of IR' efforts during the HSAF process were therefore expended on closely monitoring and raising awareness of the process itself by advocacy means and capacity-building at local level by engaging with its local partners to ensure they had the capacity to assess critically the draft protocol, its shortcomings and the risks it potentially entailed (e.g. through workshops). The action taken included the sending of letters to the IHA and information to IR' network partners. The IR representative interviewed said that media attention was low-profile while the HSAF was taking place (2008–2010), as compared to the period when the new Protocol was eventually adopted (2010–present), although IR also contacted a few journalists to express its concerns about the HSAF.

Zachary Hurwitz, IR' Policy Program Coordinator, stressed in his interview that his organisation had not adopted an official position on the HSAF. However, though critical of the lack of broad-based participation, he noted that IR was not interested in becoming a member of the HSAF. Furthermore, when the Governance Council asked IR in 2012 to join one of the chambers currently being populated for the further assessment of the final protocol (the WWF representative mentioned during his interview that they had asked IR to join the environmental chamber), it had declined. It instead engages in capacity-building to enable

[15] The protocol's scoring method was initially criticized and regarded as a problem during the first and second consultation phases (see IHA Association 2009). During the second consultation phase, various stakeholders considered the subjective and qualitative nature of the scoring and the wording of scores to be a basic flaw (see IHA 2010b).

dam-affected people and civil society organisations "to speak for themselves and to participate in the forum consultation phase".

IR was not therefore motivated to participate in the forum because that might have been construed by its network partners as legitimizing the process. Hurwitz also pointed out that IR was and continues to be "cautious", primarily because of the dominant part played by the industry in the initiative, which cast serious doubt on IR' ability to change things from within. He also questioned the industry's credibility as a genuine supporter of sustainable hydropower: was it "not just using it as a way to improve [its] public relations record?". In this, he was referring to the clear "conflict of interest" that lies in the hydropower industry's very *raison d'être*: to build dams. The participation in the forum of such NGOs as the WWF was, for instance, perceived both negatively, as a way of lending credibility, and positively, as a means of influencing some of the wording of the protocol. However, the sustainability outcomes of NGO participation will need to be assessed in the future (interview with Zachary Hurwitz).

After the adoption of the final IHA protocol, IR announced its official position (IR 2011). Six months later, Hurwitz attended the IHA Congress held in Iguaçu/Brazil on 16 June 2011, where the protocol was officially launched, on behalf of IR to express the NGOs' concern about the new IHA protocol and to urge the principal stakeholders to resign and not to support the protocol. He was supported at the Congress by the Brazilian movement *Xingu Vivo Para Sempre*, which is deeply concerned about the present construction of the Belo Monte Dam. IR therefore engaged more openly with the media to draw attention to the Protocol. In a subsequent letter to the IHA Congress, it upheld its organisational position and that of its supporting network (including *Xingu Vivo*) on the protocol, seeing it as "a purely voluntary assessment tool". It again emphasized the non-binding, non-enforceable nature of the Protocol and its concern that, rather than contributing to hydropower sustainability, the HSAP (the IHA Protocol) neglected and would further weaken existing standards, and shift the control over assessments from governments to the hydropower industry. Hence, as IR doubts the credibility of the hydropower industry's efforts, it believes that leaving dam developers, for instance, to "self-assess" sustainability is dangerous because of the lack of objectivity. It is like "putting the fox in charge of the hen house", according to Hurwitz. Unless provision is made for punitive measures and for the protocol to be more binding and unless the dam industry can actually demonstrate that it has not built destructive dams because of negative IHA Protocol trial assessments, IR will continue to oppose the current 2010 IHA Protocol (interview with Zachary Hurwitz).

Meanwhile, IR' website still generally refers to the Protocol as a (voluntary) *scorecard* "that rewards hydropower companies and financiers with a green-washed stamp of approval; it does not represent a true step towards the actual practice of sustainability in the sector". It is further characterized as a *points-based reward system* "similar to a frequent flyer program: accumulating a higher amount of points would allow a developer to claim that its projects are sustainable" (Hurwitz 2011; see also IR 2010). Now that it has been adopted, recent criticism of the new IHA Protocol has opposed its legitimacy as a WCD follow-up and pointed to the risk of the fundamental flaws in the protocol resulting in hydropower being *greenwashed*. IR' strategy has not changed fundamentally, but continues to seek to raise awareness and build capacities at operational level, but with the obvious goal of increasing media coverage. It has thus generally integrated its criticism of the IHA Protocol into its advocacy campaign *Stop Dam Greenwashing*.

In the context of this campaign, such initiatives as the 2010 IHA Protocol are seen as a mere WCD follow-up "attempt to circumnavigate the more robust requirements of the WCD while paying lip service to sustainability" (Imhof and Lanza 2010). The campaign is currently focused entirely on the IHA Protocol and its forthcoming trial assessments, but more generally highlights the risks of "voluntary private-led policy initiative(s)" (interview with Zachary Hurwitz).

According to Hurwitz, IR also supports the global water justice movement's demands for the full implementation by national governments of UN Resolution 64/292 on the human right to water and sanitation and for the rejection of the current "corporate-driven and market-based" development model that deregulates and privatizes water and sanitation services worldwide rather than ensuring supplies of water as a common good, managed in a "public or community-owned, social, cooperative, participatory, equitable and not for profit" manner.[16] In fact, IR staff members officially participated in protests at the 6th World Water Forum in Marseilles in March 2012, where protestors called for greater public participation, discredited the World Water Forum and the current IHA Protocol, and denigrated, for instance, the justification of unsustainable development projects as being consistent with the "green economy".

IR continues to monitor the application of the 2010 Protocol and now calls on its network to "publicly scrutinize the results of protocol sustainability assessments" and to "hold developers, governments, and financiers accountable to the real standards that define best practices in the industry" (Hurwitz 2011), not because they reject, for example, other NGOs' intentions to work with the private sector to help improve their performance, but because of what the Protocol turned out to be and the concern about what it could have been IR fears that the voluntary approach to dam and hydropower sustainability may replace a more binding approach and not be effective at all.

6.5 Divergent NGO Strategies for Sustainable Dams

The theoretical and conceptual framework of this article suggests that NGOs are generally seeking social or institutional change. The aim has been to determine how NGOs go about effecting this change and to explain why some NGOs choose a more or less collaborative stance towards a specific sector of industry, namely

[16] See Civil Society Declaration at the 6th World Water Forum, Marseilles 2012, http://www.internationalrivers.org/resources/civil-society-declaration-at-6th-world-water-forum-marseille-2012-4278. Accessed 02 October 2012.

the hydropower industry. The work of Den Hond and De Bakker (2007) led to the assumption that NGOs' general ideological differences—the beliefs and attitudes they advance in justification of their reactions to problematic issues (Den Hond and De Bakker 2007, p. 903)—may result in their pursuing different strategies towards the industry: symbolic gain and symbolic damage strategies.

The WWF's involvement in the HSAF as one of the main initiators from the very beginning of the process, its continuing support during the establishment of the governance structure in the post-HSAF phase, and its general endorsement of the 2010 Protocol underline its collaborative approach to the hydropower industry. Nor, when the forum dynamics during the HSAF process itself are analysed, does the WWF's attitude appear to be particularly different from those of other NGO members. All were perceived to be essentially "constructive" and consensus-oriented. This is not surprising, however, if, as was suggested during various interviews, one of the criteria from the outset was that constructive NGOs should be selected. The article identifies the WWF as a reformative NGO likely to employ a symbolic or material gain strategy that focuses on the replacement of the hydropower industry's norms and standards in cooperation with that industry, cooperation that may subsequently have a positive reputational impact on the industry through the adoption of new sustainability practices.

Empirical evidence suggests that the WWF had many reasons for backing the IHA's setting up of the HSAF, among them a desire to enhance its own credibility vis-à-vis funders and regulators as it pushed its own agenda, including the industry-wide embodiment of WWF Water Security Initiative concerns, and to lend credibility to a process it thought was a good approach. However, WWF rationale also included the belief that the dam industry is not ready to adopt or accept any binding standards, thus positively acknowledging and therefore backing an initial voluntary approach to sustainable industry performance. This line of argument indicates that the WWF's justification of its participation in the HSAF was based primarily on its views on the role played by the hydropower industry in the development of sustainable hydropower.

The WWF welcomed the IHA's interest, but also highlighted its earlier efforts to develop protocols and the final consensus-based 2010 Protocol, and gave an assurance that it would continue to collaborate with the IHA on a future certification system, for example. It believed that the IHA and its members might "lead by example" and that it has set a clear international benchmark for progress towards hydropower sustainability, one that does not have to hide behind the WCD process, as Dr Jian-hua Meng said when interviewed. In addition, as the IHA drew up both previous documents (IHA 2004, 2006) on its own, it lacked the general approval of such other stakeholders as NGOs, which had a fundamental effect on its credibility. Consequently, it was in the IHA's own best interests to continue collaborating with such NGOs as the WWF. However, the WWF's rationale and resource-based incentive to cooperate with the IHA was that it also offered the WWF a way to improve its credibility in the eyes of funders and regulators when bringing forward its own sustainability agenda and improving its image as a "constructive" NGO willing to compromise in a market-driven environment that called on NGOs to be more efficient and consumer-oriented.

However, such non-members of the forum as IR generally regarded supporting the IHA in its endeavours as a way of lending credibility to a process they thought was essentially flawed. IR was earlier identified as an advocacy NGO. Empirical evidence has shown that it followed a clear, but initially low-profile media strategy aimed at raising awareness of a misguided way of endorsing a voluntary approach to hydropower sustainability that handed over more control to a subjective dam industry. IR' clear lack of interest in joining the forum and its refusal to join one of the chambers were also intended to deny the Protocol legitimacy, since it would otherwise lose credibility in the eyes of its network of activists. Its symbolic damage strategy against the hydropower industry argued at both a consequential and a structural level: first, because the industry's failure, for example, to include broader sectors of civil society was "half-hearted" (questioning its credibility and legitimacy), and second, because the very reason for the hydropower industry's existence is to build dams, and its good will was therefore in doubt.

On that basis, IR opposed the process and Protocol by writing numerous letters and reports and tried to build the capacity of the people affected at local level to assess the HSAF and the current protocol critically. At present, it is continuing to address the Protocol trial assessments in its advocacy campaign against dam greenwashing. The hydropower industry's economic interests are seen as a dominant incentive and constraint that does not match IR' vision of dams. It promotes a development model that centres on the water and energy needs of the people above all else and is not merely economic and market-based. It emphasizes a human-rights-based approach to the rural, rather than urban, dam-affected population. This ideological difference from the WWF's philosophy was also partly evident from the WWF representative's comparison of "dams and cars".

In this regard, we also see that, strictly speaking, neither IR nor the WWF opposes dams; they both endorse the WCD's priorities and principles and hence, with some nuances, pursue the same objective of binding standards for truly sustainable hydropower. It can be argued that IR' interests essentially differ from those of the WWF in having a far broader human rights and social issue bias, but Oxfam International, while clearly sharing some of those interests, still decided to participate in the HSAF process. I should point out that ideological stances alone may not account for an NGO's decision to participate (or not) in a forum like the HSAF. Other organisational dimensions, such as funding and capacity in terms of time and expertise, as Meng commented when interviewed, may also be determining factors in this respect, although, in ideological terms, they are more or less likely to justify a collaborative approach to industry.

We also mentioned that NGOs' strategies change as a function of the political or cultural constraints they encounter, a situation analysed by Keck and Sikkink (1998) and others. The tactics employed by the WWF's national offices at operational level may differ from those specified in its international strategy and encourage both confrontational campaigns against a single dam project with the aim of "shaming" the practices of dam builders or financiers, and such collaborative action as cooperation with companies or government agencies.

As for the general perception of the divergent strategies analysed in the chapter, one interviewee regarded IR as an actor that that does not hesitate to broach painful subjects. The hydropower sector deals very carefully with sustainability issues "because you always know that, when something goes wrong, IR will be there and proclaim it to the world". Nonetheless, a more radical and less compromise-based position may reduce an NGO's credibility in the eyes of other stakeholders and lead to reluctance when it comes to establishing collaborative partnerships or multi-stakeholder dialogues like the HSAF. On the other hand, even the WWF strategy has its downside. Today, NGOs that are willing to collaborate with industry have been accused of *supping with the devil* (see also Wilson 2002).

6.6 Conclusion

The aim of this chapter was to analyse and explain how and why NGOs choose different strategies for dealing with the dam industry. Our comparative case study has clearly shown that IR' and the WWF's approaches to the dam industry differ essentially in the strategies and tactics employed. The theoretical framework and empirical evidence suggest that ideological differences, and the different visions of development that go with them, explain why NGOs choose more or less collaborative strategies and approaches. The WWF clearly engaged with industry to push forward the "reinstitutionalization", that is the replacement of status quo practices in the hydropower industry with new norms and practices aimed at achieving hydropower sustainability—on a voluntary basis and as part of longer journey with binding standards as its destination. It showed, for example, far less concern about the dam industry's underlying incentives than IR did. In contrast, IR remained focused on supporting existing and introducing new binding sustainability practices in the hydropower sector, thus both discrediting existing "unsustainable" practices in the dam industry and pushing for their replacement with new practices and norms, but in a binding, non-voluntary way and involving independent assessments and punitive measures. We explained this difference on the basis of IR' lack of confidence in the industry to be objective enough to represent the rights of dam-affected and riverine communities appropriately, seeing the industry primarily as consisting of economically driven actors whose primary goal is to build dams. It is therefore perceived not as part of the solution but as part of the problem. As the hydropower sustainability debate has still not come to an end, potential empirical research may well show—as Van Huijstee and Glasbergen (2010) suggest—that these two parallel approaches will reinforce each other in the long term.

Furthermore, research is still needed into the different outcomes of divergent strategies and to identify constraints that exist in the corporate environment or in multi-stakeholder initiatives with a view to overcoming some of the main concerns associated with these new forms of governance, such as their democratic legitimacy, social accountability and the general tendency of decision-making to shift from public to private actors. However, in this regard, a criticism recently voiced in the stakeholder literature (Van Huijstee et al. 2008, p. 85) claims that much of the analysis of partnerships and dialogues is concentrated on the process itself, on selection and on the outcome of strategies, but not on actual outcomes "in terms of sustainability". Clearly, it is too early, and was not my intention, to assess the IHA Protocol and its consequences for the sector's sustainability standards. I consider it, nonetheless, to be the most promising task for future research since, as once before in the WCD's case, it remains to be seen if the Protocol has a positive influence on practices in the hydropower sector.

6.7 Interviews

Interview 1: Zachary Hurwitz, Policy Program Coordinator (International Rivers)

Interview 2: Dr. Jian-hua Meng, Water Security Initiative (WWF International, WWF Germany); currently member of the Protocol's Interim Governance Committee

Interview 3: Kirsten Nyman, HSAF forum member on behalf of BMZ; Policy Advisor on Sustainable Hydropower (GIZ)

Interview 4: Michael Simon, HSAF forum member on behalf of Oxfam International; Advocacy Manager and Lead—People, Infrastructure and Environment Program (Oxfam Australia)

Interview 5: lead member of an international organisation (anonymous) in the form of a questionnaire via e-mail exchange from 20 to 27 August 2012.

References

Ähltröm J, Sjöström E (2005) CSOs and business partnerships: strategies for interaction. Bus Strategy Environ 14:230–240

Austin JE (2000) Strategic collaboration between nonprofits and businesses. Nonprofit Voluntary Sector Q 29:69–97

Atzl A (2012) Transnational NGO-networks campaign against the Ilisu Dam, Turkey. In: Scheumann W, Hensengerth O (eds) Evolution of dam policies. Evidence from the big hydropower states. Springer, Heidelberg

Bosshard P (2009a) A giant step back from current rights and standards. A critique of the key components document of the hydropower sustainability assessment forum, March 2009. http://www.internationalrivers.org/resources/critique-of-the-hsaf-approach-4308. Accessed 23 Aug 2012

Bosshard P (2009b) The dam industry's brave new world. http://www.internationalrivers.org/blogs/227/the-dam-industry%E2%80%99s-brave-new-world. Accessed 24 Aug 2012

Brunnengräber A (2001) Offensive des Lächelns. Die NGOs und das Dilemma der Klimaverhandlungen – Vom konflikt- zum konsensorientierten Akteur. Der Freitag, 22 January 2001. http://www.freitag.de/autoren/der-freitag/offensive-des-lachelns. Accessed 17 July 2012

Brühl T (2009) Nichtregierungsorganisationen als Akteure internationaler Umweltverhandlungen. Ein Erklärungsmodell auf der Basis der situationsspezifischen Ressourcennachfrage. Campus Verlag, Frankfurt, New York:

Den Hond F, De Bakker FGA (2007) Ideologically motivated activism: how activist groups influence corporate social change activities. Acad Manag Rev 32:901–924

Dobner P (2010) Wasserpolitik. Suhrkamp Verlag, Berlin

Doh JP, Newburry WE, Teegen H (2003) Cooperative strategies in environmental nongovernmental organisations. In: Doh JP, Teegen H (eds) Globalization and NGOs. Transforming business, government, and society. Praeger, Westport, CT, pp 65–80

Doh JP, Teegen H (2003) Preface: globalization and NGO—why should we care? In: Doh JP, Teegen H (eds) Globalization and NGOs. Transforming business, government, and society. Praeger, Westport, CT

Finnemore M, Sikkink K (1998) International norm dynamics and political change. Int Org 52(4):887–917

Florini AM (2000) (ed) The third force: the rise of transnational civil society. Carnegie endowment for international peace. Washington, DC

Glasbergen P, Groenenberg R (2001) Environmental partnerships in sustainable energy. Eur Environ 11:1–13

Glasbergen P (2007) Setting the scene: the partnership paradigm in the making. In: Glasbergen P, Bierman F, Mol AP (eds) Partnerships, governance and sustainable development. Reflections on theory and practice. Cheltenham, UK, pp 1–25

Hogenboom B (2003) Cross-border activism and its limits: Mexican environmental organizations and the United States. Cuadernos del CEDLA 13. CEDLA (Centre for Latin American Research and Documentation), Amsterdam

Hurwitz Z (2011) Hydropower industry needs standards, not scorecards, to be sustainable. http://www.internationalrivers.org/blogs/258/hydropower-industry-needs-standards-not-scorecards-to-be-sustainable. Accessed 24 Aug 2012

IHA (International Hydropower Association) (2004) IHA sustainability guidelines. International Hydropower Association, London

IHA (International Hydropower Association) (2006) IHA sustainability assessment protocol. International Hydropower Association, London

IHA (International Hydropower Association) (2009) Forum response to consultation 1 issues, March 2009. http://www.hydrosustainability.org/IHAHydro4Life/media/PDFs/PDF%20docs/consultation%20phase%201/HSAF-Phase_1_Consultation_Issues-HSAF_Response_23_Mar_2009.pdf. Accessed 1 Aug 2013

IHA (International Hydropower Association) (2010a) IHA sustainability assessment protocol, London

IHA (International Hydropower Association (2010b) Hydropower sustainability assessment forum. Phase 2 consultation outcomes report. Ove Arup & Partners Ltd, London. http://www.hydrosustainability.org/Protocol/Protocol-development/Consulation-Phase-2.aspx. Accessed 1 Aug 2013

Imhof A, Lanza GR (2010) Greenwashing hydropower: the problems with big dams. World Watch Mag 23(1):8–17

IR (International Rivers) (no year) http://www.internationalrivers.org/. Accessed 18 July 2012

IR (International Rivers) (2003) Who's behind the World Water Forums? A brief guide to the world water mafia. http://www.internationalrivers.org/resources/who-s-behind-the-world-water-forums-2644. Accessed 26 July 2012

IR (International Rivers) (2008) Social and environmental standards for large dams. Comparing the strategic priorities and policy principles of the world commission on dams, the sustainability guidelines and sustainability assessment protocol of the international

hydropower association, and the performance standards of the world bank's international finance corporation. http://www.internationalrivers.org/node/3959. Accessed 24 Aug 2012
IR (International Rivers) (2009) Letter to IHA. Sent on 09-01-12. http://www.internationalrivers.org/resources/letter-to-iha-regarding-lack-of-civil-society-participation-in-hsaf-3147. Accessed 24 Aug 2012
IR (International Rivers) (2010) Voluntary approach will not resolve dam conflicts. A critique of the international hydropower association assessment protocol. http://www.internationalrivers.org/resources/voluntary-approach-will-not-resolve-dam-conflicts-4286. Accessed 26 August 2012
IR (International Rivers) (2011) Civil society statement on the launch of the hydropower sustainability assessment protocol at the congress of the international hydropower association in Foz do Iguaçú Brazil on 16 June 2011. http://www.internationalrivers.org/resources/civil-society-statement-on-the-launch-of-the-hydropower-sustainability-assessment-protocol. Accessed 24 Aug 2012
IR (International Rivers) (2012) Infrastructure for whom? A critique of the infrastructure strategies of the Group 20 and the World Bank. http://www.internationalrivers.org/resources/infrastructure-for-whom-7454. Accessed 20 July 2012
Islam S, Susskind LE and associates (2013) Water diplomacy: a negotiated approach to managing complex water networks. Routledge, New York
Jordan L, Van Tuijl P (2000) Political responsibility in transnational NGO advocacy. World Dev 28(12):2051–2065
Keck M, Sikkink K (1998) Activists beyond borders: advocacy networks in international Politics. Cornell University Press, Ithaca
Khagram S (2000) Toward democratic governance of sustainable development: transnational civil society organizing around big dams. In: Florini AM (ed) The third force: the rise of transnational civil society. Carnegie Endowment for International Peace, Washington, DC, pp 83–114
Khagram S, Riker JV, Sikkink K (2002) From Santiago to Seattle: Transnational advocacy groups restructuring world politics. In: Khagram S, Riker JV, Sikkink K (eds) Restructuring world politics. Transnational social movements, networks and norms. University of Minnesota Press, Minneapolis, pp 3–23
Locher H, Hermansen GY, Johannesson GA, Xuezhong Y et al (2010) Initiatives in the hydro power sector post-World Commission on Dams—the hydropower sustainability assessment forum. Water Altern 3(2):43–57
Locher H, Hermansen GY, Johannesson GA, Xuezhong Y et al (2011) Hydropower sustainability assessment forum. Statement of conclusion
Martínez-Alier J (2002) The environmentalism of the poor: a study of ecological conflicts and valuation. Edward Elgar Publishing, Northampton, Cheltenham
Mathews JT (1997) Power shift. Foreign Aff 76:50–66
Mayntz R (2001) El Estado y la sociedad civil en la gobernanza moderna. Revista del CLAD, Reforma y Democracia (21):7–22
McCully P (2001) The use of a trilateral network: an activist's perspective on the formation of the world commission on dams. Am Univ Int Law Rev 16(6):1453–1475
Nadelman E (1990) Global prohibition regimes: the evolution of norms in international society. Int Org 44:479–526
Nelson PJ (1997) Conflict, legitimacy, and effectiveness: who speaks for whom in transnational NGO networks lobbying the world bank? Nonprofit Voluntary Sector Q 26(4):421–444
O'Brien R (2005) Global civil society and global governance. In: Ba AD, Hoffmann MJ (eds) Contending perspectives on global governance. Coherence, contestation and world order. Routledge, London, New York, pp 213–230
Ottaway M (2001) Corporatism goes global: international organizations, nongovernmental organization networks, and transnational business. Glob Gov 7:265–292

Parker AR (2003) Prospects for NGO collaboration with multinational enterprises. In: Doh JP, Teegen H (eds) Globalization and NGOs. Transforming business, government, and society. Praeger, Westport, CT, pp 81–106

Rhodes RAW (1996) The new governance: governing without government. Polit Stud XLIV:652–667

Rohrscheider R, Dalton RJ (2002) A global network? transnational cooperation among environmental groups. J Polit 64(2):510–533

Rowley TJ, Moldoveanu M (2003) When will stakeholder groups act? an interest- and identity-based model of stakeholder group mobilization. Acad Manag Rev 28:204–219

Sikkink K (2002) Restructuring world politics: the limits and asymmetries of soft power. In: Khagram S, Riker JV, Sikkink K (eds) Restructuring world politics. Transnational social movements, networks and norms. University of Minnesota Press, Minneapolis, London, pp 301–318

Simmons PJ (1998) Learning to live with NGOs. Foreign Policy 112:82–96

Van Huijstee M, Glasbergen P (2007) The practice of stakeholder dialogue between multinationals and NGOs. Corp Soc Responsib Environ Manag 15:298–310

Van Huijstee M, Francken M, Leroy P (2008) Partnerships for sustainable development: a review of current literature. Environ Sci 4(2):75–89

Van Huijstee M (2009) Interactions between business and nongovernmental organizations. New social practices for sustainable development. Ökologisches Wirtschaften 1:39–42

Van Huijstee M, Glasbergen P (2010) NGOs moving business: an analysis of contrasting strategies. Bus Soc 49:591

Weiss TG, Gordenker L (eds) (1996) NGOs, the UN and Global Governance. Lynne Rienner Publishers, London

Williams PB (1998) An historic overview of IRN's Mission. Unpublished manuscript

Wilson J (2002) The sponsorship scam. In: Lubbers E (ed) Battling big business: countering greenwash, front groups and other forms of corporate bullying. Common Courage Press, Monroe, ME, pp 44–52

Winston M (2002) NGOs strategies for promoting corporate social responsibility. Ethics Int Aff 16(1):71–88

World Bank (2010) Involving nongovernmental organizations in bank-supported activities. http://go.worldbank.org/0WT7SICZY0. Accessed 30 January 2013

WWF (World Wide Fund for Nature) (no year) Dams initiative. http://wwf.panda.org/what_we_do/footprint/water/dams_initiative/. Accessed 18 July 2012

WWF (World Wide Fund for Nature) (2003) WWF's Dams initiative: hydropower in a changing world. http://wwf.panda.org/what_we_do/footprint/water/dams_initiative/news/. Accessed 18 July 2012

WWF (World Wide Fund for Nature) (2005) WWF's Dams initiative: to dam or not to dam? Five years on from the world commission on dams. http://wwf.panda.org/what_we_do/footprint/water/dams_initiative/news/. Accessed on 18 July 2012

WWF (World Wide Fund for Nature) (2007) WWF's Dams initiative: rivers at risk. Dams in the future of freshwater ecosystems. http://wwf.panda.org/what_we_do/footprint/water/dams_initiative/news/. Accessed 18 July 2012

Zald MN, McCarthy JD (1980) Social movement industries: competition and cooperation among movement organizations. Res Soc Mov Conflicts Change 3:1–20

Chapter 7
Transnational NGO Networks Campaign Against the Ilisu Dam, Turkey

Andreas Atzl

Abstract Large dams provide renewable energy, increased irrigation and food, better access to clean drinking water and flood control. However, they have been criticized since the 1990s by civil society organisations for their environmental, social, cultural and economic impacts. Dam proponents and dam opponents are engaged in a "discursive battle" over norms and standards for large projects of this kind. Under pressure from transnational advocacy networks, the World Commission on Dams was set up in 1998. Two years later it presented its report entitled *Dams and Development. A New Framework for Decision-Making*. The reactions to this report ranged from agreement to rejection. Nevertheless, the debate on norms and standards for large dams entered a new stage. Since then other actors, among them the World Bank, private banks and the OECD countries' export credit agencies, have established new or revised existing standards. This chapter analyses the role of transnational networks in the application of norms for large dam projects. Taking the example of the Ilisu Dam on the River Tigris in Southeastern Turkey, it shows the pathways these networks have taken to influence decision-makers. As their efforts to influence Turkish decision-makers had little impact, the anti-Ilisu campaign members addressed European decision-makers and export credit agencies. The chapter considers the actors involved in the anti-Ilisu campaign, their origin and goals and the means they have employed, which comprise information politics, symbolic politics, leverage politics and accountability politics. The chapter concludes by discussing the implications of the anti-Ilisu Dam campaign for the debate on large dams and assesses the campaign's success and the strategy it has adopted to ensure compliance with norms and standards for large dam projects.

A. Atzl (✉)
Karlsruhe Institute of Technology (KIT), Karlsruhe, Germany
e-mail: atzl@kit.edu

W. Scheumann and O. Hensengerth (eds.), *Evolution of Dam Policies*,
DOI: 10.1007/978-3-642-23403-3_7,

7.1 Introduction

For more than 20 years a debate has raged over the benefits and risks of large dam projects. Newly industrialized countries, such as China, India, Turkey and Brazil, are developing their hydropower capacities for socio-economic development, and China has even become an important financier of large dam projects abroad (Briscoe 2010, p. 412).

During the past two decades, non-governmental organisations (NGOs) have become new political actors and drawn attention to the problems associated with large dams. Linked through transnational advocacy networks (TANs), civil society organisations have acted as norm entrepreneurs, trying to induce policy and behavioural changes at international and domestic levels (Finnemore and Sikkink 1998, 892 ff.).

The dam debate of the last two decades has centred on two issues: the right of states and civil society to make or influence decisions that affect a large number of people, and the degree to which norms and standards have been observed in large dam projects.[1] Depending on their background, authors approach these questions from different perspectives. Dam opponents tend to refer to the recommendations of the World Commission on Dams (WCD). Others stress the relevance of World Bank standards and draw attention to other dam-related processes, such as those initiated by the International Hydropower Association (IHA).

This chapter contributes to the debate by assessing the *strategies* pursued by civil society to gain influence. Such *strategies* consist of two elements: (*a*) *pathways* taken to reach decision-makers and (*b*) the *means* used. Keck and Sikkink (1999) argue that TANs use a set of different *means* (known as politics) from which they select the most suitable, adapting them as a function of the *pathways* they take to influence decision-makers. The question to be answered here is *whether*, *by what means* and *on what pathways* TANs lobby for compliance with international standards. The specific *strategy* used by the anti-Ilisu Dam campaign will also be assessed in the context of the ongoing debate on large dams.

The analysis of the anti-Ilisu campaign's strategies and success is based on network analysis which follows the networks-as-actors approach. This approach assumes that a network itself is a corporative actor, but one that is formed through the intentional interaction and cooperation (connection) of individual actors (nodes) with individual characteristics, interests and resources (Sikkink 2009, p. 228). The nodes in this network represent NGOs and campaigning sub-networks (Kahler 2009, pp. 3–4). The connections reveal the type of interaction and cooperation between these actors, with due regard for the quantity and quality of interaction (Lake and Wong 2009, p. 129). The network, with its actors and

[1] See for these two issues the papers by the dam proponent John Briscoe (2010) and the dam opponent Robert Goodland (2010) on the WCD in a special issue of Water Alternatives. The fact that Briscoe and Goodland have both worked for the World Bank shows the complexity of interests in the dam debate. A more analytical view is taken by Dubash (2009 and 2010).

interactions, is projected as an illustration of the NGOs' thematic interests (Fig. 7.3) and as a geographical map (Fig. 7.4). These two projections permit the illustration of existing thematic and geographical focuses, differences in types of cooperation and actor within the network and the relevance of network patterns to the campaign's strategy and success.

This chapter is based on empirical research on the Ilisu Dam project in Turkey for a thesis (Diplomarbeit) at the University of Mainz, Germany, which was undertaken with the support of the German Development Institute in Bonn (Atzl 2009). Data were collected in more than twenty qualitative interviews with decision-makers and NGO activists in Germany, Turkey and other European countries. The information was triangulated through participant observation and analysis of official documents, press articles and campaign documents.

7.2 Emergence of Global Norms for Sustainable Dam Practices

In the early 1990s, NGOs began to protest against the dams on the River Narmada in India, thus putting the demand for norms for large dams on the political agenda (Fink and Kramer 2008, p. 35). Following these protests, consultations between civil society and the World Bank led to the creation of the WCD in 1998. When the WCD presented its report (WCD 2000), reactions ranged from total agreement to complete rejection. The WCD became part of a "discursive battle" (Dubash 2010, p. 418) in which civil society actors were better organized than dam proponents, as the latter readily admit. The former World Bank staff member John Briscoe (2010, p. 404) argues that "[NGO] veterans, acutely aware of the unwritten rules of the game with international financial institutions," caused dam proponents to retire from the WCD process.

A section of the dam industry drew conclusions from its weak position in the WCD process. In 2007 the IHA set up the Hydropower Sustainability Assessment Forum (HSAF) together with the World Wide Fund for Nature (WWF), Oxfam and other moderate actors from civil society, development agencies and financial institutions.[2] According to the IHA, the forum's objective is the enhancement of the IHA Hydropower Sustainability Assessment Protocol (HSAP) (IHA 2010). However, the forum participants did not agree on the process before it began: while the IHA regards its own guidelines as a basis for the process, civil society organisations see the WCD recommendations as the point of departure (Bosshard 2010, pp. 64–65).

By the late 1990s multilateral financial institutions were already becoming more proactive, revising or establishing operational policies and standards for

---

[2] See http://www.hydropower.org/sustainable_hydropower/HSAF_forum%20membership.html. Accessed 09 December 2010.

large dams. In 1998, after the debate on the River Narmada projects, the World Bank implemented new safeguard policies on environmental and social impacts. In addition, the International Finance Corporation (IFC) established performance standards (IFC 2012a) and an environmental and social review procedure.[3] Banks in the private sector launched the Equator Principles (EPFI 2006), which explicitly refer to the IFC performance standards as benchmarks for the assessment of social and environmental impacts. Nevertheless, the World Bank reacted cautiously to the WCD recommendations, supporting their core values and strategic priorities, but judging the guidelines to be impracticable (World Bank 2001).

Export credit agencies (ECAs) also came under pressure from civil society in the 1990s. In a transnational campaign NGOs mainly from Europe and North America demanded reforms that would lead to the application of social and environmental standards during project approval procedures for granting export credit guarantees. As a consequence, the Organisation for Economic Co-operation and Development (OECD) governments agreed in 2001 on the *Common Approaches to the Environment and Officially Supported Export Credits* (hereinafter "the Common Approaches") (Knigge et al. 2003, p. 14). The Common Approaches are not legally but morally binding, they are meant to preclude "negative downward competition among ECAs" (Knigge et al. 2003, p. 14). The OECD's Common Approaches have been revised several times, most recently in 2007 (OECD 2007). They use the World Bank's safeguard policies and the IFC's performance standards as benchmarks for the assessment of projects (OECD 2007, p. 5).

Despite these reforms, it is for national governments to take the final decision to build a dam, and their policies determine whether or not they apply a specific set of standards to such projects. In proliferating related norms, civil society organisations have played a proactive role (see, for instance, Scheumann 2008; Schneider 2005; Fujikura and Nakayama 2009). Their access points are financial and export credit institutions committed to social and environmental standards (Schoene-Alaluf et al. 2011). Of this the anti-Ilisu Dam campaign is a clear example.

7.3 The Two Anti-Ilisu Campaigns

The Ilisu Dam will be the second largest dam and one of the key parts of the hydraulic infrastructure of the Southeast Anatolia Project in Turkey. It will have a capacity of 1,200 MW and the potential to produce 3,833 GW h of electricity per annum (CoE-R 2009, p. 6). The reservoir will cover an area of 313 km^2 (DSI 2006, p. 5). The dam is situated about 70 km North-West of the Turkish-Syrian border, immediately north of the village of Ilisu (see Fig. 7.1). The public authority responsible for planning and implementing the Ilisu Dam project is the Turkish General Directorate of State Hydraulic Works (DSI).

---

[3] Since 2006 the IFC Sustainability Framework (IFC 2012b).

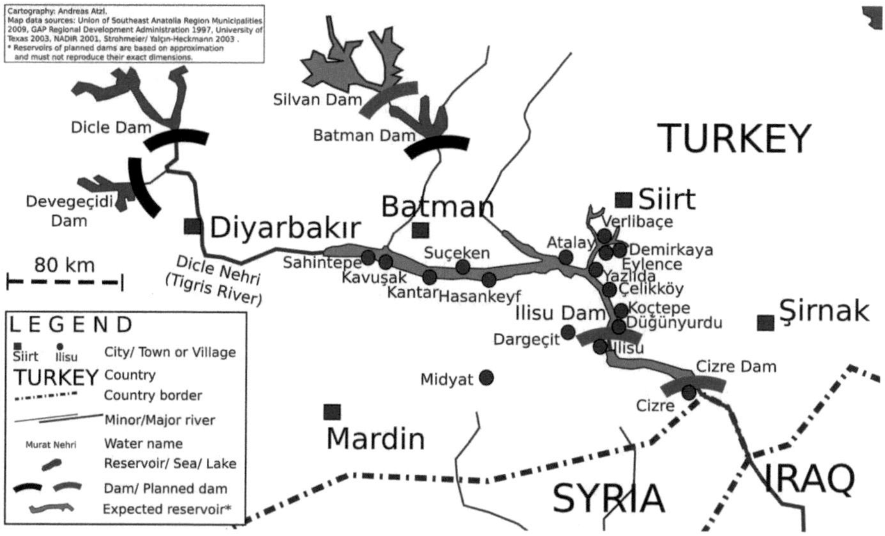

Fig. 7.1 Map of the upper Tigris region showing the location of the Ilisu Dam (author's own compilation)

7.3.1 The First Campaign

A consortium of private companies and banks was established in 1997. It consisted of Austria's Sulzer Hydro (today VA Tech Hydro/Andritz AG), Switzerland's ABB (today Alstom Switzerland), the Union Bank of Switzerland, the UK's Balfour Beatty, Italy's Impregilo, Sweden's Skanska and Turkey's Nurol, Kiska and Tefken (Scheumann 2008, p. 71). This consortium applied to the ECAs of Austria, Germany, Italy, Japan, Portugal, Sweden, Switzerland, the UK and the USA for export credit guarantees (Kibaroglu and Scheumann 2009).

Led by the British NGOs Kurdish Human Rights Project (KHRP), The Corner House and Friends of the Earth (FoE), the first anti-Ilisu campaign claimed that the dam would have negative social, environmental and cultural impacts.[4] The campaign also included the NGOs World Economy, Ecology and Development (WEED) (Germany), ECA Watch (Austria) and Berne Declaration (Switzerland), which had been part of the ECA reform campaign network.

As a consequence of the campaign, the ECAs presented the Turkish government with four conditions to be met if export credit guarantees were to be granted: the project must include a resettlement programme, upstream water treatment

---

[4] Where no explicit sources are quoted, information on the campaigns, their networks and NGOs in this and the following chapter was obtained during interviews with decision-makers and civil society actors.

plants to ensure water quality, the maintenance of the water flow downstream and the preservation of the archaeological heritage (Hilyard et al. 2000, pp. 1–2).

In October 2000, the KHRP undertook a fact-finding mission in the region to gauge the compliance of the Ilisu Dam project with these four conditions.[5] At roughly the same time, the Swedish company Skanska and the UK's Balfour Beatty withdrew from the project. Italy's Impregilo and the Swiss Union Bank followed in 2001 and 2002.

The Turkish government reacted negatively to the Ilisu Dam campaigners, accusing them of ignoring Turkish efforts to safeguard the archaeological heritage and the Ilisu Dam Lake Area Subregional Development Plan, which had been initiated by the Regional Development Administration of the Southeast Anatolia Project in the 1990s (Ünver 1997, p. 456). In particular, Turkey accused the members of the KHRP fact-finding mission of publishing false information. However, criticisms of Ilisu came not only from NGOs, but also from such other actors as former World Bank staff (Cernea 2006). The end of the first consortium was celebrated by the NGOs as a victory, and the debate surrounding the Ilisu Dam went quiet for about 2 years.

7.3.2 The Second Campaign

In late 2004, the Turkish government relaunched negotiations for a second consortium, which was established in 2005. The partners were Austria's VA Tech Hydro/Andritz, Switzerland's Alstom, Stucky, Colenco and Maggia, Germany's Ed. Züblin AG and four Turkish companies. The Bank of Austria, Züricher Kantonalbank (ZKB), Société Générale and two Turkish banks were to provide export credits (Walder 2008) (see Fig. 7.2 for the actors in the second consortium and their relationships).

The Turkish government, represented by the DSI, signed a contract with the consortium for the construction of the Ilisu Dam. For the investment the consortium needed export credits from private banks, which demanded export credit guarantees against the loss of credit. The consortium therefore applied to the German, Austrian and Swiss ECAs (hereinafter "the European ECAs") for guarantees. Having agreed to respect the OECD's Common Approaches, the ECAs had to consider the environmental and social aspects in their decision. Consequently, they negotiated with the Turkish government on how to improve the Ilisu project, which resulted in their signing an agreement with the Turkish government. These terms of reference (ToR) provided for the creation of a Project Implementation Unit and three Committees of Experts on the Environment, Resettlement and the Cultural Heritage and an External Monitoring Group on

[5] See the report of the fact-finding mission in Hilyard et al. (2000).

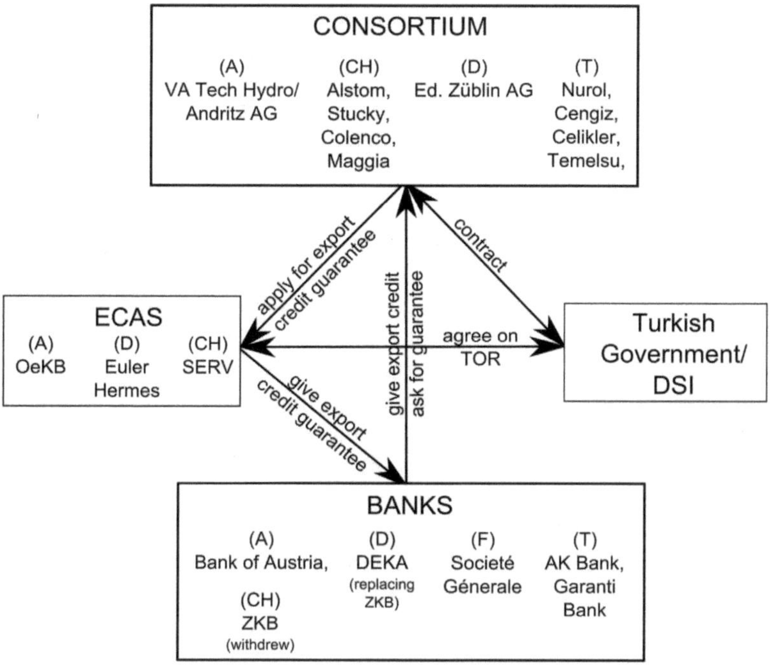

Fig. 7.2 The actors in the second Ilisu consortium and their interrelationships (author's own compilation based on Scheumann 2008 and Walder 2008)

Resettlement. The ToR were meant to influence the project's impact on the environment, the cultural heritage and resettlement. They also referred to the World Bank's standards. By agreeing on the ToR, the Turkish government and European ECAs established a legally binding framework for the improvement of standards for the Ilisu Dam. Turkey also improved its legislation on Environmental Impact Assessments and on expropriation and resettlement. Thus the formal and legal environment for applying international standards at the beginning of the second anti-Ilisu campaign was already better than at the time of the first campaign.

In response to the formation of the second consortium, the NGOs involved in the first Ilisu campaign reactivated their contacts and relaunched their campaign. The second campaign was led by three NGOs from Germany (initially World Ecology, Economy and Development, later CounterCurrent), Austria (ECA Watch) and Switzerland (Berne Declaration), because these countries' ECAs were about to grant export credit guarantees.

At the same time, local groups in and around Diyarbakır, Batman and Hasankeyf (provinces which would be affected by the project) began organizing activities against the new plans. In early 2006 they founded the Initiative to Keep Hasankeyf Alive (Hasankeyf'i Yaşatma Girişimi) (hereinafter "the Initiative"). Contacts between activists in the region and Europe were also reactivated.

As envisaged by the OECD's Common Approaches, the European ECAs set a deadline for the submission of formal objections to the Ilisu project. Civil society groups collected information and encouraged people affected by the project to submit complaints. After the deadline had expired in February 2006, the European ECAs and their governments discussed the applications for export credit guarantees and decided in December 2006 to approve in principle the export credit guarantees under the ToR.

Meanwhile, the construction of the dam had officially begun in August 2006. Throughout 2007, the Initiative and the European NGOs organized a number of activities. Among these was the publication of a report claiming that Turkey was failing to comply with the Terms of Reference (The Cornerhouse and KHRP 2007). In December 2007 and May 2008, the Committees of Experts visited the Ilisu site. Their reports[6] stated that, despite recognizable progress, the Ilisu project did not yet comply with the mutually agreed conditions.

In October 2008 the European ECAs therefore set Turkey a target of 60 days to overcome the problems (Gaserow 2008a). In December 2008 the Committees of Experts again visited the site. Their reports[7] again acknowledged that some progress had been made, but referred to continuing resettlement problems. This led the European governments to suspend the guarantees on 23 December 2008 and to set a final deadline of 6 months thence, after which they would withdraw their export credit guarantees (Gaserow 2008a).

In May and June 2009 the experts' final assessment read as follows[8]:

- As regards resettlement "almost all of the 'Measures to be completed prior to beginning of construction' as agreed with the ECAs have been completed with only minor issues remaining" (Ilisu External Monitoring Group 2009b, p. 2).
- As regards the environmental terms, "the conditions for items to be carried out during the suspension period are fulfilled. The required activities have started. Ongoing activities are under way and have produced satisfactory results" (CoE-E 2009a, p. 7).
- As regards the cultural heritage, "it must, however, clearly be stated that the goals described in autumn 2008, necessary for a start of major construction work for the project, in major fields have not been reached yet" (CoE-CH 2009, I).

However, NGOs continued to criticize Turkey's resettlement policy for failing to comply with the World Bank's standards. This position was supported by Robert Goodland, who had been involved in their development (Drillisch and Eberlein 2009, pp. 1–2).

6 See CoE-E (2008a), CoE-CH (2008a), CoE-R (2008a) for the first field visit reports and CoE-E (2008b), CoE-CH (2008b), CoE-R (2008b) for the second visits.

7 See the report of the External Monitoring Group on Resettlement (Ilisu External Monitoring Group 2009a).

8 CoE-E (2009a), CoE-CH (2009), CoE-R (2009) and Ilisu External Monitoring Group (2009b).

Shortly after the experts' field visits and 2 weeks before the deadline ended on 6 July 2009, the German newspaper *Frankfurter Rundschau* reported that the German government was about to back out of the Ilisu Dam project (Schwab 2009). On 7 July, the three ECAs announced the withdrawal of the export credit guarantees (SERV, OeKB and Euler Hermes 2009). On the same day, the Société Générale, the Bank of Austria and the DEKA Bank withdrew their export credits (DEKA Bank, Bank Austria, and Société Générale 2009).

While NGOs welcomed this decision, attributing it to their campaign (GegenStrömung and Hasankeyf Initiative 2009), Andritz, the lead company in the consortium, regretted the decision and said that it would await further decisions by the Turkish government before deciding on its future involvement (Andritz 2009). This contrasted with Züblin's and Alstom's withdrawal. In late 2009, the Turkish government presented a new financing concept with the support of two Turkish banks, Akbank and Garantibank (Die Presse 2010). According to information from the campaign, work on the construction of Ilisu has continued since May 2010. VA Tech Hydro/Andritz have remained interested in the project (ECA Watch 2010).

7.4 The Ilisu Campaign Network and its Actors

7.4.1 The NGOs' Interest in the Ilisu Dam Project

The actors in the Ilisu campaign network can be characterized according to their thematic interest in dams and their geographical origin. Figure 7.3 shows the Ilisu campaign network and categorizes its actors in four sectors, each representing a thematic core interest: (a) the *ECA reform campaign* (upper right sector), (b) *human rights issues* and the *situation of the Kurdish people* (upper left sector), (c) *environmental protection* (lower left sector) and (d) actors with *no or other predominant interests*, such as chambers,[9] universities and actors with a general interest in river issues (lower right sector). The (sub-)campaigns are placed on the inner circle, organisations involved in the campaigns on the outer circle. Actors outside the circle form part of the campaign's organisational environment and will be described later.

[9] Meaning such professional associations as chambers of lawyers, chambers of engineers, chambers of architects and chambers of physicians. In Turkey, they often pursue a political goal and engage in political and societal discussions.

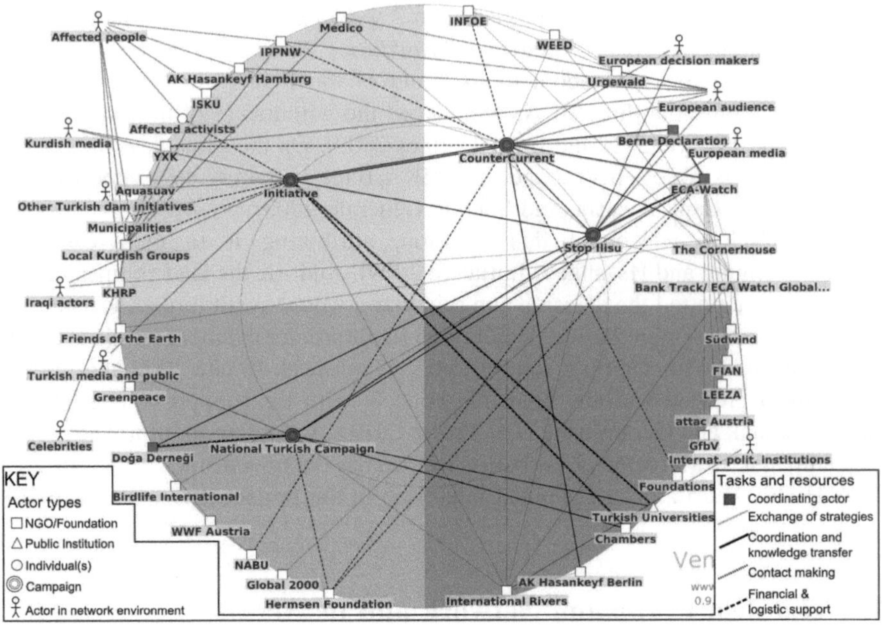

Fig. 7.3 Breakdown of the anti-Ilisu campaign network by thematic interest (author's own compilation)

7.4.2 Actors in the ECA Reform Campaign

The NGOs participating in the ECA campaign were Cornerhouse (UK), ECA Watch (Austria), Berne Declaration (Switzerland), WEED, Urgewald and INFOE[10] (all Germany). Major transnational NGOs included Bank Track and ECA Watch Global.

7.4.3 Human Rights Activists

When approached by Kurds in Southeastern Anatolia who were affected by the dam, the UK's KHRP launched the first campaign against Ilisu in Britain in 2001. Later on, the Kurdish community in Germany became active through the Kurdish Students Association (YXK) which made contact with CounterCurrent and groups in Southeastern Anatolia, including the Initiative. Other pro-Kurdish actors included AK Hasankeyf and Information Point Kurdistan (IPK) in Hamburg and Aquasuav in Italy.

[10] Institut für Ökologie und Aktions-Ethnologie, an NGO that supports ethnic minorities.

Such organisations as the Society for Threatened Peoples (GfbV), Medico International and International Physicians for the Prevention of Nuclear War (IPPNW) in Germany do not focus specifically on the Kurds. While the Society for Threatened Peoples is an international human rights organisation, Medico and International Physicians were founded by physicians. Both organisations have maintained, independent of the Ilisu issue, contacts in the region for many years.

The Initiative to Keep Hasankeyf Alive organizes several associations of people affected by the project, chambers, university institutes and municipalities in the region where the large cities of Diyarbakır and Batman are located.

7.4.4 Environmental Protection Activists

The first environmental organisation involved in the Ilisu campaign was FoE UK. Environmental NGOs play a minor role in the European campaign. The only German environmental NGO actively participating is the German section of Birdlife International (NABU), and two Austrian environmental NGOs are also members of ECA Watch. Although the environment is an important issue in dam projects and can be media-effective, few environmental organisations were initially involved.

This changed in 2008, when the Turkish nature association Doğa Derneği launched a second, *nationwide campaign in Turkey*. Doğa Derneği, the Turkish section of Birdlife International, had been interested since 2005, but had made no more than sporadic contributions to the campaign. In response to personal contacts, it launched the new campaign with Austria's ECA Watch and Germany's Manfred Hermsen Foundation, a private body that supports environmental activities and has been involved in the European campaign since 2006.

7.4.5 Actors with Other Backgrounds

The main interest of some NGOs on the periphery of the Ilisu campaign is globalization. Some members of ECA Watch work in this field, and the campaign has occasionally received support from anti-globalization activists. Other organisations in the network, such as Turkish chambers and university staff supporting the campaign, do not have a clear thematic focus.

Other NGOs focus on water issues in general, the most prominent being International Rivers (IR), which can be seen as the transnational anti-dam NGO. It maintains contact with all the European and both Turkish campaigns. A further actor is AK Hasankeyf Berlin, which includes members of the "critical water NGO scene" in Berlin, Kurds, environmentalists and other activists interested in the Ilisu dam project.

7.4.6 The Spatial Distribution of NGOs: From Europe to Southeast Anatolia

Figure 7.4 shows that actors in the anti-Ilisu campaign network come from four regions: Southeastern Anatolia, where the dam is to be built, Western Turkey,[11] Europe and the German-speaking countries.[12] Some *transnational actors* are also involved in the campaign. They are shown outside the map.

Within the *German-speaking countries* actors are involved in all four thematic areas, the majority focusing on human rights and ethnic issues. The Berne Declaration is practically the only Swiss actor involved in the campaign, whereas it has the support of some fairly large networks of NGOs in Germany and Austria.

The main NGOs in the rest of *Europe* are the three British NGOs involved in the first campaign. As other organisations and individuals in Italy, France and Belgium play only a minor role, they are not all shown.

In *Turkey*, the Initiative and Doğa Derneği differ somewhat in their focus: while actors in Istanbul, where Doğa Derneği is headquartered, have an environmental focus, the Initiative includes actors with a variety of interests, particularly in the human rights field. A specific characteristic of the Turkish campaigns is the involvement of public institutions and political actors. Municipalities, universities and chambers participate in or maintain contact with the campaign.

Of the actors who could not be assigned to a region owing to their transnational nature, IR is the most prominent, since it networks with the various sub-campaigns and actors involved in water-related issues.

Both Turkish campaigns have a dense network of contacts with actors in Europe, although many of these links existed before the anti-Ilisu campaign began. Among the new links is that between the Initiative and Doğa Derneği. While the link between the Turkish campaigns is tenuous (owing to their differing interests) compared to the links between each campaign and European campaigns, their relationship became closer when the Turkish and European network organisations converged on the Ilisu issue.

7.5 Information, Symbols and Normative Pressure: Means Used by the Anti-Ilisu Campaign

The anti-Ilisu campaign is now analysed using Keck and Sikkink's (1999) four types of politics employed by advocacy networks: *information politics*, *symbolic politics*, *leverage politics* and *accountability politics*.

---

[11] Mainly from the two large cities of Istanbul and Ankara, but also from other dam-affected regions in Turkey.

[12] Actors in these countries joined in conducting a single German-language campaign against Ilisu.

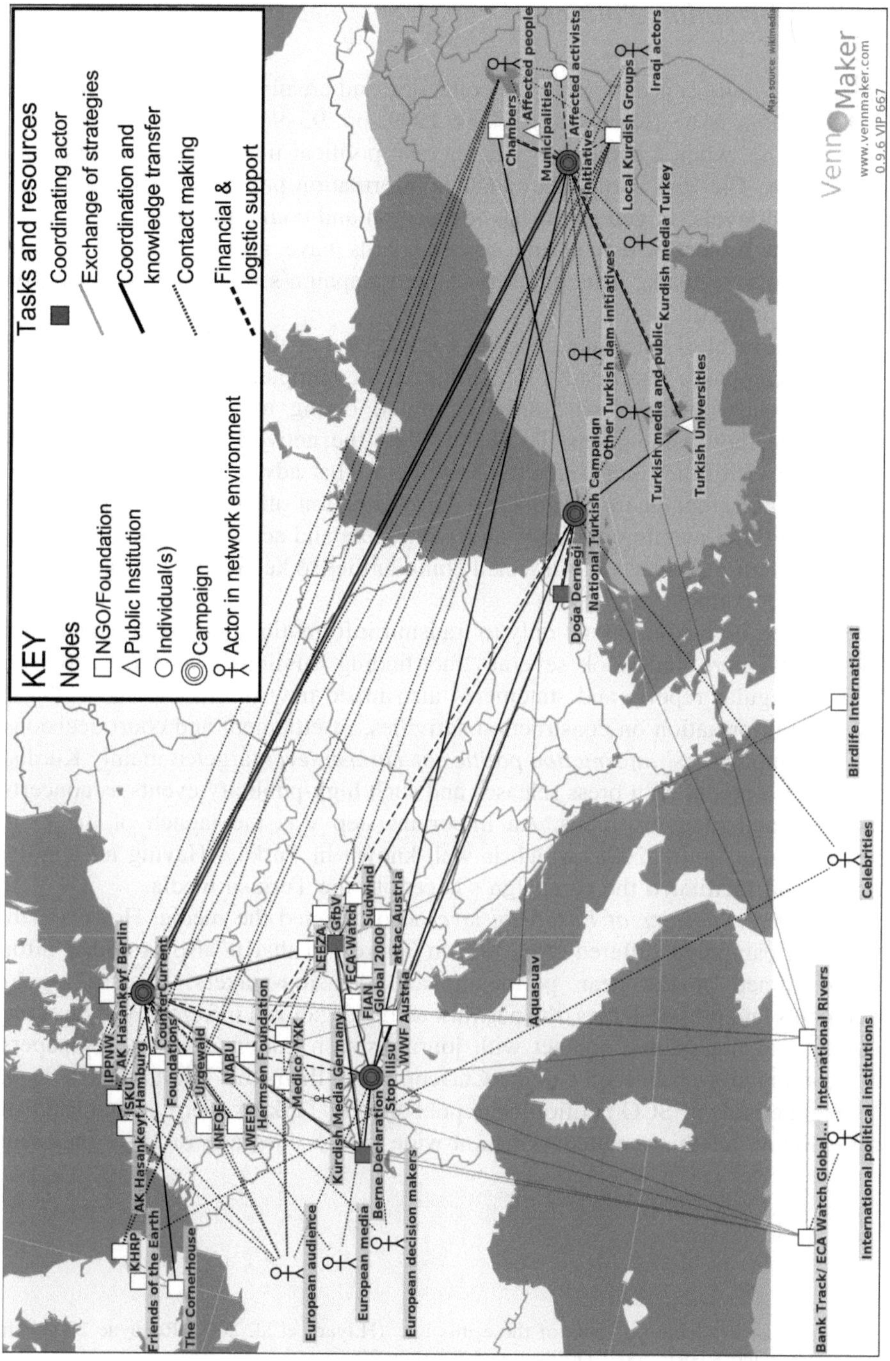

Fig. 7.4 Breakdown of the anti-Ilisu campaign network by spatial distribution (author's own compilation)

7.5.1 Information Politics

Information politics is defined as the collection and creation of information related to a campaign issue (Keck and Sikkink 1999, pp. 95–96). One aspect of this is *issue framing*, when a campaign adds its own political interpretation to objective information. The Ilisu campaign conducts information politics at different spatial and societal levels. In general, it has focused on and communicated those aspects of the Ilisu project where norms and standards have not been respected, and ignored progress made. This has served the campaign's objective of halting the Ilisu Dam project.

The first level of information politics is directed at the *people affected*, which gives the campaign legitimacy. According to the campaigners, support from the people directly affected was initially limited owing to the small amount of information they were given. To change this, the network arranged for people affected to visit other dam areas to show them what adverse effects resettlement had had. The campaigners also opened an information office in Hasankeyf, which can be used as a meeting place for affected people and activists. In addition, local campaign staff regularly visits affected communities to keep them informed and to collect information.

However, it is essential not only to transmit information but also to generate it. The campaigners undertook several "fact-finding missions" between 2000 and 2007.[13] Regular reports and statements also made new information available,[14] including information on construction activities, resettlement and court decisions.

The campaigners' *information politics at Turkish level* targeted mainly Kurdish and Turkish media with press releases and such high-publicity events as concerts, symposia and demonstrations. An important step was the launch of the Doğa Derneği campaign in 2008, which is well-known in Turkey. Having no Kurdish background facilitated the campaign's access to the Turkish media.

Information politics at European level also targeted the media. However, the European campaign differed from that in Turkey in that it also provided information aimed at European politicians and decision-makers. Besides regular statements and press releases, the network actors organized information events and protests and maintained contact with journalists on major German newspapers. The campaign websites of CounterCurrent, Stop Ilisu, the Initiative and Doğa Derneği and a UNESCO petition were published in Turkish, German and English. In Germany, all this resulted in nation-wide media coverage of Ilisu, including

[13] Reports are available on some of these missions (Hilyard et al. 2000; Ronayne 2005; The Cornerhouse and KHRP 2007; Drillisch and Eberlein 2009).

[14] See, for instance, KHRP, Bar Human Rights Committee and The Cornerhouse (2003), Setton and Drillisch (2006), Doğa Derneği (2006), KHRP (2007) and ECA Watch et al. (2008).

reports in major newspapers (*Der Spiegel*, *Süddeutsche Zeitung*, *Frankfurter Rundschau*)[15] and on television.

In addition, the campaigners contacted and briefed politicians. For example, photographs showing that the construction of the dam was continuing despite the suspension of the export credit guarantees in 2008 were passed from the local to the German campaign and from there to the Green Party, whose delegates presented them to the German parliament. First and foremost, this kind of information politics reached those politicians who had already been critical of Ilisu, including members of the left-wing and Green parties and those whose remit included human rights, the environment and the cultural heritage. To convince other politicians, the campaigners organized four visits to Austria, Germany and Switzerland for delegations of mayors and other representatives of towns which would be affected by the project. These visits took place in May 2006, February 2008, October 2008 and May/June 2009. While the local campaigners helped to identify members of the delegations, their European counterparts arranged meetings with parliamentarians and representatives of ministries, ECAs and the companies involved.

The Ilisu network thus acted as a relay station, transferring information from the site to decision-makers in Europe. To increase pressure on European decision-makers, the campaigners used European media. At the beginning of the campaign, information politics targeted audiences and politicians in Germany, Austria and Switzerland because it was their governments who had to approve export credit guarantees. With the new Doğa Derneği campaign, the NGOs tried to strengthen another pathway: the Turkish public.

7.5.2 Symbolic Politics

Such symbols as places, objects, events, dates and celebrities were used to strengthen the campaign's stance, the most important being the city of Hasankeyf itself. Being part of the Initiative's name, it stands not only for a large number of endangered archaeological sites, but also for the people who have to be resettled. Furthermore, the city is a national and international tourist destination. The campaign used the picturesque image of Hasankeyf with its minaret, ruins and cave dwellings and started a petition in 2009 aimed at having it declared a UNESCO World Cultural Heritage Site.

Doğa Derneği organized two loyalty trains from Istanbul to Hasankeyf in 2005 and 2007, and German activists arranged a street theatre performance on the history of Ilisu, which was staged in Hamburg and Bremen. Tree planting activities near Hasankeyf, some attended by Bianca Jagger, symbolized other options for the use of the region's natural resources. German-Turkish pop star Tarkan visited

[15] See Elger (2008), Spiegel (2000, 2006), Strittmatter (2007), Süddeutsche (2008), Gaserow (2008a, b), Schwab (2009).

Hasankeyf, participated in other events arranged by the anti-Ilisu campaign and produced the song Uyan (Wake Up) in cooperation with Doğa Derneği to protest against environmental destruction.

An international event with a high symbolic value was the 5th World Water Forum held in Istanbul in March 2009, which was criticized by civil society for being organized by the water industry and for not representing all stakeholders equally. Although the Ilisu campaigners were able, with some difficulty, to hold their own event during the Forum, they were more heavily involved in the alternative forum that took place in Istanbul at the same time. Activities associated with this forum included an award ceremony for local activists and the Istanbul Water Tribunal, where civil society organisations symbolically sued the Turkish government for continuing with the Ilisu project.[16]

The withdrawal of the European ECAs has symbolic relevance in itself since it can be seen as a sign of Turkish non-compliance with international standards. When the Turkish banks stepped into finance the project in 2009, the campaign criticized them for assisting a project that had lost European support. Moreover, *Ilisu itself* is a form of symbolic politics since it is seen by many activists as a precedent for the commitment of governments and financial institutions to international standards.

Symbolic politics is closely related to information politics and the interpretation of information. Consequently, the anti-Ilisu campaign has used symbolic politics on much the same pathways as information politics, namely European audiences. In addition, symbols have allowed the campaign to introduce values and emotions into the debate. Compared to information politics, the campaign has made greater use of symbolic politics to influence Turkish decision-makers, involving Turkish celebrities and organizing protests. It has also used such symbols as the *World* Water Forum and the *World* Cultural Heritage to demonstrate the global relevance of the Ilisu project.

7.5.3 Leverage Politics

Leverage politics describes the attempt to gain influence through persuasion and pressure to change policies. While *material leverage politics* tries to link two issues that are not directly connected (e.g. in the form of economic sanctions), *moral leverage politics* tries to shame actors that "place a high value on international prestige" (Keck and Sikkink 1999, p. 97).

The most obvious form of material leverage politics in the case of Ilisu is the campaign's attempt to cut the credit lines for the dam by pressurizing the ECAs to withdraw from the project.

---

[16] See http://www.boell.de/oekologie/ressourcen/ressourcenpolitik-oekologie-6363.html. Accessed 17 July 2013.

When the campaign lacked material leverage politics after the withdrawal of the ECAs, the arguments for moral leverage politics increased. The high symbolic meaning of the *European withdrawal* was used to expose Turkey and to denounce the companies remaining in the consortium for continuing to support a project that did not meet international standards.

Leverage politics was the most important tool used by the anti-Ilisu campaign until the European withdrawal in 2009, which can be seen as a success for the campaign, but it also cut the campaign's most important pathway.

7.5.4 Accountability Politics

Accountability politics describes the means used to put pressure on actors who fail to comply with their own values (e.g. laws or international treaties signed by their government). At the beginning of the campaign, accountability politics targeting the European governments referred mainly to the *OECD Common Approaches*. Later, the campaign cited the *Terms of Reference* concluded between the government of Turkey and the European ECAs, although publicly the NGOs did not consider these standards sufficient. Another tool for accountability politics used particularly against Germany consisted in the *critical statements* on Ilisu made by the German Ministry for Economic Cooperation and Development (BMZ) and in the Ministry's official support for the WCD's recommendations. However, it remains unclear how far the campaign "used" the Ministry and how far the Ministry "used" the campaign to strengthen its own position in relation to the other German ministries involved in the export credit approval process.

For accountability politics targeting the Turkish government the campaign referred mainly to *Turkish law*. Between 1999 and 2006, four court actions were initiated in Turkey. Two cases concerned formal errors in the tendering process and the legal basis for expropriations. Another referred to Turkish legislation on the protection of historical monuments. The fourth case, opposing a new law on dams and archaeological sites, was brought by the Initiative in cooperation with other anti-dam campaigns. A fifth legal action was initiated by activists before the European Court of Human Rights, claiming that a cultural heritage site (Hasankeyf) with importance for the history of humankind would be destroyed. Although the campaign was partly successful, it was able to halt neither the expropriation of land nor the construction of the dam.

The campaign's main strategy of bringing influence to bear directly on Turkish decision-makers falls into the category of accountability politics under Turkish law. Furthermore, NGOs used accountability politics at European level with reference to the ToR and the OECD Common Approaches. In this way, the campaign endeavoured to gain indirect influence on Turkish decision-makers. However, this was possible only because of the ECAs' policy learning process and the emergence of environmental and cultural heritage laws in Turkey in recent decades.

7.6 The Campaign's Influence on Decision-Makers

The strategies mentioned so far have targeted specific groups that are capable of influencing decisions on Ilisu. The degree of success differs as a function of the pathways used by the campaign.

7.6.1 Points of Access to Decision-Makers

The main access points used by the *European campaign* were the ECAs, the governments of Germany, Austria and Switzerland and, to a lesser degree, European companies and banks. In Germany, strategic partners of the transnational advocacy network were the two *opposition parties* the Greens and the Left, and *individual politicians* from the then governing parties, the Social Democrats (SPD) and the Conservatives (CDU). These actors were in regular contact with the German campaign, put the topic on the political agenda and helped to pass information from the Ilisu region to the German parliament and ministries.

The *BMZ* was another strategic partner. It took a critical view of the Ilisu dam and maintained close contact with the campaign. However, the German government as a whole, along with the banks and companies, supported the dam until June 2009, while acknowledging the campaign's right to criticize the decision and to discuss the issue. The fact that the consortium attempted to maintain a dialogue with the anti Ilisu-campaign also shows that the NGOs' information politics had been taken seriously by the European actors.

In contrast, the campaign had very little contact with Turkish decision-makers. It hoped to stop the construction of the dam by using the pathway via European decision-makers. This can be interpreted as strategic support from European NGOs for the Turkish campaign.

This pattern reveals existing asymmetries within the anti-Ilisu campaigning network. Unlike Keck and Sikkink (1998), Dobner (2010) argues that asymmetrical relations are often to be found in TANs. Referring to Rohrschneider and Dalton (2002) and Hogenboom (2003), she detects "internal power asymmetries and differing formal and informal positions in the network"[17] and assumes that the "unequal distribution of resources and differences of opinion on strategic short-term objectives or what constitutes correct tactical behaviour give rise to tensions"[18] (Dobner 2010, p. 278; author's own translation) within TANs. This is true of the anti-Ilisu network: one indicator is the central role played by European

[17] "(...) interne Machtasymmetrien und unterschiedliche formelle und informelle Positionen im Netzwerk (...)" (Dobner 2010, p. 278).

[18] "(...) und ungleiche Ressourcenverteilungen sowie Differenzen über strategische Nahziele oder richtiges taktisches Verhalten Spannungen produzieren (Rohrschneider and Dalton 2002; Hogenboom 2003)" (Dobner 2010, p. 278).

NGOs in the network. As we have seen, further asymmetries exist with respect to both the current objectives and the access points used by the members of the network. This asymmetry of objectives, resources and access points may, but need not, imply asymmetries in the network's power relations. In the Ilisu campaign, such power differentials became obvious in only a few respects.

One example of this asymmetry is the role played by "Kurdish issues" in the campaign. During research in the region directly affected by the dam and the reservoir, a high degree of mistrust between sections of the Kurdish population and the Turkish administration was evident.[19] Subsequently, Kurdish issues also played a role in the local campaign. Despite this, little importance was attached to them in the strategies pursued by the international campaign. On the contrary, the majority of European actors in the anti-Ilisu network consciously avoided the "Kurdish question". Hence, despite the high relevance of the Kurdish issue for the local campaign, neither the international nor the national campaign in Turkey referred to it. Although European NGOs did not openly try to influence their partners in Turkey, this avoidance of certain issues in the transnational campaign shows how strongly European actors influenced the campaign's overall strategy.

On the other hand, the aforementioned mistrust within Turkish-Kurdish society, combined with a fundamentally different approach to development, explains, the campaign's limited influence on decision-makers by the direct pathway in Turkey. The Turkish government and the DSI see the dam as necessary for development. For them, the problems are caused not by the dam, but by the protests against it. In addition, given the Turkish government's and DSI's long tradition of non-participatory water management, they lack the organisational capacities and personnel to conduct a dialogue with the campaign.[20] The campaign's approach is, however, similarly uncompromising. This is evident from the attempt to have Hasankeyf declared a UNESCO World Heritage Site. Such a declaration would block the possibility of any dam, including the solution involving three smaller dams proposed in a study by the Middle Eastern Technical University in Ankara (Yalcin and Tigrek 2010).

7.6.2 The Network's Success in Europe and Turkey

The Ilisu debate is also influenced by the "discursive battle" on dams that is taking place at international level. The campaign's success at Turkish, European and international level therefore needs to be assessed. Taking Sikkink's (2009, p. 235) classification of network effectiveness, the analysis distinguishes three areas in

[19] For a good overview of the local political and societal environment in South-Eastern Turkey and its implications for the Ilisu dam project see Fliesser 2010, 15 ff. and 33 ff.

[20] This view was also expressed by German politicians in the ruling party when they were interviewed.

Table 7.1 Overview of the success of the anti-Ilisu and previous anti-dam campaigns (author's own compilation)

Type of success	Principal objective	Specific objectives by level		
		International	Europe	Turkey
(a) Agenda-setting and information provision	Raise awareness	Former and Ilisu campaigns: success	Former and Ilisu campaigns: success	Ilisu campaign: success
	Initiate a basic discussion on dams	Former campaigns: success (WCD, DDP, HSAF)	No assessment	No success yet
(b) Policy and discursive change	Implementation of standards in policies	Former campaigns: success (WCD, OECD Common Approaches, Safeguard Policies, Performance Standards)	Former and Ilisu campaigns: success (OECD Common Approaches, ToR, Committees of Experts)	Ilisu campaign: partly successful, lapsed owing to European withdrawal (improvements to resettlement plans)
(c) Behavioural change in key actors	Practical change of paradigms and behaviour in water and dam management	No success yet (full implementation of standards in dam projects)	Ilisu campaign: success (export credit guarantees stopped)	Ilisu campaign: partly successful (improved planning, Ilisu dam halted, alternative assessment)

Success **Partly successful** **No success yet**

which a transnational advocacy network can be successful: agenda setting and information provision; policy and discursive change; and behavioural change in key actors. Table 7.1 gives an overview of the different objectives of the campaign, categorized by the level and type of success. The overview includes former campaigns, such as the one which brought about ECA reform.

The anti-Ilisu campaign has been most successful at *European level.* It succeeded in putting Ilisu on the political agenda and provided European decision-makers and media with information (type a). The anti-Ilisu and previous campaigns also influenced the decision of the Turkish and European governments to agree on the ToR and set up Committees of Experts (type b). A number of German politicians stated during the research that, without the public attention drawn to it, the project would probably have been accepted unconditionally by the Europeans. Finally, the campaign was at least partly successful in stopping European participation in the project (type c). Like the ToR, the withdrawal of European ECAs, banks and companies from the project can be at least partly attributed to the campaign. However, one European company—VA Tech Hydro/Andritz—still remains in the consortium.

This success at European level is also the result of former campaigns at *international level*, which had already resulted in a certain degree of agenda-setting (type a) and policy change (type b) in the form of the WCD recommendations and the OECD Common Approaches. In addition, the Ilisu campaign can be seen as a successful contribution to the objectives pursued by these international campaigns of raising awareness (type a) and changing the paradigms in water and dam management (types b and c).

The network's success at *Turkish level* is difficult to assess. The campaign has been successful in bringing the issue to the attention of the Turkish media and in providing information about the dam (success type a). As none of the political parties in the Turkish national parliament opposes large dam projects, the network was unable to initiate a basic discussion on dams (type a).

However, the anti-Ilisu campaign has been able to achieve some changes in Turkey's policies (type b). Turkey did not meet all the ToR, but it did improve stakeholder participation and the resettlement process. Although it has not been bound by the ToR since the European withdrawal, there has been at least some internal policy learning and an improvement in legislation for future dams. However, behavioural change (type c) has so far been limited. While project implementation has improved, there has been no general review of the Ilisu dam project itself.

7.7 Conclusion

The aim of this article was to analyse *whether*, by what *pathways* and by what *means* a transnational advocacy network is influential in ensuring compliance with international standards. The aim of the Ilisu campaign was not to ensure that

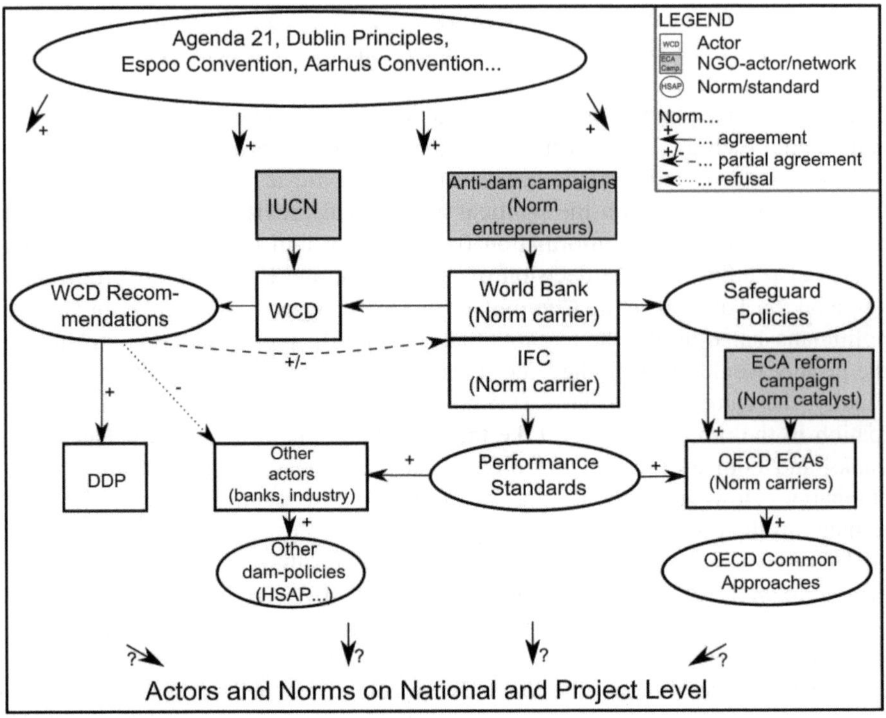

Fig. 7.5 The role of actors in norm dissemination for large dams: the international level (author's own compilation)

standards were applied, but to prevent the construction of the dam (although, for the campaign, stopping Ilisu would be the same as ensuring the application of those standards). Although this objective has not been achieved, the campaign has influenced the application of standards, as shown in Figs. 7.5 and 7.6. The diagrams link dam-related actors (squares) to dam-related norms and standards (ovals). Figure 7.5 shows the international level of norm dissemination, Fig. 7.6 the dissemination of norms in the case of the Ilisu dam.

The actors and NGOs driving the early dam campaigns in the 1980s and 1990s were *norm entrepreneurs* because they successfully established a dialogue between the World Bank and civil society actors, which culminated in the establishment of the WCD (see Fig. 7.6).[21] As a result, the World Bank became an important *norm carrier* for dam-related norms. However, actors differed in their understanding of the WCD recommendations or proposed more practical alternatives. The most frequently used standards to date are those of the World Bank.

[21] The idea is based on the concept of the norm life cycle by Finnemore and Sikkink (Finnemore and Sikkink 1998, 895 ff.).

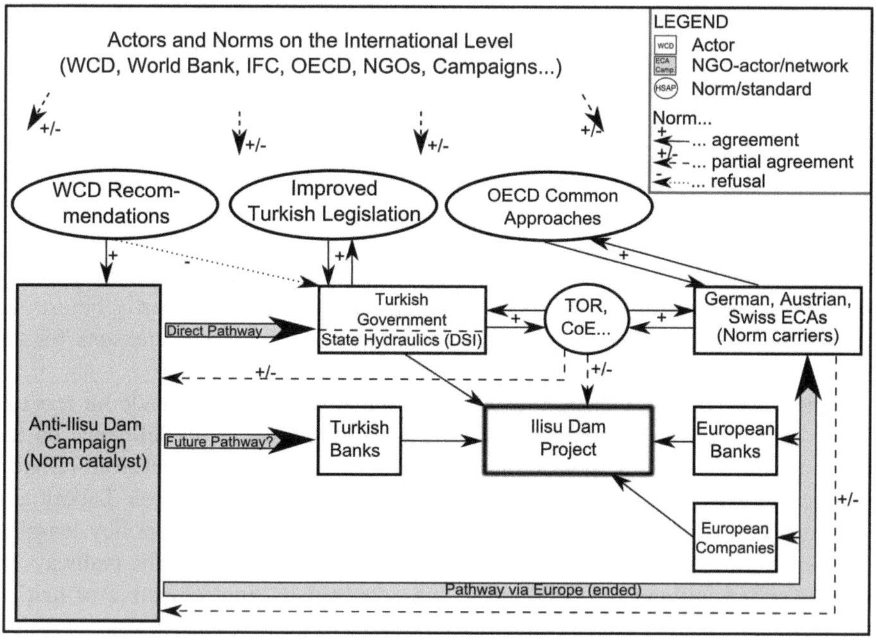

Fig. 7.6 The role of actors in norm dissemination for large dams: the case of the Ilisu Dam (author's own compilation)

Under pressure from the ECA reform campaign, the ECAs of the OECD countries developed the OECD Common Approaches (which referenced the World Bank standards). They also became *norm carriers*. In this process, the ECA reform campaign assumed the role of a *norm catalyst*, because it accelerated the emergence of the OECD Common Approaches.

When Turkey began negotiating on the establishment of an international consortium for the construction of Ilisu, the ECA reform campaign used the project. From this emerged a separate anti-Ilisu campaign (see Fig. 7.6). The end of the first Ilisu consortium was celebrated by civil society actors as a success. It meant that the Turkish government had to agree on social and environmental standards when it began negotiations for a second consortium. Of its negotiating partners, the German, Austrian and Swiss ECAs were also interested in compliance with social and environmental standards because they were bound by the OECD Common Approaches. The OECD ECAs thus became *norm carriers* when the ToR and the Committees of Experts were established.

However, the campaign did not consider the improvements made by the Turkish government to be sufficient. It continued to use predominantly the *pathway* via *European actors* to cut the credit lines for the Ilisu Dam project. This indicates asymmetrical approaches as between Turkish and European NGOs in their access to decision-makers and their ability to have issues put on the political agenda. While this is one way for local campaigns to have their concerns put on international agendas, the risk is that local issues and interests will not be adequately reflected in international discourses. The anti-Ilisu campaign used information, symbolic, accountability and leverage politics. As a consequence of the campaign, European ECAs withdrew their support for Ilisu. The Ilisu network therefore successfully triggered a behavioural change in norm carriers. The question is, however, why the European ECAs withdrew their support despite recognizable improvements in Turkish legislation and major improvements in compliance with the Terms of Reference. It suggests that the high public pressure and the intensification of the campaign's activities in Europe shortly before the date of the last deadline for Turkey in July 2009 were the main reasons for the European ECAs' withdrawal.

This can be regarded as problematic, given that the campaign made far less use of the remaining *direct pathway to Turkish decision-makers*. Its influence on the application of standards and on behavioural change in Turkey therefore remains indirect. However, improved legislation and the agreement between Turkey and the European ECAs on the Terms of Reference indicate that the policy learning process in Turkey is ongoing. But with the European withdrawal, the pathway via Europe ended without the network having achieved its final objective of halting the dam.

As a consequence, the current challenge for the anti-Ilisu campaign is to find *new pathways* and access points to influence decision-makers in Turkey. The campaign has already made some initial attempts: the aim of the Doğa Derneği campaign is to communicate the issue to a wider public in Turkey. Another attempt consists in the emerging network between Ilisu and other dam campaigns in Turkey. In June 2009, dam activists from different regions met in Ankara to demonstrate against large dams. When the Turkish Garanti Bank and Akbank joined the project in spring 2010, the network launched a campaign against them.

However, the campaign's *strategy* has led it into a situation where it has lost its most important pathway of influence. In these circumstances, it is unlikely that the campaign will achieve its primary objective of stopping the Ilisu Dam project. Yet the anti-Ilisu campaign has to be seen as another contribution to the process of norm dissemination. From this perspective, which is likely to be shared mainly by European and international civil society actors, it is of minor importance for the debate on large dams whether the campaign finally succeeds in stopping the Ilisu dam. For the local campaign and the people affected by the Ilisu dam project, the actual legal and behavioural changes in Turkey are more important. Whether the campaign could have achieved more by adopting a less fundamental approach to the dam project is an open question.

Acknowledgements Thanks go to my supervisors at the University of Mainz, Prof. Peter Preisendörfer, Dr. Jürgen Schiener, Prof. Rolf-Dieter Wilken and to Dr. Waltina Scheumann from the German Development Institute for her valuable comments and supervision of my study.

References

Andritz (2009) ANDRITZ bedauert Ausstieg der Exportkreditversicherungen aus dem türkischen Wasserkraftwerksprojekt Ilisu. http://www.andritz.com/de/ANONIDZ0BA5D16E6AB3A104/news?id=56101&count=2. Accessed 10 Aug 2009

Atzl A (2009) The role of transnational NGO-networks in the implementation of social and environmental standards. The Ilisu Dam Project in Turkey. Diploma thesis in Organisational Sociology, Johannes Gutenberg University, Mainz

Bosshard P (2010) The dam industry, the World Commission on Dams and the HSAF process. Water Altern 3(2):58–70

Briscoe J (2010) Overreach and response: the politics of the WCD and its aftermath. Water Altern 3(2):399–415

Cernea M (2006) Comments on the resettlement action plan for the Ilisu Dam and HEPP Project. Prepared for the Berne Declaration, Switzerland and the Ilisu Dam campaign Europe

CoE-CH (Committee of Experts on Cultural Heritage) (2008a) Ilisu Hydropower Project, Turkey. Report of the Sub-Committee on Cultural Heritage. First site visit 2–11 Dec 2007. http://www.designconsult.com/ilisu/themes/blue_style/images/force-download.php?file=./archae/coe/coe_cultural_heritage.pdf. Accessed 09 Aug 2009

CoE-CH (Committee of Experts on Cultural Heritage) (2008b) Ilısu Hydropower Project, Turkey. Report of the Sub-Committee on Cultural Heritage. Second site visit 10–21 Aug 2008. http://www.designconsult.com/ilisu/themes/blue_style/images/force-download.php?file=./archae/coe/Ilisu_CoE_Rep_Site_Visit_CH2_end.pdf. Accessed 10 Aug 2009

CoE-CH (Committee of Experts on Cultural Heritage) (2009) Ilisu Hydropower Project, Turkey. Report of the Sub-Committee on Cultural Heritage. Forth site visit 18–23 May 2009. http://www.designconsult.com/ilisu/themes/blue_style/images/force-download.php?file=./archae/coe/Ilisu_CoE_Rep_SV4_VE2.pdf. Accessed 08 Aug 2009

CoE-E (Committee of Experts on Environment) (2008a) Ilisu Hydropower Project, Turkey. Sub-Committee on Environment: Final Report. First site visit 2–11 Dec 2007. http://www.designconsult.com/ilisu/themes/blue_style/images/force-download.php?file=./umwelt/coe/coe_environment.pdf. Accessed 09 Aug 2009

CoE-E (Committee of Experts on Environment) (2008b) Ilisu Hydropower Project, Turkey. Sub-Committee on Environment Report. Biodiversity, EMP and related aspects. Second site visit 15–25 May 2008. http://www.designconsult.com/ilisu/themes/blue_style/images/force-download.php?file=./umwelt/coe/Report_CoE_E_2nd_Site_Visit_Zwahlen.pdf. Accessed 10 Aug 2009

CoE-E (Committee of Experts on Environment) (2009a) Ilisu Hydropower Project. Subcommittee on Environment Report. Biodiversity, EMP and related aspects. Fourth site visit 4–12 June 2009. http://www.designconsult.com/ilisu/themes/blue_style/images/force-download.php?file=./umwelt/coe/sub_committee_on_env_16_02_09.pdf. Accessed 09 Aug 2009

CoE-R (Committee of Experts on Resettlement) (2008a) Ilisu Hydroelectric Dam Power Plant Project. Report on the first field visit of the Committee of Experts-Resettlement. 29 Nov–11 Dec 2007. http://www.designconsult.com/ilisu/themes/blue_style/images/force-download.php?file=./umsiedelungen/coe/coe_resettlement02.pdf. Accessed 09 Aug 2009

CoE-R (Committee of Experts on Resettlement) (2008b) Ilisu Hydroelectric Dam and Power Plant Project. The Committee of Experts—Resettlement. Institutions and capacity building for resettlement in Ilisu. Report on the second field visit of the Committee of Experts—Resettlement. 10–19th March 2008. http://www.designconsult.com/ilisu/themes/blue_style/images/force-download.php?file=./umsiedelungen/coe/Report_CoE_R_2nd_Site_Visit_Cernea.pdf. Accessed 10 Aug 2009

CoE-R (Committee of Experts on Resettlement) (2009) Ilisu Dam and Hydroelectric Power Plant Project. Report on the field visit and Evaluation Workshop of the Committee of Experts—Resettlement. 7–12 June 2009. http://www.designconsult.com/ilisu/themes/blue_style/images/force-download.php?file=./umsiedelungen/coe/Final_COE_R_4th_report_in_20090706.pdf. Accessed 08 Aug 2009

DEKA Bank, Bank Austria, and Société Générale (2009) Bank Austria, DekaBank and Société Générale will no longer support the export financing of the Ilisu project. July 7. http://csr.socgen.com/content/download/928/16850/version/1/file/PR+Ilisu+070709+EN.pdf. Accessed 10 Aug 2009

Der Spiegel (2000) Ärger um Hermes. Der Spiegel 12

Der Spiegel (2006) Am türkischen Tropf. Der Spiegel 44

Derneği Doğa (ed) (2006) Review of the Environmental Impact Assessment Report submitted for the Ilisu dam and the hydro-electric power project. Doğa Derneği, Ankara

Die Presse (2010) Türkei: Ilisu-Staudamm wird ab April gebaut. Die Presse, Februar 17. http://diepresse.com/home/wirtschaft/eastconomist/540217/index.do. Accessed 24 Aug 2010

Dobner P (2010) Wasserpolitik. Suhrkamp, Berlin

Drillisch H, Eberlein C (2009) Ilisu trip report from 18 to 24 May 2009. http://www.gegenstroemung.org/drupal/sites/default/files/Report%20NGOs%20Ilisu%20May2009-p_0.pdf. Accessed 26 Aug 2009

DSI (Turkish General Directorate of State Hydraulic Works) (2006) Ilisu Dam and HEPP project amendments on Updated Resettlement Action Plan (URAP). http://www.dsi.gov.tr/ilisu/ilisu_yyp_eke.pdf. Accessed 17 Aug 2010

Dubash NK (2009) Global norms through global deliberation? reflections on the World Commission on Dams. Glob Gov 15(2):219–238

Dubash NK (2010) Reflections on the WCD as a mechanism of global governance. Water Altern 3(2):416–422

ECA (Export Credit Agency) Watch (2010) Bauarbeiten in Ilisu wieder aufgenommen—aktuelle Bilder. StopIlisu.com. May 14. http://m-h-s.org/ilisu/front_content.php?idcat=143&idart=635. Accessed 24 Aug 2010

ECA (Export Credit Agency) Watch, WEED (World Economy, Ecology and Development), EvB, Hasankeyf Initiative, Manfred-Hermsen-Stiftung (2008) Europas Schmutzige Hände. Stopp Ilisu. Umstrittenes Staudammprojekt am Tigris. Factsheet distributed by the Ilisu campaign

Elger K (2008) Die Macht des Wassers. Der Spiegel 52(6)

EPFI (Equator Principles Financial Institutions) (2006) The "Equator Principles". http://www.equator-principles.com/documents/Equator_Principles.pdf. Accessed 09 Aug 2010

Fink M, Cramer A (2008) Towards implementation of the World Commission on Dams Recommendations. Experiences and reflections after 5 years. In: Scheumann W, Neubert S, Kipping M (eds) Water politics and development cooperation. Springer, Berlin, pp 33–53

Finnemore M, Sikkink K (1998) International norm dynamics and political change. Int Org 52(04):887–917. doi:10.1162/002081898550789

Fliesser L (2010) Governance of the Ilisu Dam Project. Stakeholders—International Institutions—Compliance—Legitimacy. Tectum, Marburg

Fujikura R, Nakayama M (2009) Lessons learned from the World Commission on Dams. Int Environ Agreem Polit Law Econ 9(2):173–190

Gaserow V (2008a) Berlin schickt Ankara Blauen Brief. Frankfurter Rundschau, Oktober 9. http://www.fr-online.de/in_und_ausland/politik/aktuell/?em_cnt=1610246&. Accessed 10 Aug 2009

Gaserow V (2008b) Abschied vom Mega-Projekt. Frankfurter Rundschau, Dezember 22. http://www.fr-online.de/in_und_ausland/politik/aktuell/?em_cnt=1649224&. Accessed 10 Aug 2009

GegenStrömung, Hasankeyf Initiative (2009) Europäische Beteiligung offiziell beendet. July 7. http://www.gegenstroemung.org/drupal/de/node/51. Accessed 10 Aug 2009

Goodland R (2010) The World Bank versus the World Commission on Dams. Water Altern 3(2):384–398

Hilyard N, Tricarico A, Eberhardt S, Drillisch H, Norlen D (2000) "If the river were a pen…"—The Ilisu Dam, the World Commission on Dams and export credit reform. The final report of a fact-finding mission to the Ilisu Dam region 9–16 Oct 2000. KHRP, London

Hogenboom B (2003) Cross-border activism and its limits: Mexican environmental organizations and the United States. Cuadernos del CEDLA 13, Amsterdam

IFC (International Finance Cooperation) (2012a) IFC Performance Standards on Environmental and Social Sustainability. Effective January 1, 2012. http://www.ifc.org/wps/wcm/connect/c8f524004a73daeca09afdf998895a12/IFC_Performance_Standards.pdf?MOD=AJPERES. Accessed 06 Oct 2013
IFC (International Finance Cooperation) (2012b) IFC Sustainability Framework. Effecitve January 1, 2012. http://www.ifc.org/wps/wcm/connect/b9dacb004a73e7a8a273fff998895a12/IFC_Sustainability_+Framework.pdf?MOD=AJPERES. Accessed 06 Oct 2013
IHA (International Hydropower Association) (ed) (2010) Hydropower sustainability assessment protocol (HSAP). http://www.hydrosustainability.org/getattachment/7e212656-9d26-4ebc-96b8-1f27eaebc2ed/The-Hydropower-Sustainability-Assessment-Protocol.aspx. Accessed 06 Oct 2013
Ilisu External Monitoring Group (2009a) External Monitoring Group Report Ilisu Dam and HEPP Project. March 2009. http://www.ilisu-wasserkraftwerk.com/themes/blue_style/images/force-download.php?file=./umsiedelungen/coe/EMG_Report_March%202009.pdf. Accessed 08 Aug 2009
Ilisu External Monitoring Group (2009b) External Monitoring Group Report Ilisu Dam and HEPP Project. June 2009. http://www.ilisu-wasserkraftwerk.com/themes/blue_style/images/force-download.php?file=./umsiedelungen/coe/EMG_June_2009_Report.pdf. Accessed 08 Aug 2009
Kahler M (2009) Networked politics: agency, power and governance. In: Kahler M (ed) Networked politics: agency, power, and governance. Cornell University Press, Ithaca, pp 1–20
Keck ME, Sikkink K (1998) Activists beyond borders. Advocacy networks in international politics. Cornell University Press, Ithaca
Keck ME, Sikkink K (1999) Transnational advocacy networks in international and regional politics. Int Soc Sci J 51(159):89–101. doi:10.1111/1468-2451.00179
KHRP (Kurdish Human Rights Project) (2007) The Ilisu Dam Project. A flawed plan is revived unchanged. KHRP Briefing Paper. June 18. http://www.khrp.org/khrp-news/human-rights-documents/2007-publications/doc_download/203-the-ilisu-dam-project-a-flawed-plan-is-revived-unchanged.html. Accessed 26 Aug 2010
KHRP (Kurdish Human Rights Project), Bar Human Rights Committee, The Cornerhouse (eds) (2003) Meaningful consultation and the Ilisu Dam: the trial of human rights defenders. Trial Observation Report. KHRP, London http://www.khrp.org/khrp-news/human-rights-documents/2007-publications/doc_download/203-the-ilisu-dam-project-a-flawed-plan-is-revived-unchanged.html. Accessed 26 Aug 2010
Kibaroglu A, Scheumann W (2009) Transboundary water politics in the Euphrates-Tigris rivers system. In: Kibaroglu A, Scheumann W, Kramer A (eds) Turkey's water policy: national frameworks and international cooperation. Springer, Heidelberg
Knigge M, Görlach B, Hamada AM, Nuffort C, Kramer RA (2003) The use of environmental and social criteria in export credit agencies' practices. A study of export credit agencies' environmental guidelines with reference to the World Commission on Dams.GTZ and Ecologic (on behalf of BMZ), Eschborn
Lake DA, Wong W (2009) The politics of networks: interests, power, and human rights norms. In: Kahler M (ed) Networked politics: agency, power, and governance. Cornell University Press, Ithaca, pp 127–150
OECD (Organisation for Economic Cooperation and Development) (2007) Revised council recommendation on common approaches on the environment and officially supported export credits. http://www.olis.oecd.org/olis/2007doc.nsf/LinkTo/NT00002B8E/$FILE/JT03228987.PDF. Accessed 03 Aug 2009
Rohrschneider R, Dalton RJ (2002) A global network? transnational cooperation among environmental groups. J Polit 64(2):510–533
Ronayne M (2005) The cultural and environmental impact of large dams in Southeast Turkey: fact-finding mission report. National University of Ireland and KHRP, Galway, London. http://www.khrp.org/khrp-news/human-rights-documents/doc_download/2-the-cultural-and-

environmental-impact-of-large-dams-in-southeast-turkey-fact-finding-mission-report.html. Accessed 26 Aug 2010

SERV, OeKB, Euler Hermes (2009) Projekt Ilisu/Türkei: Exportkreditgarantien enden. Pressemitteilung. http://www.agaportal.de/pdf/presse/pm_ilisu_07072009.pdf. Accessed 10 Aug 2009

Scheumann W (2008) How global norms for large dams reach decision-makers. In: Scheumann W, Neubert S, Kipping M (eds) Water politics and development cooperation. Springer, Berlin, pp 55–80

Schneider AK (2005) From commitment to implementation. The report of the World Commission on Dams after five years. International Rivers Network, Berkeley

Schoene-Alaluf E, Bittner V, von Waldenfels P (2011) Global standards for officially supported export credits. Zeitschrift für Wirtschafts- und Unternehmensethik 12(1):123–138

Schwab T (2009) Deutschland lässt Mega-Projekt fallen. Frankfurter Rundschau, June 19. http://www.fr-online.de/top_news/?em_cnt=1803702&. Accessed 15 July 2013

Setton D, Drillisch H (2006) Zum Scheitern verurteilt. Der Ilisu-Staudamm im Südosten der Türkei. WEED, Berlin

Sikkink K (2009) The power of networks in international politics. In: Kahler M (ed) Networked politics: agency, power, and governance. Cornell University Press, Ithaca, pp 228–247

Strittmatter K (2007) Warum bringt ihr uns um? Süddeutsche Zeitung, 26 March 2007

Süddeutsche (2008) Ilisu-Projekt vor dem Aus. Süddeutsche Zeitung, 08 October 2008

The Cornerhouse, Kurdish Human Rights Project (eds) (2007) Ilisu Dam Downstream Water Impacts and Iraq. Report of Fact Finding Mission to Iraq, 29 March 2007. http://www.thecornerhouse.org.uk/sites/thecornerhouse.org.uk/files/IlisuDownstream.pdf. Accessed 24 Aug 2010

Ünver İHO (1997) Southeastern Anatolia Project (GAP). International Journal of Water Resources Development 13(4):453-484. doi:10.1080/07900629749575

Walder C (2008) Und macht euch die Erde Untertan (documentary film). ECA-Watch

WCD (World Comission on Dams) (2000) Dams and development. A new framework for decision-making. Earthscan Publications, London

World Bank (2001) World Bank Official Response to Dams & Development—a new framework for decision-making. http://www.dams.org/report/reaction/reaction_wb2.htm. Accessed 28 June 2009

Yalcin E, Tigrek S (2010) Investigation of alternative solutions to Ilisu Dam and HEPP, Ankara

Chapter 8
Interaction of Chinese Institutions with Host Governments in Dam Construction: The Bui Dam in Ghana

Oliver Hensengerth

Abstract The study analyses the role of Chinese companies and financing institutions and Ghanaian governmental agencies in planning and constructing the Bui Dam. The analysis focuses on the division of responsibilities between Sinohydro and China Exim Bank on the one side and the Ghanaian government on the other side. The findings show that environmental and resettlement planning for Bui was commissioned and financed by the Government of Ghana without Sinohydro's involvement. The obligation of the firm is to abide by the environmental regulations that are monitored by the regulatory authorities. The role of Sinohydro consists in building the dam, maintaining the construction site, contracting workers, and providing for workers' health and safety. The firm has no role in resettlement, which is carried out by the Bui Power Authority. While there is clear evidence that the Bui Power Authority does not follow the recommendations of the Resettlement Planning Framework, Sinohydro appears to abide largely by the conditions set out in the Environmental Impact Assessment study whose implementation is monitored by the Ghanaian Environmental Protection Agency and the Ghanaian Water Resources Commission.

8.1 Introduction

The launch of China's Going Out Strategy in 2003 marked the beginning of concerted and large-scale Chinese government assistance for investment by Chinese firms in developing countries in Latin America, Africa and Southeast Asia. The Organisation for Economic Co-operation and Development (OECD) world reacted negatively and fearfully to this owing to a mix of three interrelated concerns: that cheap Chinese competition would be unfair, that Chinese companies

O. Hensengerth (✉)
Department of Social Sciences and Languages, Northumbria University, Newcastle, UK
e-mail: Oliver.Hensengerth@northumbria.ac.uk

W. Scheumann and O. Hensengerth (eds.), *Evolution of Dam Policies*, 229
DOI: 10.1007/978-3-642-23403-3_8,

would eat away at markets hitherto dominated by Western companies (many of the Europeans among them being very active in their former colonies), and that Chinese companies and financiers would operate without regard for the social and environmental effects of infrastructure investment.

While the academic discourse has become highly diversified, the "fear of China" discourse has remained a favourite theme in political and non-governmental organisation (NGO) circles. However, in an apparent change of strategy, some NGOs such as International Rivers have now begun to highlight the positive effects of Chinese investment and the environmental reform processes that are going on in China and also affect Chinese companies with a sizeable foreign presence (such as Sinohydro).

Avoiding the futile question whether Chinese investment is good or bad, the present study seeks to examine the role of Chinese companies in foreign hydropower investment, focusing particularly on the division of responsibilities between companies and government agencies in the planning (technical design, environmental impact assessment and resettlement) and construction of a hydropower dam. The specific case studied is the Bui Dam in Ghana.

Examining the processes that occur in the collaboration between foreign entities (here: Sinohydro and the China Export Import Bank) and national government agencies (here: Ghana's government departments and regulatory bodies) during planning and implementation makes it possible to identify the processes over which a government retains control, the areas of responsibility delegated to the company and the mechanisms for ensuring the company's compliance with national legislation and international standards.

Data for this study were gathered during a 2-week field study in Ghana's capital, Accra, and in the Bui Dam area. Interviews were held with government agencies, NGOs, researchers, and the operators of the Akosombo and Bui Dams.

The following sections discuss the history, planning and implementation of the Bui Dam. The first section outlines Ghana's present energy situation and the Bui Dam's role in the country's energy and economic development. The second section briefly reviews the history of the Bui Dam project. This is followed by an analysis of the financing, planning and implementation processes with a particular focus on the sharing of responsibilities between the Chinese contractor, Sinohydro, and Ghanaian government agencies and on the role of participatory procedures in the environmental and resettlement planning and implementation processes.

8.2 The Role of Hydropower in Ghana's Energy/Electricity Sector

Ghana's electricity supply is highly dependent on hydropower. Ghana currently has two large hydroelectric dams: Akosombo, which was commissioned in 1965 and has a capacity of 1,012 megawatt (MW), and Kpong, which was commissioned in 1982 and has a capacity of 160 MW. Bui will be the third large

Table 8.1 Installed capacity (current and under development) in Ghana (Malgas 2008, p. 9, Table 1; Volta River Authority no year)

Plant	Type	Owner	Installed capacity in MW
Currently installed			
Akosombo	Hydro	Volta River Authority	1,020
Kpong	Hydro	Volta River Authority	160
Takoradi	Thermal	Volta River Authority	330
Takoradi II	Thermal	TAQA and Volta River Authority	220
Tema I	Thermal	Volta River Authority	110
Mines Reserve Plant	Thermal	Volta River Authority	80
Currently under development			
Bui	Hydro	Bui Power Authority	440
Takoradi Expansion Project T3	Thermal	Volta River Authority	132
Tema II	Thermal	Volta River Authority	49.5
Kpone	Thermal	Volta River Authority	220

hydroelectric dam. The Strategic National Energy Plan for the period 2005–2025 refers to four hydroelectric projects: Bui (to be built by 2012), Hemand and Juale (to be built by 2015) and Pwalugu (to be built by 2020). The Strategic National Energy Plan also proposes an increase in generation capacity at Akosombo and Kpong "by a combined 160 MW" (IWP&DC 2006). In addition, a further hydro project on the Pra River is under discussion (IWP&DC 2006).

Hydropower is supplemented by thermal power. The most important thermal power plants are the 330 MW Takoradi plant at Aboadze, which became operational in 1997, and the 220 MW Takoradi II plant run by the Takoradi International Company (TICO), which has been in operation since 2000. 90 % of Takoradi II was initially owned by the US energy company CMS Energy and 10 % by the Volta River Authority. In 2007 CMS sold its shares to Abu Dhabi's TAQA (Malgas 2008, pp. 8–9 and p. 9, note 7). The governments of Ghana and Canada have also begun work on a 132 MW plant, the Takoradi Expansion Project T3. The Volta River Authority also operates thermal plants at Tema. These include Tema I (110 MW) and the Mines Reserve Plant (80 MW), both commissioned in 2008; Tema II (49.5 MW), to be commissioned by the end of 2010; and a plant at Kpone, near Tema (220 MW), which is currently being developed (Volta River Authority s. a.). Table 8.1 summarises currently installed capacity and capacity under development.

Before the Takoradi thermal plant was commissioned, electricity generation was almost exclusively by hydropower, as Fig. 8.1 shows.

The installation of thermal power plants from 1997 must be seen against the backdrop of two severe droughts in 1983/1984 and 1998, which seriously affected hydropower generation at the Akosombo and Kpong dams (for the 1998 drought see also Sect. 8.3 below on "Planning and Construction of Bui"). In addition, electrification in Ghana is unevenly distributed, with a focus on the southern parts

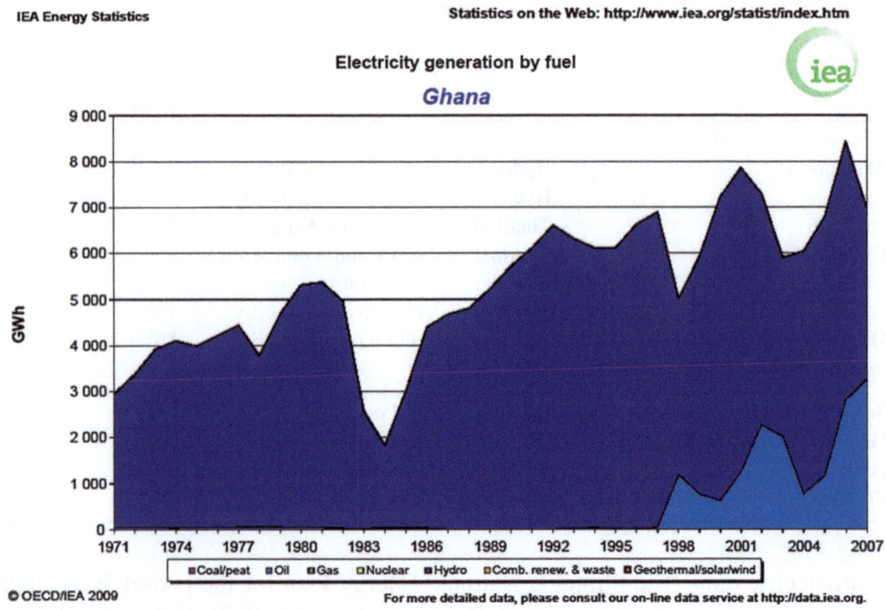

Fig. 8.1 Electricity generation by fuel (IEA 2009)

Table 8.2 Summary of the project characteristics of the Bui Dam (Bui Power Authority no year g)

Purpose	Reservoir volume	Dam height	Installed capacity	Average annual output	Number of relocatees	Beginning of construction
Multi-purpose	12.57 billion m^3	108 m	400 MW	969 GWh	1,216	December 2008 (river diversion)

of the country (IWP&DC 2008). Bui will also be equipped with a facility designed to irrigate 30,000 ha of land. Table 8.2 summarises the main project characteristics of the Bui Dam.

8.3 Planning of the Bui Dam

The history of the planning of the Bui Dam begins in 1925, when Albert Kitson discovered the Bui Gorge and found it suitable for a hydroelectric dam. For decades, Bui has seen its fortunes rise and fall. After the construction of two dams, Akosombo in the 1960s and Kpong in the 1980s, both of which are run by the Volta River Authority, energy demands continued to rise sharply and new power plants were debated. By 1978 plans for Bui had reached an advanced planning stage, with Australian and World Bank involvement. However, four coups d'état in Ghana made implementation impossible, and Bui remained dormant until 1992 (Fink 2005, p. 69).[1]

In 1992, with demand for energy surging, the project was revived and the Ghanaian government commissioned the French consulting company Coyne et Bellier to draw up a feasibility study. In 1998 droughts and the subsequent decline in water levels in the Akosombo and Kpong reservoirs caused a power crisis. The water level in the Akosomobo reservoir fell by 74 m, reducing power generation capacity from 912 MW to 300 MW (IWP&DC 2006). In response, the Ghanaian government decided to import electricity from Ivory Coast, to construct new thermal power plants in the longer term, to improve generation capacity at Akosombo and Kpong and to construct new hydropower projects (IWP&DC 2006, 2008).

In 1999 the government set up a Bui Development Committee along with a Bui Development Secretariat to look for suitable candidates for the development of the Bui project. Later that year the Committee announced that its search had been successful, and the government, represented by the Volta River Authority, signed a Memorandum of Understanding with Halliburton Brown and Root. The Volta River Authority then commissioned consulting firm BKS Acres in 2001 to conduct an Environmental Impact Assessment (EIA). This, however, was never completed. Again in 2001, the government cancelled the Memorandum of Understanding with Halliburton Brown and Root "for reasons that remain unclear" (Fink 2005, pp. 71–72). In the same year, the government cancelled plans for Bui and dissolved the Bui Development Committee, announcing that it would be less expensive to opt for such thermal power sources as natural gas.

Then, in 2002, the government set up a new Bui Development Committee, and the search began anew with the issue of an international call for tenders for the preparation and implementation of the Bui project. However, only one company submitted a bid. As a consequence of the unsuccessful tendering procedure, President Kufuor then asked China and Russia for assistance. In 2004 Ghana's Energy Commission published the Strategic National Energy Plan for the period 2005–2025, which stated that an "investment decision for Bui hydro project will be taken before the end of 2006" and that construction would take 5 years.

In 2005 the Energy Commission suggested that the government should invest US$ 2–3 million in a feasibility study "which would make the project bankable and enable BDS [Bui Development Secretariat] to obtain firm commitment from developers for the project" (Fink 2005, p. 71). Subsequently, Coyne et Bellier updated their earlier feasibility study. In 2005 the National Development Planning Commission approved Bui "as compliant with overall Ghanaian development policies." Later that year the Ministry of Energy commissioned UK consulting firm Environmental Resources Management (ERM) to undertake a new EIA.

---

[1] Unless otherwise indicated, this historical account follows the PhD thesis of Michael Fink (Fink 2005, pp. 69–72).

Ghana's Environmental Protection Agency commented that a new scoping study would be necessary before the actual EIA process could be started (Fink 2005, p. 71).

Even later in 2005 the government announced that the Chinese government had expressed willingness to extend funding for Bui and that Sinohydro had agreed to build the dam. The World Bank reported that the Ghanaian government had "accepted an unsolicited bid from the Sinohydro Corporation of China to finance and construct" the Bui hydroelectric scheme (World Bank 2008, p. 108). The Ministry of Energy and Sinohydro then signed a Memorandum of Understanding on the construction of Bui; funding was expected to come from the China Exim Bank (Fink 2005, p. 71). During the 2006 summit of the Forum on China-Africa Cooperation in Beijing, Prime Minister Hu Jintao announced the Chinese government's commitment to fund the Bui Dam. In 2007 the Ghanaian parliament passed Act 740 establishing the Bui Power Authority and dissolving the Bui Development Committee and Bui Development Secretariat. The Bui Power Authority set to work and was given full responsibility for planning, executing and managing the Bui project (Zigah 2009, p. 25). As a result, the Volta River Authority lost its responsibility for the Bui Dam and was reduced to overseeing the Akosombo and Kpong Dams and managing the Volta Lake. Henceforth, the Bui Power Authority was to be the Bui project owner.

Since the Ghanaian government managed to obtain an agreement in principle on construction and financing from Chinese agencies in 2005 and 2006, the planning and construction of the Bui Dam have been governed by five key documents. Three of them govern the relationship between Ghanaian and Chinese organisations:

- the 2005 Memorandum of Understanding (MoU) between the Ministry of Energy and Sinohydro,
- the 2007 Engineering, Procurement and Construction (EPC) turnkey contract between the Ministry of Energy and Sinohydro and
- the 2007 and 2008 loan agreements between the Ministry of Finance and Economic Planning and the China Exim Bank.

Two further documents, contracted by Ghanaian institutions, guide the governmental decision-making process and also contain norms and procedures with which Sinohydro as contractor has to comply:

- the 2006 feasibility study by French consulting company Coyne et Bellier and
- the 2007 Environmental and Social Impact Assessment by UK-based consulting firm Environmental Resources Management.

Table 8.3 reviews the main steps in the planning and construction of Bui.

Table 8.3 Timeline of Bui Dam reports and developments (author's own compilation based on ERM 2007b; Bui Power Authority no year a, b, c, d, e; Ghana News Agency 2007; Fink 2005)

Year	Events/Reports	Actors/Report authors
1925	"The possibility of Bui Gorge as the site of hydroelectric station", in Gold Coast Geological Survey Bulletin No. 1	Albert E. Kitson
1964	Bui Hydroelectric Station on the Black Volta River, Republic of Ghana (Moscow: USSR State Committee on Power Engineering and Electrification)	S. J. Zhuk Hydroproject (Moscow)
1966	Bui Resettlement Study: A Report for the Ministry of Fuel and Power (Kumasi: Faculty of Architecture, University of Science and Technology)	D. Butcher and L. Huszar
1976	Bui Hydroelectric Project, Feasibility Study	Snowy Mountains Engineering Corporation
1985	Ghana Generation Planning Study	BKS Acres International Limited
1995	Bui Hydroelectric Development Feasibility Study Update	Coyne et Bellier
1999	Establishment of first Bui Development Committee along with Bui Development Secretariat	Government of Ghana
1999	Memorandum of Understanding	Halliburton Brown and Root
2001	Bui Hydroelectric Power Project, Environmental Impact Assessment, Scoping Report	BKS Acres and Volta River Authority
2001, November	Constitution of second Bui Development Committee	Government of Ghana
2005, October	Memorandum of Understanding	Ministry of Energy and Sinohydro
2005	Inclusion of Bui in the 2006–2009 planning cycle	National Development Planning Commission
2005, December	Contract for EIA study concluded with Environmental Resources Management	Ministry of Energy
2006, October	Bui Hydroelectric Power Project Feasibility Study Update	Coyne et Bellier
2007, January	Environmental and Social Impact Assessment of the Bui Hydropower Project: Final Report	Environmental Resources Management and Bui Development Secretariat
2007, February	Letter of approval of ESIA study to Bui Development Secretariat	Environmental Protection Agency
2007, April 19	EPC contract signed	Ministry of Energy and Sinohydro
2007, July 30	Approval of US$ 622 million financing package	Parliament

(continued)

Table 8.3 (continued)

Year	Events/Reports	Actors/Report authors
2007, August	Bui Power Authority created, Bui Development Committee and Bui Development Secretariat dissolved. Volta River Authority loses responsibility for the Bui project	Government of Ghana
2007, October	Beginning of field investigations for construction at dam site	Sinohydro
2007, November	Construction of basic infrastructure (roads, camps etc.)	Sinohydro and Bui Power Authority
2008, May	Beginning of Resettlement Phase A	Bui Power Authority
2008, September 4	Concessional loan agreement signed	Ministry of Finance and Economic Planning and Ministry of Commerce through China Exim Bank
2008, December	River closure and river diversion with diversion channels and coffer dams completed	Sinohydro
2009, January–November	River bed excavation and foundations	Sinohydro
2009, December	Beginning of construction of main dam	Sinohydro
2010, June	Beginning of Resettlement Phase B	Bui Power Authority

8.4 Project Planning and Clearance Procedures

Step One: Inclusion of Bui in National Development Planning

The planning of dams officially begins at the National Development Planning Commission. The Commission includes hydropower projects in its planning cycle by identifying demand and supply gaps in the energy sector. So far, the National Development Planning Commission has been responsible for three planning cycles: from 2003–2005 (Medium-Term Development Plan under the Growth and Poverty Reduction Strategy, or GPRS I), 2006–2009 (Medium-Term Development Plan under the Growth and Poverty Reduction Strategy, or GPRS II) and the current Medium-Term Development Plan for the period 2010–2013. The Commission included Bui in the 2006–2009 planning cycle.[2]

The identification of a demand-supply gap in the energy sector is followed by the formulation of a policy for addressing it. If it is decided to build a dam, then the policy must indicate why a dam is necessary and how much energy it is to

[2] Interview A02072010.

generate. There can be no dam construction without that specific justification. In Bui's case, the Commission decided that the dam should be built to increase power generation and so overcome power shortages.[3]

However, it seems that the question was not if Bui should be built to redress Ghana's energy shortages. After decades of unsuccessful attempts to fund the dam and temporary decisions in favour of thermal power, the government of President John Agyekum Kufuor (2001–2009) again prioritised the project. Therefore, the National Development Planning Commission's decision to include Bui in its planning cycle was based on a dominant government policy that has been continued by current President John Atta-Mills (since 2009).

The National Development Planning Commission monitors adherence to the plan. Once included in the planning cycle, the dam—like any other project endorsed by the Commission—is followed up by annual monitoring and evaluation reports. The focus of these activities is on the implementation of planned activities, with particular importance attached to the timetable. Other problems concerning delays in wage payments or service delivery, for example, are not the Commission's responsibility.[4]

Step Two: Environmental Clearance

The first regulatory hurdle the Bui Power Authority as project owner had to overcome was to obtain the Environmental Permit required by the Environmental Protection Agency. Before an Environmental Permit can be issued, an Environmental and Social Impact Assessment (ESIA) must be carried out. The ESIA was commissioned and paid for by the Government of Ghana on behalf of the Bui Development Secretariat (the predecessor of the Bui Power Authority), because the Environmental Protection Agency requires the Environmental Permit to be issued before the loan agreement with the China Exim Bank can be signed.[5] This conforms to the Exim Bank's *Guidelines for Environmental and Social Impact Assessments of the China Export Import Bank's Loan Projects* of August 2007. Article 13 of the *Guidelines* requires the project owner to submit government approval documents, including environmental and social impact assessments, to the Bank before the loan agreement can be signed. The environmental approval documents then become part of the loan application that is reviewed by the China Exim Bank (China Exim Bank 2007). This is indeed what happened, as the loan procedure above has shown. It was claimed that this procedure, i. e. environmental clearance before the loan agreement, is also used for World Bank-financed projects[6] and that the Exim Bank would not have given its approval for the loan if the Environmental Protection Agency had not issued the Environmental Permit.[7]

3 Interview A02072010.

4 Interview A02072010.

5 Interview A21062010-3.

6 Interview A21062010-3.

7 Interview A30062010-2.

It was in fact confirmed in an interview that the Environmental Permit was issued to the Bui Power Authority during the loan negotiations, but the loan agreements were signed only after the Environmental Permit had been issued.[8]

Step Three: Licensing and Permitting Process

When the Environmental Permit is issued, the Energy Commission convenes a Siting Clearance Committee to approve the dam site.[9] In Bui's case, this process appears to have been nothing more than a formality, given that the planning process for Bui has essentially been in place since the 1960s, when the Bui Gorge was certified as a site for power generation.[10] Consequently, even though the Environmental Protection Agency is not yet represented on the Siting Clearance Committee, it probably would not have made any difference to the outcome of the Siting Clearance process.[11]

After the Energy Commission had approved the site, the permit and licensing process continued. As with the Environmental Permit, it was the Bui Power Authority which had to apply for all the permits and licences described in the following, and it is responsible for monitoring Sinohydro's compliance with the permit and licence conditions.

To begin the river works, the Bui Power Authority had to apply to the Water Resources Commission for a permit, which is needed by anyone wanting to use, divert, store or abstract water. It issues three permits: Diversion Permits, Construction Permits and Water Use Permits. This means that, when the project proponent wishes to begin building the dam, the river must be blocked and diverted. Before the river can be diverted, a Diversion Permit is needed. Once actual construction is due to begin, the project proponent needs a Construction Permit. When the dam is completed, but before it is commissioned, the project proponent needs a Water Use Permit. Project proponents may apply for all three permits at the same time, or for one after the other as required while the building of the dam proceeds.[12]

The Water Resources Commission is normally invited by the Environmental Protection Agency to form part of the technical review team for the ESIA study, but this did not happen in the case of Bui.[13] Furthermore,

> *the government sometimes jumps steps. They may inform the Water Resources Commission that certain works have began, but they do not wait for the Commission's approval. [...] For Bui, the river had already begun to be diverted when the Bui Power Authority applied for the Diversion Permit. What's more, parliament had already approved the EPC contract with Sinohydro and the loan agreement with China Exim Bank, and so there was*

[8] Interview A14072010.

[9] Interview A14072010 and Energy Commission (no year).

[10] Interview A14072010.

[11] Interview A14072010.

[12] Interviews A23062010-3 and A29062010.

[13] Interviews A23062010-3 and A29062010.

nothing anymore that we could do. This is a government project, and negotiations are between two governments. The plan was already there and so the permits couldn't be refused. Therefore for Bui, the Commission addressed its concerns in the Permit conditions.[14]

Interestingly, the contention that Ghana's regulatory authorities cannot refuse permits was strongly confirmed in an interview with a government agency official. The interviewee claimed that,

had it not been a government project, the dam probably would not even have gotten the Environmental Permit. There were concerns, particularly related to the Bui National Park and the hippopotamus population. [...] Also, there was the issue of the transboundary nature of the river: although the medium-size wall height was chosen from the various dam designs, there is still danger that when it rains a lot the river might flood into Ivory Coast. Had it been a private project, the government probably would not have allowed it. In Ghana it is difficult to regulate projects completely.[15]

Since permits for Bui could not be refused, they were issued with mitigation requirements. The permit conditions applied by the Water Resources Commission concern water availability, flooding, regulated flows downstream or water pollution during construction. For Bui, the Diversion Permit required:

- a specific daily downstream flow to be recorded and reported daily;
- the relocation of the dam a short distance from the original site to prevent chips of rocks brought in for the construction of the dam and then broken up from falling into the water and so making the channel too shallow (this condition was also included in the Construction Permit);
- mitigation measures for sedimentation;
- an assurance that there would be no upstream flooding during rainfall;
- the construction of a diversion channel the same size as the original river.[16]

As procedures now stand, the Water Resources Commission becomes involved after the Environmental Protection Agency issues the Environmental Permit and when the project proponent wants to begin diverting the river. This will change once the National Dams Safety Unit is installed. This Unit is designed to ensure that dams operate to certain standards, the aim being to prevent dam breaks, for example. As it will have the authority to examine dam design, it will oversee the planning process from the design stage onwards. The National Dams Safety Unit will include representatives of all the major agencies concerned with dam planning: the Bui Power Authority, the Volta River Authority, the Ghana Water Company, the Ghana Irrigation Development Authority, the Water Resources Commission, the Environmental Protection Agency and the Ministries of Justice,

---

[14] Interviews A23062010-3 and A29062010.

[15] Interview A29062010. On a related topic see also the section on resettlement below for information on the non-implementation of resettlement recommendations contained in the ESIA study and the Resettlement Planning Framework.

[16] Interviews A23062010-3 and A29062010.

Local Government and Water Resources, Works and Housing. However, the relevant legislation has not yet been passed.[17]

As for monitoring, the Water Resources Commission has limited capacity and is unable to lay down a tight timetable that would allow frequent visits to the dam site. While the Commission has confirmed that the Bui Power Authority and Sinohydro seem to be observing the permit conditions whenever Water Resources Commission staff make unannounced site visits, it is far from certain that that is the case between visits. Furthermore, the Water Resources Commission has been in operation for only a few years and is not yet fully staffed.[18]

The Energy Commission, Ghana's regulatory body for the energy industry, then steps in. Apart from being responsible for the Siting Clearance process, it issues licences for the operation of energy facilities, including the wholesale supply licence for the installation and operation of energy facilities to produce electricity (Energy Comission 2006, p. 12).

To approve the licence application for Bui, the Energy Commission needed an Environmental Permit from the Environmental Protection Agency, Diversion and Construction Permits from the Water Resources Commission (it is unclear if the Energy Commission also needs a Water Use Permit, which has not yet been issued, or if this step can be conveniently skipped), the MoU between the China Exim Bank and the Ministry of Finance and Economic Planning, the EPC contract between Sinohydro and the Ministry of Energy and the Power Purchase Agreement between the Ministry of Energy and the Electricity Company of Ghana. According to the Energy Commission, it has not yet issued licences for Bui, but they are ready and will be issued once the Bui Power Authority pays the licence fees. Licences are granted for a maximum of 25 years. They can be revoked and are issued under conditions which are monitored by the Energy Commission.[19]

For monitoring, the Energy Commission's Inspectorate liaises with the relevant departments of the Energy Commission. When a large hydroelectric dam is to be monitored, it cooperates with the Power Division. It was claimed by one interviewee that, because of this separation of responsibilities, the Energy Commission does not regard large hydropower stations as generating renewable energy.[20] However, the author could not confirm whether this view is generally held in the Energy Commission or in the Ghanaian government. Asked what thresholds are applied to determine what qualifies as a large dam, the Volta River Authority responded that it applied the definition of the International Commission of Large Dams (ICOLD), which defines a large dam as being 15 m or higher or between 10 and 15 m if it meets at least one of the following conditions: a crest length of not

[17] Interviews A23062010-3 and A29062010.

[18] Interviews A23062010-3 and A29062010.

[19] Interview A24062010-3.

[20] Interview A24062010-2.

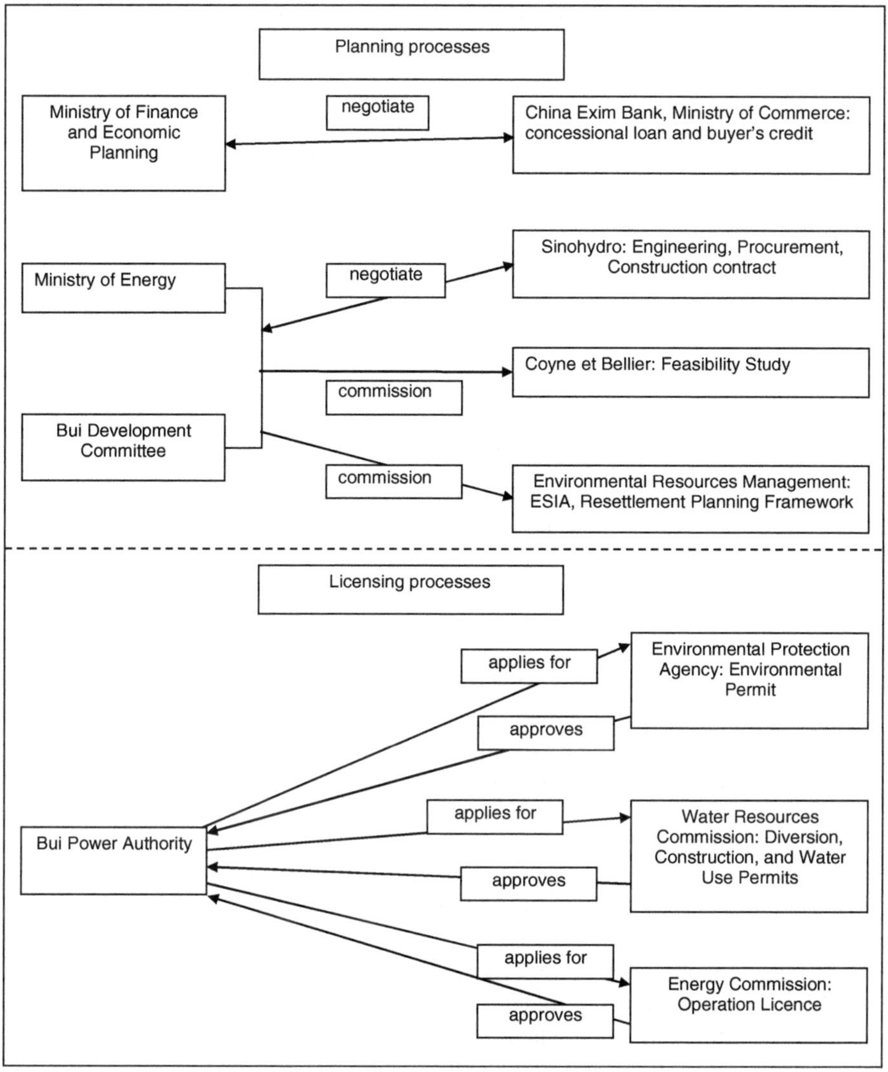

Fig. 8.2 Planning and licensing processes (author's own compilation)

less than 500 m; a spillway discharge potential of at least 2,000 m^3 per second; and a reservoir volume of not less than one million cubic meters.[21]

All licences must be granted before construction may begin.[22] Figure 8.2 summarises the planning and licensing processes.

[21] Personal communication, 26 July 2010.

[22] Interview A24062010-3.

8.5 Ghanaian Environmental Impact Assessment Regulations

8.5.1 Role of International Norms in Ghana's Environmental Impact Assessment Process

Government discourse in Ghana on environmental issues began with the 1972 Stockholm Conference on the Human Environment. In 1973 the government set up the Environmental Protection Council, which advised on how to improve environmental protection. However, it took until 1985 for the Council to set up a committee to examine ways of introducing an EIA system in Ghana. In 1989 the Council established the National Environmental Action Plan with the aim of forming an EIA regime for all new projects that might harm the environment (Appiah-Opoku 2001, pp. 60–61).

Around the same time, a series of environmental hazards attracted public and media attention. They consisted specifically in epidemics following the toxic pollution of rivers by gold-mining operations in the Ashanti Region. In the Western Region, villagers took up arms and threatened to take action against polluting mining operations near their village, which had destroyed crops and caused houses to collapse following the use of dynamite (Ofori 1991 and Graham 1993 quoted in Appiah-Opoku 2001, p. 61).

These developments eventually led to the 1994 Environmental Protection Agency Act, which set up the Environmental Protection Agency in succession to the Environmental Protection Council (Appiah-Opoku 2001, p. 61). The Environmental Protection Agency was tasked with the development of an EIA regime, which came into effect in 1999 as the *Environmental Assessment Regulations*. These *Regulations* have since formed the legal basis of environmental planning in Ghana.

The *Regulations* specify two main project schedules: Schedule 1 projects must be registered with the Environmental Protection Agency and an Environmental Permit must be obtained before construction, but an ESIA is not mandatory. Schedule 2 projects are required to undergo an ESIA process before the Environmental Permit can be granted. Schedule 2 covers "power generation and transmission" projects, which specifically include "dams and hydroelectric power schemes" (Environmental Protection Agency 1999). Whether these actually need to undergo ESIA depends, however, on the size of the dam. A medium-sized hydroelectric dam, for example, may need no more than a preliminary ESIA. It is "at the discretion of the Environmental Protection Agency" to make this categorisation.[23]

[23] Interview A21062010-3. At the time of writing (July 2010) there was a draft EIA guideline on the energy sector with specific reference to dams. The final version was said to be published before September. The draft guideline had been reviewed by the International Association for Impact Assessment. Its comments were being incorporated at the time of writing (interview A21062010-3).

The Environmental Protection Agency is also the main authority for setting environmental protection standards in Ghana.[24] These standards are established "in conformity with international standards."[25] This was confirmed by a Ghanaian regulatory authority official, who claimed that standards set by the Environmental Protection Agency are based on international standards, particularly those developed by the World Bank, and standards of the International Organisation for Standardisation (ISO).[26] Water resources engineers confirmed that the technical standards they apply in their work include those of funding agencies, e.g. the World Bank and International Finance Corporation (IFC), plus those of ICOLD, the International Hydropower Association, and ISO 14001.[27]

The Bui ESIA study specifically refers to World Bank and Ghanaian regulations:

> *Ghanaian Environmental Impact Assessment Procedures require that a project proponent prepares an environmental impact statement that presents a clear assessment of the impacts of the proposed undertaking on the environment, and set out four steps to be followed in the conduct of an Environmental Impact Assessment: Project Registration, Project Screening, Scoping Report, and Environmental Impact Statement. [...] In addition, the ESIA of the Bui Hydroelectric Power Project has been carried out in accordance with international standards, as reflected in the policies, safeguard procedures, and guidance of the World Bank Group* (ERM 2007a, p. i).

As regards public consultation, the ESIA states that it observes Ghanaian regulations and the consultation guidelines of "international financing institutions, specifically the World Bank Group". Ghanaian regulations contained in the 1999 *Environmental Impact Assessment Regulations* are:

- First, "advertisement of the scoping notice": to "relevant ministries"; "in at least one national and one local newspaper if there is any"; and "copies of the scoping report available for inspection by the general public in the locality of the proposed project".
- Second, the "Environmental Impact Statement must be made available to the general public, relevant public agencies, organisations, NGOs, metropolitan, Municipal and District Assemblies and local communities [... so that] they may make any comments and suggestions on any matters therein".
- Third, the "Environmental Protection Agency must conduct a public hearing if there appears to be an adverse public reaction to the project; the undertaking will involve resettlement; the undertaking could have extensive effects on the environment" (ERM 2007a, pp. 149–150).

Writing in 2001, Appiah-Opoku (2001, p. 65) comments that a "lack of public input" is a constraint on the realisation of environmental regulation in Ghana. We

---

24 Personal communication with Ghanaian water resources engineers in July 2010.

25 Personal communication with Ghanaian water resources engineers in July 2010.

26 Interviews A23062010-3 and A29062010.

27 Personal communication with Ghanaian water resources engineers in July 2010.

will consider throughout the study whether this is still relevant or whether the 10 years or so that have passed since then have led to proper account being taken of affected people's concerns in the ESIA procedure.

8.5.2 Status of Social and Environmental Impact Assessment in the Project Cycle

Within the project cycle, ESIA forms part of the feasibility study. In Bui's case, however, this was not so, because Coyne et Bellier simply had to update the 1976 and 1995 feasibility study reports.[28] This means that the feasibility study and ESIA were separate processes. The company tasked with the ESIA was the UK-based environmental consulting firm Environmental Resources Management. It submitted the ESIA study in January 2007.

Specifically, the environmental planning procedure follows the steps outlined below:

First, project proponents must register their dam projects with the Environmental Protection Agency by completing an EIA registration form. In the case of Bui, the proponent is the Bui Power Authority. Sinohydro is only the contractor.[29]

Second, the Environmental Protection Agency requests a scoping exercise, which includes preliminary consultations with interested and affected parties. The scoping report with the draft Terms of Reference (ToR) was submitted by the Bui Power Authority to the Environmental Protection Agency.[30] The ToR called for the development of a Resettlement Planning Framework (RPF) and a Monitoring Plan, for a comprehensive assessment of impacts (physical, biological, social, public health, cultural, cumulative and international), for the discussion of alternatives, for the establishment of a Mitigation and Environmental Management Plan and for public consultations (ERM 2007b, pp. A3–A17).

The ToR also identified stakeholders to be interviewed and the methodology to be applied in the ESIA study. It appears that Environmental Resources Management did not hold public consultations for the scoping phase since they had already been held by BKS Acres. The introduction to the ToR states specifically that the ToR are "essentially the same" as those drawn up by BKS Acres for their Scoping Report, which was drafted in 2001 but never submitted (ERM 2007b, p. A1). The 2007 ESIA is therefore largely based on the 2001 Scoping Report.

Third, the Environmental Protection Agency set up a technical review committee, that is, a special cross-sectoral team made up of representatives of—among others—the Wildlife Division of the Forestry Commission, the Hydropower Division of the Ministry of Works and Housing and the Environmental Protection

[28] Interview A21062010-3.

[29] Interview A21062010-3.

[30] Interview A21062010-3.

Agency itself. This committee was asked to consider whether consultations for the scoping report had been conducted appropriately and would be adequate for the ESIA. The review committee was also asked to determine whether the impacts of the dam on wildlife and the Bui National Park were fully understood. Importantly, one interviewee explained that the scoping study needed to focus on certain key points: the impact on downstream communities and downstream flows to the Akosombo and Kpong Dams, resettlement issues and other impacts on such things as public health. When the review committee found the scoping study acceptable, the ESIA study was allowed to go ahead.[31]

The ESIA study was required to present comprehensive baseline data on social conditions, natural resources, cultural and economic factors and the land tenure situation. The project description was required to include such details as the design, construction schedule, location of facilities and material requirements and an assessment of alternatives. In Bui's case, there was a reference to earlier studies of the effects of differences in dam height on power capacity, the communities and the Bui National Park and of the extension of the reservoir into Ivory Coast. Eventually, the decision was made to opt for medium wall height to prevent the reservoir from extending into Ivory Coast. The study had to provide a plan for mitigating and monitoring significant impacts. A provisional environmental management plan was also required, as was a decommissioning proposal. Finally, the ESIA study had to specify levels of consultation, who was consulted and how, and if and how the comments were considered in the report.[32]

In compliance with the public participation requirements, hearings were held in Accra in April 2006 and in Bamboi, Banda Nkwanta, Banda Ahenkro and Bungasi in August 2006. The Accra meeting was organised by the Bui Development Committee and the Environmental Protection Agency, the local hearings by the local offices of the Environmental Protection Agency with the assistance of Environmental Resources Management, and the persons invited to the August meetings were selected by the local offices of the Environmental Protection Agency (ERM 2007a, p. 151; ERM 2007c, p. 112). Key themes of the consultation rounds and the concerns expressed by the people attending are summarised in the ESIA. They relate mainly to compensation and losses of livelihood (ERM 2007a, pp. 151–156). On the question of how these comments were incorporated in the ESIA study, the latter states that:

"All comments raised during the consultation carried out during preparation of this ESIA have been fully taken into account in preparing the Environmental and Social Management Plan and the Resettlement Planning Framework that have been prepared as companion volumes to this ESIA, and issued in January 2007" (ERM 2007a, p. 156).

We will return to this point when discussing resettlement below.

---

[31] Interview A21062010-3.

[32] Interview A21062010-3.

The fact that none of the hearings were held at the dam site drew criticism from NGOs, given that current road conditions make it difficult to travel from the villages and that the meeting held in Accra was of no interest to and beyond the control of affected people.[33] The fact that people from neighbouring villages not affected by the dam were allowed to attend the local meetings was also criticised on the ground that the involvement of non-affected people might dilute the strength of criticism voiced by the affected communities and might be used by the project proponents as an excuse for ignoring affected peoples' concerns.[34] However, the results of hearings are not legally binding and hearings were "rather like public forums: everybody in the project area is allowed to attend."[35] Interestingly, the national-level hearing in Accra was attended by 121 individuals, among them representatives of Sinohydro, government agencies, local chapters of international NGOs, academia and the media (ERM 2007b, Annex N).

The contents of the ESIA submitted in January 2007 were then evaluated by the Environmental Protection Agency. It began by setting up a technical review committee. Simultaneously, the report was opened to public inspection for 21 working days in the Agency's library, on its website, at its regional offices in Northern and Brong Ahafo regions, in the affected districts of Tain and Bole and at other, similar locations. The availability of the report was advertised in national newspapers.

As mentioned above and as pointed out by one of the interviewees, the 1999 *Environmental Impact Assessment Regulations* require hearings to be held on ESIAs in three cases: first, if the project entails the resettlement and relocation of peasants; second, where there are serious environmental concerns (this being determined at the discretion of the Environmental Protection Agency); third, if the project arouses intense public concern.[36] The Bui project met all three requirements.[37]

For the hearings on the ESIA report, the Environmental Protection Agency put together an independent panel consisting of the presiding members of the district assembly, a member of the Ghana Institute of Engineers and a member of the Wildlife Division of the Forestry Commission. First, it was the panel's responsibility to have the Bui Power Authority, as project proponent, present the project in the local language, explaining what it was planning to do and how, what the impact on communities and the environment would be and how it intended to manage this. Second, the panel asked the general public to voice their concerns. Questions were collected (four to five at a time) and then answered by the proponent. The panel then summarised the issues and presented them to the Environmental Protection Agency as input into the technical review.[38]

[33] Interviews A22062010-1 and A22062010-2.

[34] Interviews A22062010-1 and A22062010-2.

[35] Interview A21062010-3.

[36] Interview A21062010-3 and ERM (2007a, pp. 149–150).

[37] Interview A21062010-3.

[38] Interview A21062010-3.

Next, the Environmental Protection Agency sent the technical and public review comments to the Bui Power Authority and asked it to have Environmental Resources Management revise the ESIA study: as regards the impact on the Bui National Park, the Environmental Protection Agency stressed the need to compensate for the loss of 21 % of the Park; it also required the Bui Power Authority to add a management plan for the Park and a rescue plan for the hippopotamus population and to detail how resettlement was to be managed and what impact Bui would have on the Akosombo Dam downstream.[39]

In line with these comments, the Bui Power Authority revised the study and returned it to the technical review committee. The committee then accepted the revised ESIA study, and the Environmental Protection Agency issued the Environmental Permit to the Bui Power Authority, on the basis of predictions of the environmental effects outlined in the revised ESIA study. The Environmental Permit was issued subject to a number of conditions to be observed by the Bui Power Authority:

- the submission of a schedule of reports on water quality, resettlement, etc.;
- the submission of an annual environmental report to the Environmental Protection Agency;
- the presentation of an Environmental Management Plan within 18 months of the commencement of construction (the interviewee remarked that this time had come for Bui, and that he therefore had to check whether the Environmental Management Plan had already arrived at the Environmental Protection Agency);
- the fulfilment of the commitments outlined by the Bui Power Authority in the ESIA report, regarding, for example, the management plan for Bui National Park;
- the reporting of changes (to construction, flow regime, etc.);
- the beginning of construction within 18 months of the Environmental Permit being issued. The Environmental Permit expires if construction does not begin within 18 months of its issue;
- an Environmental Certificate must then be obtained; it replaces the Environmental Permit and is based on actual performance. It must be renewed and the Environmental Management Plan must be revised every 3 years.[40]

To monitor compliance, the regional offices of the Environmental Protection Agency visit the site.[41]

[39] Interview A21062010-3.

[40] Interview A21062010-3.

[41] Interview A21062010-3.

8.6 Ghanaian Regulations on Compensation and Resettlement

Resettlement is carried out under a Resettlement Action Plan, which is based on the RPF submitted by Environmental Resources Management along with the ESIA study in January 2007.[42] According to one interviewee, the Resettlement Action Plan was drawn up by a Ghanaian agency contracted by the Bui Power Authority.[43]

While the Resettlement Action Plan is not accessible, the RPF is. From this and the situation of the Gyama Resettlement Site—the only resettlement site so far—this section will draw some conclusions regarding the implementation of recommendations for public participation and livelihood reconstruction spelt out in the RPF.

A total of eight villages have to be resettled. Four have so far been relocated. Table 8.4 shows the current state of resettlement.

Numbers of affected communities given in the ESIA study differ significantly and are more detailed (see ERM 2007a, p. 52, Table 4.2). The first difference concerns the numbers of people and households in villages that will be inundated. Here, the table compiled by Environmental Resources Management shows 1,710 people to be resettled (Bui Camp is not included) as against 1,216 people in the table compiled by the Bui Power Authority. The second difference is that, while the Bui Power Authority table lists only villages that have to be resettled, the table by Environmental Resources Management shows in greater detail how villages that will not be resettled are affected: some will lose some of their farmland; others are situated along proposed transmission lines and therefore will at least be affected by the noise and air pollution due to the erection of those lines; thirdly, some villages will be affected by changes in the flow regime of the Black Volta River due to upstream electricity generation and will therefore be subject to a flow that depends on the Bui Dam operation schedule.

Resettlement is divided into three phases: A, B and C. During Phase A, which began in May 2008, four dam site villages were resettled: the villages of Brewohodi, Agbegikuro, Dam Site and Lucene were relocated to Gyama Resettlement Township. Phase B began in June 2010. The inhabitants of Bui Village, Bator Akanyakrom and Dokokyina are being resettled at a location between Bongase and the dam site (Bui Power Authority (no year a and b)). Phase C has not yet commenced (at the time of writing).

[42] Interview A08072010.

[43] Interview A14072010.

Table 8.4 Bui resettlement as at June 2010 (Bui Power Authority 2010)

Affected communities	Number of households	Number of people	Status
Brewohodi	10	48	Resettled
Dam Site	6	36	Resettled
Agbegikuro	22	107	Resettled
Lucene	4	26	Resettled
Bui Village	42	297	Ongoing
Bator Akanyakrom	63	437	Ongoing
Dokokyina	36	165	Ongoing
Bui National Camp	36	100	Not yet begun
Total	219	1216	

8.6.1 Legal Basis and Planning Recommendations

To identify compensation and resettlement procedures for Bui, the RPF uses World Bank guidelines and Ghanaian law as points of reference: the World Bank Operational Policy on Involuntary Resettlement (O.P. 4.12 and its Annex A), the World Bank Involuntary Resettlement Sourcebook and Ghana's 1992 Constitution, 1986 Land Title Registration Act, 1962 State Lands Act, 1962 Administration of Lands Act, 1960 State Property and Contracts Act and 1965 Public Conveyancing Act. Of these, the State Lands Act is the main body of legislation governing compulsory land acquisition. It requires the authority wishing to acquire land to apply to the Regional Minister, who refers the claim to the Site Advisory Committee, which includes the District Chief Executive and representatives of the Ministries of Lands, Forestry and Mines, Health and Water Resources, Works and Housing, the Town and Country Planning Department, the Ghana Water Company, the Electricity Company of Ghana, the acquiring authority and the Lands Commission. The Site Advisory Committee makes a recommendation to the Regional Minister. The Regional Minister then refers the case to the Lands Commission, which prepares an Executive Instrument for acquisition and sends it to the Ministry of Lands, Forestry and Mines for execution. There the Executive Instrument is signed, and the acquisition is publicised in newspapers and the government Gazette. The acquiring authority then applies for a lease, while the former owners can submit claims for compensation to the Land Valuation Board at the Lands Commission within six months of the publication of the Executive Instrument. The Land Valuation Board determines and pays compensation and resolves conflicting claims. Compensation is also governed by the State Lands Act and by the State Property and Contracts Act: it is paid for land at the market value, for damages suffered as a result of the acquisition of the land and for reasonable expenses caused by relocation elsewhere. According to the State Lands Act, the Land Valuation Board or an agency designated by the President is responsible for relocating those who need to be resettled on suitable alternative land that supports the economic well-being and the social and cultural values of the community (ERM 2007c, pp. 56–61).

As Environmental Resources Management applied World Bank standards and Ghanaian law as guidelines for establishing a proper resettlement procedure, the RPF identifies disparities between the World Bank standards and the Ghanaian legislation. Many of these have to do with a lack of clarity of Ghanaian law regarding eligibility for compensation, supplementary assistance for vulnerable individuals, the basis of evaluation of assets or timing and location selection (ERM 2007c, pp. 62–79). However, the most striking difference is that Ghanaian legislation does not require public participation in expropriation or the development of resettlement sites; however, Ghanaian ESIA regulations require public consultations during which resettlement issues may be raised (ERM 2007c, p. 67 and 70). Minutes of local consultations held during the ESIA phase and lists of stakeholders included in the ESIA study are annexed to the RPF (ERM 2007c, Annex E and F).

To reconcile these stark differences, the RPF recommends public participation:

> *Public participation in the process of land acquisition and proposed resettlement must be promoted. Procedures or guidelines for such public consultation ought also to be clearly spelt out* (ERM 2007c, p. 78).

For the development of resettlement areas, the RPF suggests that:

> *New resettlement sites or host communities must be provided with basic infrastructure and public services to ensure that basic levels of amenities are accessible to the displaced persons and host communities. The provisions of the 1992 Constitution which requires that resettlement sites must be "suitable" having due regard for their economic well-being and social and cultural values should be adhered to* (ERM 2007c, p. 79).

According to the World Bank Involuntary Resettlement Sourcebook, the following criteria should be observed in the selection of resettlement sites: they should be "acceptable to the resettlers; have the capacity to support the incomes and living standards of the people to be resettled; provide for population growth; supply infrastructure and services better than, or at least similar to those available before resettlement; and be incorporated into the jurisdiction of local government agencies before completion of the resettlement programme" (ERM 2007c, p. 101 and Box 10.1). In addition, according to Ghanaian and World Bank standards, resettlement villages must be equivalent to or better than the original settlements (ERM 2007c, p. 101).

The RPF identifies three groups of affected persons eligible for compensation:

- Group 1, households requiring resettlement are entitled to full rehabilitation measures based on the assets and crop inventory;
- Group 2, households that lose land but do not need to be resettled will receive compensation for land (should they own it), crops, trees, grazing and forest products. Households that lose more than 20 % of their land will be captured by a Livelihood Enhancement Programme that is designed to produce livelihoods of at least the same level as before relocation;
- Group 3, host communities will receive rehabilitation measures in relation to "pressure on natural resources, public infrastructure, and social services."

- In addition, households "with limited resources" will receive additional help with relocation; households "unable to restore or improve their livelihoods" after relocation will be classified as vulnerable and will receive special assistance (ERM 2007c, pp. 83–84, 90, Table 9.3).

Compensation is of four types: cash, in-kind ("land, houses, other structures, building materials, seedlings, agricultural inputs and financial credits for equipment"), assistance (such as a "transition allowance, transportation and labour") and replacement ("of public services, infrastructure and productive assets such as fish ponds") (ERM 2007c, p. 90, Table 9.2 and p. 96).

8.6.2 Implementation of Planning Recommendations

For effective resettlement implementation, the RPF had suggested an institutional setup to address livelihood issues. Most importantly, this setup would have created a Steering Committee composed of official representatives of the Bui Power Authority, district officials, chiefs, the Land Valuation Board and NGOs. The Steering Committee would have cooperated with a Working Group made up of a Resettlement Coordinator, village representatives, NGOs (including one specifically appointed to monitor implementation) and technical personnel in charge of host site preparation (architects, agricultural experts, etc.). The Working Group would have been responsible for implementing resettlement, including pre-resettlement consultation and overseeing the process of relocation—which would have involved regular coordination with the construction contractor. After resettlement, the Working Group would have been responsible for implementing and monitoring rehabilitation measures and responding to grievances. The Working Group would have received input from relevant authorities on all levels of government and from specialised NGOs. For grievance procedures, a special mechanism would have been put in place, including a grievance officer (ERM 2007c, pp. 122–133).

This institutional setup was to have been rounded off with an internal and external monitoring and evaluation procedure: internal monitoring and evaluation would have been conducted as part of the daily resettlement process, and progress would have been measured against a set of indicators. External monitoring would have been carried out by an independent agency applying a set of indicators 3–6 months, 18 months and again 36 months after resettlement. The results would have been reported to the Steering Committee and the Working Group (ERM 2007c, pp. 141–146).

This arrangement was entirely ignored, except that the Bui Power Authority has appointed a Resettlement Officer, who is in charge of resettlement coordination. The Resettlement Officer cooperates with government agencies, and most importantly with the Land Valuation Board, which is responsible for valuing the assets of affected people before their relocation (ERM 2007c, p. 81).

The main problem seems to be that there is no legal requirement for the monitoring of resettlement—just as there is no legal requirement for public participation in resettlement planning. Whereas hearings can be conducted as part of the ESIA process, Ghanaian EIA regulations make no mention of resettlement monitoring (Otu-Tei 2009, p. 118). There is no institutionalised inclusion in the resettlement process of paramount and village chiefs, village representatives appointed by the chiefs, district and regional authorities, relevant ministries and service-delivery NGOs in such sectors as health, education and agriculture. Rather, the impression is that the Bui Power Authority management has difficulties cooperating with NGOs, since NGO criticism of the resettlement process in general and the Bui Power Authority in particular is often perceived as wholesale opposition to the dam-building process.

It should be noted that the RPF did not allocate a particular role to the construction company, Sinohydro. It was apparently to be largely uninvolved in relocation, but would be affected by resettlement insofar as the progress of construction depends on the relocation of communities. Consequently, the Bui Power Authority has to coordinate with Sinohydro when relocation is in progress.

8.6.3 Information Politics

Bui Dam resettlement is rife with rumours. Few people outside the key governmental agencies know anything definite. Not only are local chiefs and resettlement communities uncertain about exact proceedings, such as who builds the infrastructure in resettlement townships, but the affected communities have no knowledge at all of the resettlement timelines and key actors. Some affected communities claimed that the Chinese had constructed their houses,[44] while others pointed to the Volta River Authority.[45] A local community leader maintained that the Bui Power Authority had contracted local companies to do the work.[46] Ministerial officials argued that the EPC contract also contains a resettlement plan,[47] while others said that resettlement has been subcontracted to local contractors.[48] One interviewee, who is involved in land management, said that resettlement is contained in the Ghanaian contribution of US$ 60 million.[49] If correct, this would mean that not only the implementation but also the financing of resettlement is an exclusively Ghanaian responsibility.

---

[44] Interview J01072010-2.

[45] Interview W01072010-1.

[46] Interview W01072010-1.

[47] Interview A30062010-2.

[48] Interview A30062010-1.

[49] Interview A08072010.

A point of contention and much confusion is the construction of Bui City. Bui City is meant to be an entirely new city, where all resettlers are to be eventually moved. All resettlement prior to Bui City was intended to be temporary to enable dam construction and allow time for the construction of the City. However, there is no time schedule for Bui City, which so far exists only on the drawing board. People from NGOs and land planning circles described Bui City as a "myth" and a "dream" that has so far failed to materialise.[50]

In February 2007, before the signing of the EPC and loan agreements, the Ghanaian newspaper *The Statesman* ran an article saying that the construction of Bui City would involve Chinese and Ghanaian private-sector firms. Of the overall amount of US$ 622 million, US$ 25 million was said to have been estimated as resettlement costs (Benjamin 2007). However, this information could not be verified. The major document, the RPF, calculates total compensation costs at slightly over US$ 4 million (ERM 2007c, pp. 136–140).

The National Development Planning Commission explicitly refers to Bui City in Appendix II of its *Draft Medium-Term National Development Policy Framework* for the period 2010–2013. Under "Sect. 2: Special Development Areas" of "Chapter VII: Reducing Poverty and Income Inequalities" Bui City is mentioned in connection with issues pertaining to "Uncoordinated demands for special development areas [and] lack of coordination of ongoing and emerging national development opportunities and challenges." As a consequence, the National Development Planning Commission recommends action to "Coordinate the selection and development of special development areas" by enhancing "planning and coordination of the development of [...] Bui City" and other special development areas (National Development Planning Commission 2009, p. 73).

Another issue here is the release of information during local consultation. Claire Sutcliffe, who conducted extensive interviews in affected villages, has said that the recommendations in the RPF for addressing villagers' concerns about their livelihoods[51] have not been taken up during resettlement. In particular, villagers were not given a timeframe for resettlement or for compensation payments, and they were not informed of appropriate channels of communication "to make their grievances known." As a consequence, authorities ignored the Consultation Objectives in the RPF (Sutcliffe 2009, pp. 2–3), especially in respect of the following:

> *Stakeholders should be publicly informed by the relevant authorities of the details of the resettlement activities. The information made public and provided to each household will include cut-off dates for each affected group (if the cut-off date differs), entitlements, eligibility criteria, modes of compensation, complaints and grievance redress procedures* (ERM 2007c, p. 111, quoted in Sutcliffe 2009, p. 2).

---

[50] Interviews A22062010-1, A22062010-2 and A08072010.

[51] For details on issues raised by stakeholders and how they are addressed in the RPF see ERM (2007c, pp. 115–116, Table 13.1).

An interviewee involved in land management claimed that the RPF processes had not been adopted in the Resettlement Action Plan and that the two documents were very different.[52] Affected communities claimed that they were not consulted, but merely given certain information. They were what they called "partially consulted" in that they were promised that Gyama would be temporary (followed by relocation to Bui City) while rocks were being blasted and that their situation would improve significantly in the Bui City settlement, with new houses and infrastructure. Information on negative impacts was not disclosed[53] (Ghana News Agency 2009b).

8.6.4 Rehabilitation of Livelihoods

Indeed, when the recommendations in the RPF are compared with what has actually happened, a rather bleak picture emerges. To ensure the reconstruction of livelihoods after relocation, the RPF advocates the implementation of a Livelihood Enhancement Programme. This Programme adopts a two-pronged approach: the targeting of vulnerable households and the enhancement of livelihoods in resettler communities in general. The focus is on the major livelihood activities of the villages: agriculture; fishing; trading; and grazing, hunting and collection of forest products. The RPF provides for a monitoring body consisting of an NGO appointed to implement the Programme, while a committee oversees implementation. The committee is to consist of representatives of traditional authorities, the Ministry of Food and Agriculture, the project owner's resettlement coordinator, i. e. the Bui Power Authority's Resettlement Officer, and the NGO appointed to implement the Programme. Quarterly monitoring reports are to ensure the success of the Programme (ERM 2007c, p. 96).

A glance at the fishing section of the Livelihood Enhancement Programme soon shows that the recommendations it contains have not been put into effect for the relocated fishing community at Gyama Resettlement Township:

> *The key principle with respect to rehabilitation of fishermen and women is to ensure that they are given access to equivalent or improved fishing grounds. The measures […] will aim to improve fishing livelihoods to levels over and above pre-resettlement levels* (ERM 2007c, p. 97).

The measures specifically comprise the allocation of equivalent fishing grounds, the development of fishing opportunities (i. e. the establishment of fishing associations, business planning, micro-credit support, as well as storage, transport and processing refrigeration facilities) and the development of alternative livelihoods (i. e. agriculture, small service enterprises and artisan workshops and training for other livelihoods to be identified and for which there is demand)

[52] Interview A08072010.

[53] Interview A08072010 and Interview J01072010-2.

(ERM 2007c, p. 98). In addition, to compensate for lost fishing grounds, the RPF suggests providing a "transportation allowance for fishing equipment (boats etc.)" and a "transition allowance until fishing livelihood is restored" (ERM 2007c, p. 94, Table 9.6).

The main problem here is the identification of suitable alternative fishing grounds. According to the RPF, the Gyama Chief suggested that the fishing village of Agbegikuro, populated by members of the Ewe tribe, should be resettled on his land because, though of different tribes, the two communities have known each other for a long time and because Gyama Village is close to the river. The RPF also states that the Resettlement Action Plan must "ensure the viability for fishing in the light of changes in river flows during construction and subsequent operation" (ERM 2007c, p. 103 and 104, Table 10.1).

The resettled fishing community had been fishing in the river near the dam site for several generations. As the Gyama Chief had suggested, it was resettled on dry land next to Gyama Village when construction began. It was claimed that, when they were resettled, they did not receive any seeds (having to buy them at the market instead), tools or training in farming, which prompted the Gyama Chief to provide them with seeds, cassava and technical farming know-how.[54]

It was also claimed that the Bui Power Authority was slow to pay compensation. Communities and people involved in land management mentioned that compensation had not been paid, except for the construction of houses (house to house compensation).[55] But no compensation had been received for crops, economic trees or land at the time the interviews were conducted in Ghana in June and July 2010—although the basis for the claim that compensation for land had not been received is unclear, because resettlers had been given land to cultivate. However, it was claimed that the land was not fertile.

Resettlers were given three types of cash compensation: first, a one-off payment of GH¢100.00 per person regardless of age; second, a farm grant of GH¢50.00 per household to enable initial crop cultivation on the land the Bui Power Authority had acquired from the Gyama Chief (two acres per household); and third, a monthly allowance for 1 year from May 2008 to April 2009 per household as temporary income support.[56] This monthly allowance was meant to assist resettlers—some of whom were farmers, some fishermen—to travel back to their fields, 4.5 km from their new settlement, to harvest their produce while simultaneously developing their new fields at Gyama Resettlement Township (Zigah 2009, pp. 28–29; Bui Power Authority no year f). There is also hope that once the reservoir is filled, fishing opportunities will re-emerge and improve for the fishing communities at Gyama and possibly include the establishment of aquaculture

[54] Interviews J01072010-3 and J01072010-2.

[55] Interview A08072010.

[56] According to the Bui Power Authority, the calculation of the monthly allowance of GH¢100.00 was based on a study of alternative livelihoods conducted by the Faculty of Human Settlement of Kwame Nkrumah University of Science and Technology in Kumasi (Zigah 2009, p. 29).

(Otu-Tei 2009, pp. 107–108). So far, however, Otu-Tei (2009, p. 113) says that resettlers had to spend a larger portion of the monthly allowance on food than on travel, because they "were unable to access their old farms". Indeed, resettlers claimed that even though they had been told that they could return to harvest their mangoes, they were not in fact allowed to do so.[57] They also claimed that the Land Valuation Board of the Lands Commission had made mistakes when valuing the properties in the four villages and that these mistakes had not been rectified.[58]

Otu-Tei (2009, pp. 110–112) also argues that timing was the crucial mistake during the Bui resettlement, since the resettlement process was not attuned to the agricultural cycle. The Gyama resettlement occurred "immediately after the main season for planting maize and yam, the main staple foods in the area, because the resettlement was determined by the pace of the dam construction," even though the RPF had provided information on the growing seasons for all crops grown in the area. He concludes that, because of this timing criterion, "critical issues" in resettlement implementation, such as:

> *compensation, preparation of resettlement site, relocation, implementation of livelihood programs, and monitoring were ignored [...]. Completion of the dam was prioritized at the expense of sustainably restoring or improving the living conditions of the affected people* (Otu-Tei 2009, p. 116).

It can therefore be said that none of the recommendations of the Livelihood Enhancement Programme for fishing communities has so far been implemented, with the exception of some initial help with farming. Being part of resettlement Group 1, the fishing community is entitled to full compensation and may also be classified as vulnerable and therefore entitled to additional support for income restoration. The Bui Power Authority may well argue that monthly cash payments fulfil this requirement and the recommendation for a transition allowance. However, it is uncertain whether suitable new fishing grounds can be found. It is also clear that, with the expiration of the 1-year financial assistance and without their key source of food and income and thorough training for alternative livelihoods, the fishing community is finding it hard to sustain acceptable levels of livelihood comparable with or better than pre-resettlement conditions.

Gyama Resettlement Township does provide certain improvements in infrastructure: stone houses and boreholes provide secure shelter and drinking water. Overall, however, the township appeared to be in a dilapidated state, with muddy roads and a run-down structure that was featured as a community centre. Furthermore, as mentioned above, it was meant to be temporary until Bui City was completed. But at present it is unclear when Bui City will be built and how it will look. Indeed, the Bui Power Authority has reportedly found re-relocation to be too expensive and has therefore promised to improve living conditions in the Gyama

---

57 Interview J01072010-2.

58 Interview J01072010-2.

Resettlement Township, where rooms are very small and roofs are already leaking. But this has not yet happened.[59]

It is a matter of interpretation whether Gyama Resettlement Township meets the constitutional requirement of suitability for a resettlement site. Certainly, the recommendation in the RPF that resettlement sites should provide basic infrastructure has been put into effect, but whether public services have been provided as recommended is questionable. A similar assessment can be made on the basis of the requirements set out in the World Bank Involuntary Resettlement Sourcebook: Gyama Resettlement Township does "supply infrastructure and services better than, or at least similar to those available before resettlement." However, the location is clearly not "acceptable to the resettlers," and given that no alternative fishing grounds have been demarcated, the farmland is allegedly poor, and no thorough training in farming techniques has been provided, it must be doubtful that the settlement has "the capacity to support the incomes and living standards of the people to be resettled" (ERM 2007c, p. 101 and Box 10.1).

For resettlement Phase B, houses are already being built, and according to a local community representative, eleven local contractors have been engaged.[60] Representatives of the communities mentioned that they had been shown construction plans of houses and had subsequently been to the construction site to measure for themselves whether the houses matched the plans. And they did.[61] One community chief said that his village, which is to be resettled during Phase B, would receive compensation for crops, infrastructure and land. No compensation had been paid by the time of the interview, but he said that the Land Valuation Board had been to his village and valued everything.[62] The Phase B resettlements may turn out to be an improvement on the Phase A resettlements, given that Bui City is a long way from realisation.

It seems that the Gyama resettlement is a case where villagers ignorant of their legal rights and overwhelmed by official rhetoric and the power of a large bureaucracy found themselves at the mercy of a government organisation. Affected communities that are still to be resettled therefore maintained that the Gyama resettlement case, made known to them largely through the Ghana Dams Dialogue (see Box 8.1), which also covered the negative experience of the Akosombo Dam resettlement communities, helped, as a test case, to show them how to improve their own fate.[63] Importantly, there were no indications that the Bui Power Authority had set up the monitoring body recommended by the RPF to allow an independent and transparent review of the resettlement programme.

---

[59] Interviews J01072010-2, A22062010-2, A22062010-1 and A08072010.

[60] Interview W01072010-1.

[61] Interview W01072010-1.

[62] Interview W01072010-1.

[63] Interview W01072010-1.

Box 8.1: The Ghana dams dialogue

It should be noted that the dilapidated state of Gyama Resettlement Township is at least partly due to the fact that it was meant to be a temporary solution. Overall, even critics of the Bui Dam acknowledge that resettlement for Bui has so far proceeded rather better than the Akosombo resettlement; but they contend there is considerable room for improvement in local consultation, the transparency of decision-making and the time it takes to pay compensation.[64]

The improvement in the resettlement process seems to be the result of two factors: lessons learnt from the Akosombo Dam and the establishment of the Ghana Dams Dialogue. This Dialogue is funded by the German Development Cooperation (GIZ) and hosted by the International Water Management Institute (IWMI) and the Volta Basin Development Foundation (VBDF) in Accra. It has increased the awareness of the affected communities and has acted as a confidence-building platform for all major pro- and anti-Bui Dam stakeholders in government, affected communities, research organisations and civil society. During interviews with Dialogue and non-Dialogue members, government and non-government agencies acknowledged the usefulness of the Dialogue[65]; only the Bui Power Authority management was vehemently opposed to it, arguing that the importance of the Dialogue had been greatly exaggerated and that the Bui resettlement had been an improvement on the Akosombo resettlement because lessons had been learnt from the disastrous Akosombo experience and not because of the Dialogue.[66] The Ghana Dams Dialogue was not universally known, however: interviews at the Ministry of Finance and Economic Planning and the National Development Planning Commission showed that people there had not heard about it.

Asked about the immediate effects of the Ghana Dams Dialogue, one interviewee said that one of the Akosombo resettlement communities had had a problem with electricity payments: the community refused to pay the tariffs, but the Volta River Authority demanded payment. A delegation from the Ghana Dams Dialogue went to the community to mediate and solved the problem.[67]

---

[64] Interviews A08072010, A22062010-2 and A22062010-1.

[65] Interviews A22062010-2, A22062010-3, A23062010-3, A29062010, A25062010-1, W01072010-1 and J01072010-2.

[66] Interview A28062010-2.

[67] Interview W01072010-1.

8.7 The Role of Chinese Agencies: Financing and Construction Under EPC Contract Arrangements

China's involvement in the above procedures takes the form of the financing and construction of the Bui Dam under the EPC contract between Sinohydro and Ghana's Ministry of Energy and the loan agreements between the China Exim Bank and the Ghanaian Ministry of Finance and Economic Planning. Planning was the Ghanaian government's responsibility. The dam design was prepared by Coyne et Bellier in the feasibility study commissioned by the Ministry of Energy and the Bui Development Committee. According to Sylvester Zigah, the Development Programs Officer of the Bui Power Authority, Coyne et Bellier continue to act as consultants to Sinohydro, the contractor (Zigah 2009, p. 26). The Environmental and Social Impact Assessment—which includes resettlement planning—was delivered by UK consulting firm Environmental Resources Management and commissioned by the Ministry of Energy and the Bui Development Committee.

8.7.1 Sinohydro's Responsibilities

Under the EPC contract Sinohydro is the contractor, and its role is in principle to execute the dam design, abiding by the conditions attached to the permits and licences issued by Ghana's regulatory authorities. As these permits and licences are granted to the Bui Power Authority as the project owner, it is its responsibility to monitor Sinohydro's adherence to the conditions. The EPC contract complies with the international standards for EPC contracts published in 1999 by the International Federation of Consulting Engineers. However, Sinohydro can request the Bui Power Authority's approval of changes to the design.[68]Under the EPC contract, Sinohydro is responsible for

> *civil construction, fabrication and installation of metal structures, installation of power transmission lines and unit installation. Contract duration is 56.5 months. Bui Hydro-electric Project includes a 114 m high roller compacted concrete gravity dam, a 60-meter-high clay core rockfill dam, a homogeneous earth dam and a powerhouse at dam toe installed with three francis turbine-generator units having a total installed capacity of 400 MW. Bui hydropower station will be the second largest hydropower station in Ghana after its completion.*
>
> *Quantities of main works for the project include 940,000 m^3 RCC* (Roller-compacted concrete) *placement, 200,000 m^3 conventional concrete placement, 1,500,000 m^3 filling works of saddle dams, approximately 5,000 tons of metal structure (including penstocks and gates) fabrication and installation, 320 km power transmission line, and two sub-stations* (Sinohydro Bureau 2009).

Sinohydro also employs the workers (Box 8.2), but must abide by Ghanaian law, which means *inter alia* allowing the formation of a union under the 2003

[68] Interviews A23062010-2 and A28062010-2.

Labour Act. The first attempt to form a union was met by the Chinese management with "open intimidation and victimisation." The Bui Power Authority's Resident Engineer tried to intervene and take up the workers' concerns with the Sinohydro management. However, this was seen by Sinohydro as "unnecessary interference in their work." Bui Power Authority management in Accra then ordered the Resident Engineer to "stop interfering in the affairs of the Chinese."[69] This was followed by mass resignations of workers, to the point where the timetable for the construction of Bui was affected. Eventually, the Bui Power Authority called on the Sinohydro management to permit the formation of a union. This process was then launched by the Brong Ahafo Regional Secretariat of the Ghana Trades Union Congress in Sunyani and the Industrial Relations Officer of the Construction and Building Materials Workers Union at the Ghana Trades Union Congress (Baah et al. 2009, pp. 104–105).

Ghanaian law also specifies a minimum wage,[70] a maximum of 8 hours work per day or 40 hours per week, voluntary overtime (which may be made compulsory where required to ensure profitability or project viability), 15 days' paid annual leave, up to 6 months' paid sick leave, maternity leave and insurance against occupational accidents and injuries, including up to 6 months' paid sick leave (Baah et al. 2009, pp. 100–104). In addition, after 6 months' employment the status of Ghanaian workers must be changed from casual to permanent (Baah et al. 2009, p. 100).

Box 8.2: Sinohydro and disputes with workers

As Sinohydro employs the workers independently of Ghanaian government agencies, it is also required to deal with potential disputes over salary payments, working conditions, accommodation and health and safety issues. The first disputes occurred in 2008, shortly after construction began. Ghanaian workers complained about cramped 14-bed dormitories and a lack of mattresses. They claimed that their wages were just above the required minimum and in fact lower than those they had signed up for; working hours were from 7am to 6 pm; and Sinohydro would not allow workers to rest after hospital treatment, would not pay allowances during illness and treated them as casual labourers. In response, the Bui Power Authority and the Ministry of Energy called on the workers to "exercise patience and be prepared to sacrifice a little" as the Chinese workers did until the camps were fully completed and to work as hard as the Chinese, thus learning lessons from their work ethic (Ghana Broadcasting Corporation 2008; Peace FM Online 2009; Ghana News Agency 2008a, b, 2009b; Baah et al. 2009, p. 100).

[69] Baah et al. (2009, p. 104); Mohan (2009); interviews A21062010-3 and A14072010.

[70] The daily minimum wage for 2008 was GH¢2.25. For 2009, it was raised to GH¢2.655. In February 2010, the daily minimum wage for 2010 was raised to GH¢3.11 (Ghana News Agency 2009a, 2010).

Workers were also unaware that they were entitled to paid annual leave and were insured against occupational injuries and that they had a right to 6 months' paid sick leave. Chinese doctors at the dam site sometimes refuse to grant sick leave (Baah et al. 2009, pp. 102, 104).
Negotiations on the wage issue between the workers and Sinohydro began in August 2008, continued into 2009 and simmered on until April 2010. It appears that the reputational damage on the Chinese side was such that the Chinese Ambassador, Yu Wenzhe, visited the construction site in 2009 to check on the progress of the wage talks (Hall 2009). In 2010 it was reported that workers were still complaining about cramped accommodation, with five to ten people sharing one room. Workers also complained about sanitary conditions and about the fact that 80–100 workers are packed into one truck and transported between the work site and Gyama village. In addition, wages were at only GH¢2.50 for workers and GH¢3.00 for artisans. As regards the wage problem, the Chief Executive Officer (CEO) of the Bui Power Authority, Jabesh Amissah-Arthur, said that workers' welfare was the responsibility not of the Bui Power Authority but of Sinohydro and the Trades Union Congress. In late 2009 the Trades Union Congress and workers had agreed that wages should be at GH¢4.50 for labourers and GH¢5.00 for artisans, but Sinohydro management had allegedly failed to implement that decision (Appiah 2010). A compromise between Sinohydro and the Trades Union Congress was reached in April 2010 (Osei 2010).

The EPC contract stipulates that 3,000 semi-skilled and unskilled Ghanaian workers must be employed, along with around 1,000 Chinese workers.[71] As of June 2010, some 2,000 Ghanaians and fewer than 1,000 Chinese are working on the project.[72] Sixty Pakistani workers drive the heavy equipment: they count as Chinese for the purposes of the imported labour quota (Mohan 2009). Mohan and Tan-Mullins confirm that the Bui project employs 3,000 Ghanaian workers and only 700 Chinese expatriates. They conclude that, although Chinese firms "sometimes bring low-cost labour from China [...] the perception that they only employ Chinese labour is misleading" (Mohan and Tan-Mullins 2009, p. 595).[73] Indeed, while research indicates that Chinese companies in Ghana have "manipulated local laws regarding capital equipment and foreign exchange for overseas investors to earn additional foreign exchange [...] Chinese businessmen in Ghana [...] argued that Ghana is much stricter than other West African states in applying its investment code, which sets a high minimum limit for investments and its immigration laws," thus delimiting labour importation (Mohan and Tan-Mullins 2009, pp. 599–600).

[71] Interview A30062010-2.

[72] Interview A30062010-2.

[73] See also Ghana News Agency (2007).

Table 8.5 China Exim Bank credits and Ghana's contribution to the construction of the Bui Dam (Interview at the Ministry of Finance and Economic Planning in June 2010)

Type of credit	Amount (million)	Interest rate	Amortisation period (years)	Grace period (years)
Buyer's credit	US$ 292	2 % over CIRR[a]	20[a]	5
Concessional loan	US$ 270	2 %	20	5
Ghanaian government loan	US$ 60	–	–	–

Note [a] Brautigam (2009, p. 349, note 30) reports the following numbers: interest rate of 0.75 % over CIRR and a maturity period of 17 years

8.7.2 Terms of the Loan Agreements

Funding for the construction of Bui consists of three parts. Of the total project costs of US$ 622 million, US$ 60 million has been contributed by the Government of Ghana. The remaining US$ 562 million has been extended by China through the China Exim Bank in the form of a mixed financing package: a concessional loan of US$ 263.5 million, subsidised by China's Ministry of Commerce, and a buyer's credit of US$ 298.5 million provided by the China Exim Bank alone. Table 8.5 shows details of the financing arrangements.

While the concessional loan may qualify as Official Development Assistance (ODA) in OECD terms, Brautigam claims that Ghana's Ministry of Finance and Economic Planning calculated the grant element of the mixed credit at 22 %. For a concessional loan to qualify as ODA as defined by the OECD, it must contain a grant element of at least 25 % (Brautigam 2009, p. 349, note 30).

Anecdotal evidence suggests that the negotiations were rather tough. In November 2006 the *Ghana News Agency* reported that during the negotiations on the loan agreement the Chinese offered to make 50 % of the loan concessional, while the Ghanaian government pushed for 70 % (Osei Bonsu 2006). In an article for the *Statesman* in February 2007, Benjamin reports that the Ministry of Finance and Economic Planning wanted the Chinese to increase their portion of the Bui financing from 85 % of the project costs to 90 %, with 42.5 % of the loan concessional. For the concessional part, according to Benjamin, the maximum interest rate would be 2 % with a repayment period of at least 20 years. However, the government then headed by President Kufuor wanted a 1 % interest rate and a 30-year repayment period. In the case of the buyer's credit, the Chinese offer was 1.1 % over Commercial Interest Reference Rate (CIRR)—around 5.75 % at the time—with a 12-year repayment schedule and a grace period of 5 years. However, the then Ghanaian government pushed for a 15-year repayment schedule and an interest rate of 1.05 % over CIRR (Benjamin 2007). In an interview conducted by the author at the Ministry of Finance in June 2010, the interviewee reported that the Chinese offered a 17-year repayment schedule for the buyer's credit, which the Ghanaian negotiators managed to push to 20 years.

The financial arrangement is secured by cocoa as collateral.[74] According to the interviewee, the Chinese government guaranteed to purchase 30,000 tonnes of cocoa per year from the Ghanaian government at going world market prices—rather than a fixed price—until the dam was operational. The reason for this is that China wants to promote cocoa consumption in China. Until the Bui Dam becomes operational and begins generating revenue to service the loan (during the construction period there is a moratorium on the payment of the principal, which means that only interest has to be paid in that period), the proceeds from the sale of cocoa are placed in an escrow account, i. e. a trust account, held by the China Exim Bank as security for loan repayments. If not all the funds in the escrow account are needed to pay the interest, then the excess reverts to the Ghanaian government, i. e. the Ministry of Finance and the Central Bank. Once the dam is operational within 5 years of the signing of the loan agreement, the cocoa agreement lapses. By that time, the interviewee hoped, China will have developed a taste for cocoa and so become a new and important market for Ghanaian cocoa.[75]

When the Bui Dam is operational and the cocoa agreement expires, 85 % of the proceeds from sales of energy generated by the completed Bui Dam will be placed in an escrow account held by China Exim Bank to service the debt. The remaining 15 % will be used to meet the Bui Power Authority's administration costs. If not all the 85 % is needed for debt servicing, the excess will revert to the Government of Ghana, i. e. the Ministry of Finance and Economic Planning and the Central Bank. If Ghana decides not to withdraw the excess money from the escrow account, it will earn interest.[76]

The World Bank's view of the buyer's credit, and the associated repayment conditions, is that it is consistent with the Ghana Poverty Reduction Strategy II, which concluded that non-concessional funding is needed to finance the development plan:

> *... following the October 2005 Consultative Group (CG) meeting the Government communicated to the Bank that it would seek additional financing from non-concessional sources to meet the country's growth targets and achieve the MDGs. [...] This need became more pressing as the energy crisis deepened in late-2006 and early 2007. The decline in water levels at the Akosombo dam reduced power supply capacity by 25–30 %, making even more urgent the need to scale up investments to restore power supply capacity to the levels prior to the crisis and ensure that energy supply would increase in lockstep with projected real GDP growth.*
>
> *More recently, the Government's intention to access international capital markets was discussed in February 2007 and incorporated in the joint Bank-Fund Debt Sustainability Analysis (DSA) completed in April 2007. The joint DSA included the government's intention to borrow about US$ 250 million per year non-concessionally between 2007 and 2009, and US$ 350 million per year thereafter, and concluded that Ghana faced as a result a moderate risk of debt distress, albeit on the low side. The main developments since that joint DSA was completed are the Government's decision to borrow the envisioned*

[74] Interview A30062010-1.

[75] Interview A30062010-1.

[76] Interview A30062010-1.

amount from international capital markets in one single tranche, rather than spreading it over a 3 year period, and to sign the financing agreement with the China ExImbank. [...]
Moreover, the financing agreement with the China ExImbank for the construction of the hydro-electric power plant at the Bui Dam is part and parcel of the Government's efforts to keep the expansion of low cost electricity generation capacity in line with the country's projected real GDP growth (World Bank 2008, p. 6).

8.7.3 Procedures Leading to the Loan Agreements

As far as procedure is concerned, the first step leading to the loan agreement was a letter of application from the Bui Development Committee to the Ministry of Finance and Economic Planning proposing the use of funds from the China Exim Bank. Second, the Ministry of Finance and Economic Planning formally applied to the China Exim Bank and the Government of China for two lines of credit.[77]

Next, the Ministry of Energy commissioned Coyne et Bellier to conduct a feasibility study. For this, Coyne et Bellier updated their earlier study on the Bui Dam. The Bui Development Committee and Sinohydro then entered into discussions on an EPC contract. Once the EPC contract and the feasibility study had been drawn up, they were sent to the China Exim Bank for evaluation.[78]

Following this, negotiations began between the Chinese government, represented by the China Exim Bank, and the Ghanaian government, which led to draft agreements on a concessional loan and a buyer's credit facility. The Exim Bank also requested a comprehensive EIA report.[79] This was in line with the Exim Bank's *Guidelines for Environmental and Social Impact Assessments of the China Export Import Bank's Loan Projects* of August 2007, Article 13 of which requires the project owner or borrower to submit the EIA study during loan negotiations so that the environmental and social impact may be evaluated and the construction proposal amended if necessary (China Exim Bank 2007). Article 14 of the *Guidelines* stipulates that the "China EXIM Bank, if necessary, can require the inclusion of environmental and social responsibilities in the loan contract, in order to monitor and restrain the behaviour of borrowers" (China Exim Bank 2007). The submission of the EIA report to the China Exim Bank was followed by a new round of negotiations between the two sides' lawyers concerning the details of the agreements.[80]

After an agreement between the Chinese and Ghanaian sides had been reached, the Ministry of Finance and Economic Planning presented a memorandum to the Cabinet for its approval. The memorandum contained the financial details of the concessional loan and the buyer's credit facility that had been negotiated between

77 Interview A30062010-1.
78 Interview A30062010-1.
79 Interview A30062010-1.
80 Interview A30062010-1.

the China Exim Bank and the Ministry of Finance and Economic Planning. After receiving the Cabinet's approval, the loan agreements were sent to Parliament for its consent. The concessional loan agreement was signed in 2007 and the buyer's credit facility in 2008.[81] An interviewee at the Environmental Protection Agency said that the loan agreement had had to refer to World Bank guidelines and that the China Exim Bank had agreed. The loan agreement thus refers to international standards relating to dam safety, resettlement and forestry.[82] An official of the Ministry of Finance and Economic Planning interviewed in June 2010 was adamant that, as a consequence of this negotiation process, China "did not impose any loans on the Government of Ghana."[83]

To examine the financial implications for the government before the signing of the loan agreements, the Ministry of Finance and Economic Planning conducted a cost-revenue projection based on the sale of power at a certain rate and a generation capacity of 400 MW. For the projection, the Ministry calculated 4.5 US cents per kilowatt hours (kWh) (since the calculation, energy prices have risen to 7 US cents per kWh. One interviewee claimed that, once operational, the Bui project "can pay for itself, even with the lower cost projection of 4.5 cents per kWh. The Bui Dam loan agreement therefore will not impact debt sustainability." Currently, the government is paying interest and fees, but when energy generation begins, "the Bui Power Authority has to reimburse the government."[84]

According to the Bui Power Authority CEO, to monitor the spending of its money, the China Exim Bank keeps a check on loan disbursement. Articles 15–19 of the Exim Bank's environmental guidelines make Exim monitoring of its projects mandatory (China Exim Bank 2007). According to the Bui Power Authority, the China Exim Bank monitors like any other financial institution on the basis of reports submitted by the Bui Power Authority to the Exim Bank through the government (but the Bui Power Authority would not say whether the government was represented in this respect by the Ministry of Energy or the Ministry of Finance and Economic Planning). At any rate, there is no direct link between the Bui Power Authority and the China Exim Bank.[85] In June 2010 an Exim Bank team reportedly visited the Bui Dam site.[86] In the week beginning 21 June 2010 an International Review Board of Consultants (RBC) inspected the Bui Dam and briefed the CEO of the Bui Power Authority on 25 June 2010. The Bui Power Authority management did not want to disclose any information on the composition of the RBC or the outcome of the review, stating only that inviting an RBC is "standard international practice." The only response to a request for information

[81] Interview A30062010-1.
[82] Interview A21062010-3.
[83] Interview A30062010-1.
[84] Interview A30062010-2.
[85] Interview A28062010-2.
[86] Interview A30062010-1.

on the CEO's visit to China from 30 June 2010 for project meetings and manufacturing inspections was simply that the visit concerned the dam, no further details being forthcoming.[87]

8.7.4 Responsibilities of Sinohydro Under a Construction Management Plan

The ESIA study suggests that during the construction phase Sinohydro, the contractor, should be required to implement a Construction Management Plan as part of the Environmental and Social Management Plan for Bui. The Construction Management Plan would have a direct effect on the company's work. It sets out the company's responsibilities and so distinguishes them from those of the Bui Power Authority and other Ghanaian government agencies. According to the Construction Management Plan, Sinohydro is required to consider:

> *Workforce and local residents' health and safety; "good housekeeping"—site management practices (e.g. erosion control, materials storage, maintenance of silt traps and oil separators, waste management,* etc.*); flow regimes during construction (particularly during diversion); emergency response to significant accidents/pollution incidents; site access; siting of temporary structures/work locations/materials sourcing (e.g. for sand dredging); timing of certain activities (e. g. to avoid the rainy season, or the breeding seasons or migratory movements of animals); site clean-up and restoration of all areas temporarily affected by construction activities, and decommissioning of all construction-related facilities* (ERM 2007a, p. 163).

Sinohydro's responsibilities thus relate exclusively to the installation of the dam structures, including associated areas: observing permit conditions imposed on the Bui Power Authority by the regulatory authorities, contracting workers, implementing health and safety measures, timing activities according to local conditions, restoring temporarily affected areas and decommissioning construction facilities. Its responsibilities do not extend to resettlement. The ESIA study suggested that site preparation, which precedes construction, should be done by someone else: a professional logging company should be employed to clear vegetation, and wildlife officials should be responsible for rescuing the hippopotamus population from the inundation zone of the Bui National Park (ERM 2007a, pp. 162–163). However, while the hippos naturally remained the responsibility of Ghana's Forestry Commission, affected villagers claimed that Chinese workers cleared away bush and also cashew farms on expropriated land.[88] Everything other than actual construction-related work, particularly resettlement, is a government responsibility. Figure 8.3 summarises the division of responsibilities between the Ghanaian government and Sinohydro.

---

[87] Interview A28062010-2.

[88] Otu-Tei (2009, p. 83; p. 103); interview J01072010-2.

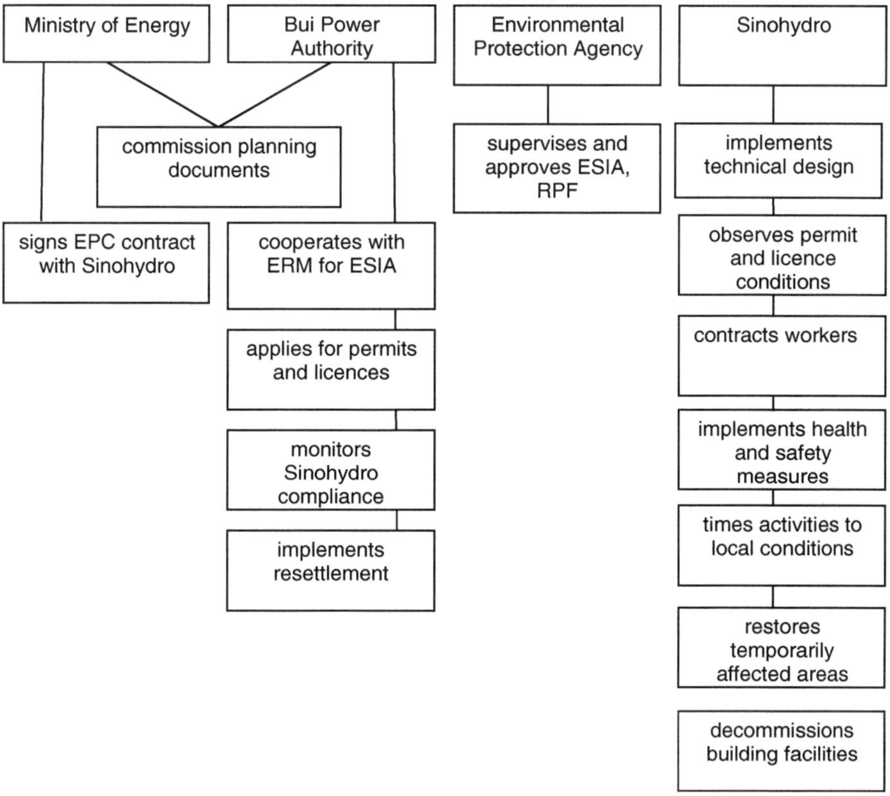

Fig. 8.3 Lines of responsibility between the Ghanaian government and Sinohydro (author's own compilation based on ERM 2007a, p. 132; pp. 157–159; p. 163)

8.8 Summary and Conclusion

This chapter has examined the role of Chinese construction companies and financial institutions in the African dam business, taking the Bui Dam in Ghana as its example. African countries have been providing raw materials for the Chinese economy and investment markets for Chinese companies. The investment activities of these companies receive political support, as evidenced by high-profile meetings at which the investments are announced to the public. The Bui Dam agreement was announced at the 2006 summit of the Forum on China-Africa Cooperation in Beijing.

The problem is not Chinese activity per se, but the nature of Chinese investment and Chinese investment projects. As the Bui Dam was a contentious investment, the Ghanaian government had difficulty finding potential investors. However, as the project had political priority, alternatives for energy production that had previously been considered were cast aside in favour of hydroelectricity. Ghana is in urgent need of an increase in energy capacity, and Sinohydro's reputation for building dams under problematical environmental conditions eventually made it the ideal candidate for the construction of the dam.

Sinohydro has a good international standing as an engineering firm, but the question of environmental sensitivity arises. At least according to Ghana's Water Resources Commission and the Environmental Protection Agency, Sinohydro seems to be abiding by the conditions attached to the various permits issued by these regulatory bodies. However, as these regulatory bodies have a relatively low standing within Ghana's ministerial hierarchy, responsibility for ensuring that provisions concerning environmental supervision are upheld ultimately rests with the Ghanaian president and the Bui Power Authority. While employees of the Water Resources Commission and the Environmental Protection Agency were clearly eager to implement environmentally sound policies, the project, being political, prevented regulatory authorities from interfering or prolonging the planning and construction process.

Importantly, Sinohydro was not involved in the planning of the Bui Dam or in the resettlement. It is responsible for constructing the dam and for such associated aspects as the contracting of workers and the implementation of health and safety measures on the basis of the planning documents and Ghanaian law. Nor, as Bui is a turnkey project, is Sinohydro the project owner, but will hand the dam over to the Bui Power Authority when construction is completed.

However, planning, implementation and monitoring are strong only on paper. The Bui Power Authority, which is responsible for resettlement, has not complied with the recommendations of the RPF. This Framework suggested an institutional setup that would have ensured proper consultation processes, which would necessarily have taken a long time to complete. Instead, the Bui Power Authority seems to have opted for a quick resettlement process to prevent the construction of the dam from being impeded by protracted negotiations with affected communities.

To preclude problems associated with careless relocation processes, the Ghana Dams Dialogue was set up (it should be noted however that the Bui Dam is only one objective of the Dialogue). The Ghana Dams Dialogue addresses the lack of public consultation by providing a cross-stakeholder platform that involves representatives of affected people, government and NGOs. The Dialogue has faced the difficulty of overcoming the suspicion of the Bui Power Authority management that it is a preventive tool in the hands of non-governmental organisations. The Bui Power Authority CEO has accordingly denied its relevance to the improvement of resettlement practices, claiming that the comparative improvement in the Bui resettlement process has been due to the experience gained from the failings of the Akosombo resettlement. However, it appears that that improvement can be attributed to the Ghana Dams Dialogue, general pressure from civil society and lessons learnt from Akosombo.

As for the environment, the low standing of regulatory authorities in this sector is a threat to strict environmental supervision and leaves room for speculation as to how closely Sinohydro is adhering to the environmental conditions attached to the permits, given that the Bui Power Authority exercises overall control over

Sinohydro's performance. If the Environmental Protection Agency and the Water Resources Commission were given an enhanced role, there would be less room for speculation and uncertainties concerning the strengths of Ghana's environmental policies and Sinohydro would be drawn more closely into a network of supervisory regulations. At the same time, this too points to the need for Sinohydro to improve its own business practices, particularly as there is a lack of Chinese regulations and compliance mechanisms regarding the approval and implementation of overseas investment by China Exim Bank and construction companies.

Acknowledgments The field study was made possible by the support of the following people and organisations: Cathleen Seeger and Kirsten Nyman of the Project of the GIZ "Policy Advice for Sustainable Hydropower"; Harriet Ludwig, the Head of Cooperation at the German Embassy in Accra; the GIZ Office in Accra; Liqa Raschid of the IWMI in Accra; Richard Twum of the VBDF; the following official Ghanaian institutions: the Ministry of Finance and Economic Planning; the Ministry of Energy; the Environmental Protection Agency; the Energy Commission; and the Water Resources Commission.

References

Appiah IS (2010) Bui dam workers in slavery. http://www.modernghana.com/news/273165/1/bui-dam-workers-in-slavery.html. Accessed 25 Apr 2013

Appiah-Opoku S (2001) Environmental impact assessment in developing countries: the case of Ghana. Environ Impact Assess Rev 2(1):59–71

Baah AY, Otoo KN, Ampratwurm EF (2009) Chinese investments in Ghana. In: Baah AY, Jauch H (eds) Chinese investments in Africa: a labour perspective. African Labour Research Network, Accra, pp 85–121

Benjamin C (2007) $25 m seed money for Bui city. http://www.thestatesmanonline.com/pages/news_detail.php?section=2&newsid=2607. Accessed 25 Apr 2013

Brautigam D (2009) The dragon's gift: the real story of China in Africa. Oxford University Press, Oxford

Bui Power Authority (2010) Bui resettlement statistics. http://www.buipowerauthority.com/resettle_main.htm. Accessed 25 Apr 2013

Bui Power Authority (no year a) Bui resettlement programme phase a. www.buipowerauthority.com/res_phasea.htm. Accessed 25 Apr 2013

Bui Power Authority (no year b) Bui resettlement programme phase b. http://www.buipowerauthority.com/res_phaseb.htm. Accessed 25 Apr 2013

Bui Power Authority (no year c) Project milestones and completion schedule. http://www.buipowerauthority.com/project_milestones.htm. Accessed 25 Apr 2013

Bui Power Authority (no year d) Project implementation. http://www.buipowerauthority.com/implement.htm. Accessed 25 Apr 2013

Bui Power Authority (no year e) Project time-line. http://www.buipowerauthority.com/timeline.htm. Accessed 25 Apr 2013

Bui Power Authority (no year f) BPA resettlement package. http://www.buipowerauthority.com/res_package.htm. Accessed 25 Apr 2013

Bui Power Authority (no year g) Project features. http://www.buipowerauthority.com/project_features.htm. Accessed 25 Apr 2013

China Exim Bank (2007) Issuance notice of the 'guidelines for environmental and social impact assessments of the China export and import bank's loan projects'. Beijing

Energy Commission (no year) Siting guidelines for the development of energy infrastructure facilities. Accra
Energy Comission (2006) Licensing manual for service providers in the electricity supply industry. Accra
Environmental Protection Agency (1999) Environmental assessment regulations. Accra
ERM (Environmental Resources Management) (2007a) Environmental and social impact assessment of the Bui hydropower project: final report. Accra
ERM (Environmental Resources Management) (2007b) Environmental and social impact assessment of the Bui hydropower project: final report, annex volume. Accra
ERM (Environmental Resources Management) (2007c) RPF (Resettlement planning framework) for the Bui hydropower project. Accra
Fink M (2005) Integrating the world commission on dams recommendations in large dam planning processes: the case of Bui. Ghana, Dortmund
Ghana Broadcasting Corporation (2008) Bui workers asked to halt complaints. http://ghanadistricts.com/districts1on1/tain/?arrow=nws&read=16894. Accessed 25 Apr 2013
Ghana News Agency (2007) Agreement on Bui dam project signed. http://mobile.ghanaweb.com/wap/article.php?ID=122740. Accessed 25 Apr 2013
Ghana News Agency (2008a) Communities within Bui hydroelectric project to receive compensation. http://mobile.ghanaweb.com/wap/article.php?ID=148215. Accessed 25 Apr 2013
Ghana News Agency (2008b) Bui workers are not on strike. http://www.ghanaweb.com/GhanaHomePage/NewsArchive/artikel.php?ID=154827. Accessed 25 Apr 2013
Ghana News Agency (2009a) Ghana minimum wage raised by 18 per cent. http://www.ghanabusinessnews.com/2009/04/03/ghana-minimum-wage-raised-by-18-per-cent. Accessed 25 Apr 2013
Ghana News Agency (2009b) Submit documents on Bui dam, minister orders. http://www.modernghana.com/news2/234587/1/submit-documents-on-bui-dam-minister-orders.html. Accessed 25 Apr 2013
Ghana News Agency (2010) New national daily minimum wage is 3.11 cedis. http://www.ghanaweb.com/GhanaHomePage/NewsArchive/artikel.php?ID=175749. Accessed 25 Apr 2013
Graham Y (1993) Not all that glitter: residents of village polluted by gold mining up in arms. West Africa Magazine 9 May 1993:736
Hall H (2009) Ghana in the balance. http://chinaafricanews.com/ghana-in-the-balance/. Accessed 25 Apr 2013
IEA (International Energy Agency) (2009) Ghana: electricity generation by fuel. http://www.iea.org/stats/pdf_graphs/GHELEC.pdf. Accessed 25 Apr 2013
IWP&DC (International Water Power & Dam Construction) (2006) Ghana balances hydro and thermal. http://waterpowermagazine.com/story.asp?storyCode=2034818. Accessed 25 Apr 2013
IWP&DC (International Water Power & Dam Construction) (2008) Moving up from Akosombo. http://www.waterpowermagazine.com/news/newsmoving-up-from-akosombo. Accessed 25 Apr 2013
Malgas I (2008) Energy stalemate: independent power projects and power sector reform in Ghana: management programme in infrastructure reform & regulation (MIR), Working Paper, Graduate School of Business, University of Cape Town, Cape Town
Mohan G (2009) Dams, development and the nation. http://www.open2.net/blogs/society/index.php/2009/05/12/title-2?blog=10. Accessed 25 Apr 2013
Mohan G, Tan-Mullins M (2009) Chinese migrants in Africa as new agents of development?: an analytical framework. Eur J Dev Res 2(4):588–605
National Development Planning Commission (2009) Draft medium-term national development policy framework: 2010-2013, appendix II. Aug 2009, version 3.0, Accra
Ofori SC (1991) Environmental impact assessment in Ghana: current administration and procedures: towards appropriate methodology. Environmentalist 11(1):45–54

Osei S (2010) Bui dam workers settle dispute with contractor. http://tain.ghanadistricts.gov.gh/?arrow=nws&read=34415. Accessed 25 Apr 2013

Osei Bonsu K (2006) China assures Ghana of financial support for Bui dam project. http://ghanaweb.com/GhanaHomePage/NewsArchive/artikel.php?ID=113252. Accessed 25 Apr 2013

Otu-Tei C (2009) Social and economic impacts of hydroelectric power dams: the Bui dam, Ghana. University of Central Washington, Ellensburg

Peace FM Online (2009) Bui dam in trouble. http://business.peacefmonline.com/news/200911/31794.php. Accessed 25 Apr 2013

Sinohydro Bureau 8 (2009) Bui hydroelectric project, Ghana. http://www.bajuintl.com/English/Web/PicView.aspx?BMID=1006&SMID=10061&NewsID=1430. Accessed 25 Apr 2013

Sutcliffe C (2009) Interviews with people to be affected by Bui dam: a field report. http://www.internationalrivers.org/files/attached-files/bui_field_report.pdf. Accessed 25 Apr 2013

Volta River Authority (no year) Power activities. http://vraghana.com/our%20mandate/index.php. Accessed 25 Apr 2013

World Bank (2008) International development association program document for a proposed credit in the amount of SDR60.8 milllion (US$100 million equivalent) to the republic of Ghana for a sixth poverty reduction support credit, Washington DC

Zigah S (2009) Speech at the 1st annual meeting of dam affected communities. In: International Water Management Institute (ed) Proceedings of the 1st annual meeting of dam affected communities, Akuse, pp 25–30

Chapter 9
Hydropower Planning in Informal Institutional Settings: Chinese Institutions and the Failure of Environmental and Social Regulation in Cambodia

Oliver Hensengerth

Abstract This study considers Chinese hydropower investment in Cambodia, examining the degree to which Chinese companies adhere to international or national environmental and social standards and analysing the roles played by and the interactions between Chinese and Cambodian actors in the building of hydropower capacity in Cambodia. The example used is the Kamchay Dam, built by Sinohydro and financed by the China Export Import Bank (China Exim Bank). An analysis of the distribution of responsibilities enables us to take a close look at interactions, processes and conflicts between the two sides in the planning and implementation of hydroelectric dams. It also enables us to identify weak points in the environmental and social regulations and to direct criticism more precisely, thus avoiding general accusations against either the Cambodian government or Chinese companies. The legal and political situation in Cambodia is made difficult by a number of factors: a neo-patrimonial, authoritarian polity in combination with elite economic interests produces a legally problematic situation in which personal interests and perceptions outweigh abstract administrative processes. This explains the failure of environmental and social regulation in Cambodia.

9.1 Introduction

Since the second half of the 1990s, the Chinese government has been increasing its political and economic profile in neighbouring countries. High-level visits, economic investments and aid have improved Chinese influence in countries across Southeast Asia, including Cambodia.

O. Hensengerth (✉)
Department of Social Sciences and Languages, Northumbria University, Newcastle, UK
e-mail: Oliver.Hensengerth@northumbria.ac.uk

W. Scheumann and O. Hensengerth (eds.), *Evolution of Dam Policies*,
DOI: 10.1007/978-3-642-23403-3_9,

One vehicle used to create influence is the holding of the annual summits of the Greater Mekong Subregion and other regional fora, at which Chinese leaders announce generous aid-investment packages that have been agreed beforehand. For example, at the Greater Mekong Subregion Summit held in Kunming in 2005, the Chinese and Cambodian leaders publicly announced that an agreement had been reached to finance and build the Kamchay Dam with an investment of US$280 million. At the time, this was the single largest foreign-invested project in Cambodia.

This study considers Chinese hydropower investment in Cambodia, examining the degree to which Chinese companies adhere to international or national environmental and social standards and analysing the roles played by and the interactions between Chinese and Cambodian actors in the building of hydropower capacity in Cambodia. The example used is the Kamchay Dam, built by Sinohydro and financed by the China Export Import Bank (China Exim Bank).

An analysis of the distribution of responsibilities enables us to take a close look at interactions, processes and conflicts between the two sides in the planning and implementation of hydroelectric dams. It also enables us to identify weak points in the environmental and social regulations and to direct criticism more precisely, thus avoiding general accusations against either the Cambodian government or Chinese companies.

The fact that all hydroelectric dams in Cambodia are concessionary build-operate-transfer projects gives the construction company a major stake in the project as it recovers its costs through the operation of the dam. This raises a number of questions: Does the government monitor the company? Does the company operate under a strict environmental supervision regime? What is the status of Environmental Impact Assessment (EIA) in the project approval process? What is the role of the company in the EIA process?

Data for the study were collected during a 3-week field trip to Phnom Penh, the Kamchay Dam area and the Kirirom III Dam area in September 2010. Interviews were conducted with non-governmental organisation (NGO) personnel, consultants and researchers, local councillors, and government officials. Due to the political sensitivity of the topic in Cambodia, all interviewees were assured anonymity. The interviews are therefore coded.

9.2 The Role of Chinese Investment in Cambodia

During the past decade, China has played a major role as investor and donor in Cambodia. The 2011 Cambodia Development Effectiveness Report shows that China has become the second largest bilateral provider of aid, with disbursements in 2010 amounting to US$138 million, or 12.9 % of total aid, up from US$46.6 million in 2005. Of the US$138 million, US$122 million went to the transport sector and US$16 million to the agricultural sector. Japan has consistently been the largest bilateral donor, with disbursements of US$146 million, or 13.6 %, in 2010, up from US$111 million in 2005 (Cambodian Rehabilitation and Development

Board of the Council for the Development of Cambodia 2011, iv; 15 Table 5; 30 Table 1.2). Chinese aid to Cambodia has thus roughly tripled.

Where approved investment is concerned, China has been the largest (2008, 2009) or second largest (2006, 2007, 2010 and the first 9 months of 2011) investor at least since 2006[1] (Council for the Development of Cambodia 2012, II-5 Table II-2-1). It should be noted that the amount actually invested—when compared to approved investment numbers—is often smaller. However, China's consistent ranking at or near the top reveals its economic importance for Cambodia.

In terms of hydropower investment, the hydropower development plan of the Department of Energy Development in the Ministry of Mines and Energy lists the construction of a number of large hydropower and coal-fired power stations by 2020. An updated table for the plan published by the Council for the Development of Cambodia in 2012 shows that, of the hydropower stations built or to be built under this plan, Chinese companies account for seven: Kamchay (operational since December 2011), Kirirom III, Stung Atay, Stung Tatay, Lower Stung Russey Chhrum, Stung Chhay Areng and Sambor. Two hydropower stations are being built by Vietnamese companies (Lower Se San II and Lower Sre Pok II). The table does not identify a developer for Lower Sre Pok III and IV (Council for the Development of Cambodia 2012, IV-2 Table IV-2-2).

Consequently, not only is China the second largest donor and the largest or second largest investor, but Chinese companies also occupy a dominant position in the hydropower sector.

Cambodia is in urgent need of physical infrastructure and energy, much of the infrastructure in large areas of the country having been destroyed in the three decades of political conflict that began with the communist insurgency (1968–1975) and continued with the "secret" bombing of Cambodia by the United States (1969–1970), the Khmer Rouge regime (1975–1979) and the civil war (1979–1991). Large portions of the country have become difficult to access, mines are still a danger in former Khmer Rouge strongholds, which were not pacified until the late 1990s, and roads are often in a dilapidated state. The railways are still being rehabilitated.

Chinese aid and investment is criticised for ignoring social and environmental standards. Indeed, one of the main criticisms is that Chinese representatives do not take part in international coordination efforts to establish benchmarks for the effectiveness of aid or in attempts to lay down social and environmental standards for investment. The China Exim Bank, for instance, has not recognised the Common Approaches for Export Credit Agencies of the Organisation for Economic Co-operation and Development (OECD), but instead published its own environmental policy in 2004 and an extended version in 2007.[2] In Cambodia,

[1] China was the largest investor in 2008 and 2009, and the second largest in 2006 (after South Korea), 2007 (after Malaysia), 2010 (after South Korea) and in the first 9 months of 2011 (after the United Kingdom).

[2] For details of the China Exim Bank's environmental policy, see China Exim Bank (2007). For an assessment of how the policy is applied in Chinese hydropower investment in Ghana, see Hensengerth (2011).

Chinese officials have attended the Cambodia Development Cooperation Forum since 2007. This body coordinates the activities of Cambodia's OECD donors. However, China has not appointed representatives to the Technical Working Groups affiliated to the Consultative Group meetings (Middleton 2008, p. 23). OECD officials have also confirmed that there is no regular exchange, formal or informal, between OECD embassies and the Chinese embassy on matters of development cooperation in Cambodia.[3]

It is this secretive behaviour of Chinese agencies that invites speculation as to the environmental and social soundness of Chinese investment. To some extent, their conduct can be seen as a consequence of the strong public criticism levelled at Chinese companies and government agencies for their high-profile engagement in major projects around the world, particularly since the Going Abroad Strategy was launched in 2003.

At the same time, the environmental bureaucracies of developing countries are rarely strong, and in Cambodia the focus on economic development has led to its environmental bureaucracy being sidelined in the investment approval process. A combination of a weak environmental bureaucracy, the hesitation of Chinese agencies to open their operations to public scrutiny and the Chinese government's policy of abiding by the norms and standards in force in the host country has resulted in the neglect of environmental and social safeguards in investment decisions and project implementation.

Increasing attention is also being paid to the developmental effects of Chinese aid and investment. The Cambodia Development Resources Institute, for example, has been engaged in research on the subject (Ouch 2010), and Kubny and Voss have conducted a similar study, taking the garment sector as their example. They conclude that Chinese investment has neither more nor less impact on development than investment from developed countries and that the social effects are not particularly detrimental, salaries in Chinese-invested firms being very similar to those in firms in which companies from developed countries have invested (Kubny and Voss 2010, p. 12).

However, when the environment is considered, the picture seems very different, at least on the surface. An interviewee in a government agency argued that Chinese garment companies in particular operate without having undergone an EIA.[4] According to the 1996 Environmental Protection and Natural Resources Law and the 1999 Sub-decree on the Environmental Impact Assessment Process, this would make the operation of these companies illegal. Yet they appear to operate without government interference. This raises a number of issues: the economic interests of Cambodia's political elite, pervasive corruption and a biased investment approval

[3] Conversations with officials from OECD embassies in Cambodia, 6 April 2010 and 10 September 2010.

[4] Interview P07092010.

process. During the following analysis we will see how this may be relevant to the hydropower sector and Chinese investment in it.

The following sections explore Chinese hydropower investment against the backdrop of Cambodia's energy situation, the relevance of environmental and social standards in Cambodia's investment approval and implementation process and the division of responsibilities between Cambodian and Chinese actors. In the final section, the findings are applied to Sinohydro's investment in the Kamchay Dam.

9.3 Cambodia's Energy Situation and the Role of Hydropower Development

Cambodia suffers from a lack of power supply. Before the Kamchay Dam became fully operational in December 2011 with an installed capacity of 194 MW, Cambodia's installed capacity was 600 MW, of which 23 MW were supplied by hydropower installations. The remainder was generated with diesel oil, heavy fuel oil and electricity imported from Vietnam, Thailand and Laos. Power is imported via transmission lines from Vietnam and Thailand (Vietnam: 220 kV; Thailand: 115 kV) and from all three countries via low-voltage lines to replace diesel oil in border communities.[5] Cambodia also imports coal from Australia and Indonesia (Council for the Development of Cambodia 2010).

For transmission, there is so far no national grid, only a special-purpose grid for energy imports from Vietnam, Thailand and Laos.[6] Provincial towns and cities have their own power stations and distribution networks. These local power stations are often fuelled by imported diesel. Prices are high, ranging from US$ 0.16/kWh to US$ 0.90/kWh in rural areas (Council for the Development of Cambodia 2010). In 2003, Cambodia had 24 isolated power systems (Ministry of Industry, Mines and Energy 2003, p. 7). The government plans to have a national grid in place by 2018.[7]

Writing in 2009, Ryder quotes the Electricity Authority of Cambodia stating that 95 % of power in Cambodia is generated with diesel fuel. Only 54 % of urban households and 13 % of rural households have access to grid-based electricity. With 85 % of Cambodians living in rural areas, only 18 % of the total population have access to grid-based electricity (Ryder 2009, pp. 21–22). This concurs with a calculation by the Council for the Development of Cambodia according to which only 20 % of the population have access to electricity mains. In the countryside, people use diesel generators and car batteries to generate electricity (Council for the Development of Cambodia 2010).

[5] Interview P10092010a.

[6] Interview P10092010a.

[7] Interview P10092010a.

At present, Cambodia produces energy from three hydropower stations: one small dam, Kirirom I, has an installed capacity of 12 MW. It was originally built by Yugoslavia in 1968 with an installed capacity of 10 MW, but stopped operating in 1970 because of the war (Ministry of Industry, Mines and Energy 2003, p. 6). In 1999, the Cambodian government tasked China Electric Power Technology Import and Export Corporation (CETIC) to rehabilitate the power station with an installed capacity of 12 MW. It was re-commissioned in May 2002 (Ministry of Industry, Mines and Energy 2003, p. 6; Sinohydro 2008, p. 13). A second small dam, O Chum II, was commissioned in 1993 with an installed capacity of 1 MW. Cambodia's first large hydropower station, on the Kamchay River, became operational in December 2009 with a capacity of 10 MW and has been operating at full capacity since December 2011.

9.4 Planning Processes for Dams

To enable rapid investment in hydropower, the government has devised a liberal investment regime that actively encourages private-sector participation in build-operate-transfer (BOT) hydropower projects.[8] Under investment regulations, Chinese companies provide the investment and financing from Chinese financial institutions, and in return the Cambodian government undertakes to purchase the power generated.[9]

It is significant in this context that Cambodia does not have the financial and technical capabilities to build hydroelectric dams on its own. All dams are therefore concessionary BOT projects; there are no turnkey dam projects, which would be handed over to the government for operation after being commissioned. However, the government regulates investment and contracts with the Chinese investor, including agreements under which the government commits itself to purchasing power,[10] as was the case with the Kamchay Dam. The government also encourages private investment in transmission lines. In 2007, for example, it granted the China Yunnan Corporation for International Techno-Economic Cooperation "a concession to finance and build a medium-voltage (230 kV) transmission line linking the country's southern transmission grid to the western grid. The concession is part of China Yunnan Corporation's Stung Atay hydropower dam concession package" (Ryder 2009, p. 31). Cambodia's 2007 Concession Law defines a concession as follows:

> any act attributable to the state whereby a competent institution entrusts to a private third party the total or partial implementation of an Infrastructure Project for which that

[8] Interview P10092010a.

[9] Interview with a representative of an international organisation in Phnom Penh, 14 September 2010.

[10] Interview P08092010.

institution would normally be responsible and for which the third party assumes a major part of the construction and/or operating risks or receives a benefit by way of compensation from government revenue or from fees and charges collected from users or customers (Royal Government of Cambodia 2007, Article 3).

According to the 2001 Electricity Law, the Electricity Authority of Cambodia is the national regulator and, as such, issues licences to producers and distributors, approves performance standards and ensures fair electricity tariffs. Independent power producers, i.e. private companies, need a licence from the Electricity Authority of Cambodia. For hydropower, these licence holders implement BOT projects and are required to sell the electricity to Electricité du Cambodge, the state-owned utility company that has the authority to generate, transmit and distribute electricity (Ryder 2009, pp. 24–30).

Within the Ministry of Industry, Mines and Energy, the Hydropower Department of the General Department of Energy is responsible for hydropower planning. It selects locations and projects, oversees construction and prepares and implements national hydropower policies and development plans.

For Sinohydro's Kamchay investment, this process meant that Sinohydro needed a licence from the Electricity Authority of Cambodia.

9.4.1 Investment Approval Process

The agency responsible for approving private-sector investment is the Cambodian Investment Board of the Council for the Development of Cambodia. The investment application for a dam submitted by the project proponent (here: Sinohydro) must contain a master plan that details dam management, water management, environmental protection mechanisms and the social benefits the project will bring for the local area. This master plan must be approved by the Ministry of Land Management, Urban Planning and Construction, the Ministry of Industry, Mines and Energy and the Ministry of Water Resources. They approve the construction plan, the water management plan and the power supply to the end user.[11] According to the Ministry of Industry, Mines and Energy, the Ministry of Land Management grants the Construction Licence, and the Ministry of Water Resources and Meteorology grants the Water Use Permit_[12] under the 2007 Water Resources Management Law. However, writing in 2008, McCallum claims that a sub-decree specifying the permit process has yet to be adopted (McCallum 2008, pp. 62). The Ministry of Environment grants the Environment Licence. The government leases the land to the project company under a land lease agreement.

[11] Interview at the Cambodian Investment Board, 8 September 2010.

[12] Interview at the Ministry of Industry, Mines and Energy, 10 September 2010. Middleton (2008, p. 20), however argues that the Ministry of Water Resources and Meteorology also grants the water construction licence.

All licences and permits are free of charge in order to prevent investment costs from rising.[13]

When it receives an application for a concessionary investment project, the Cambodian Investment Board approves investment incentives in accordance with Chapter 5 of the 1994 Investment Law, specifies all the authorisations required [i.e. EIA, construction licence, water use permit etc.] and obtains these authorisations on behalf of the applicant (Royal Government of Cambodia 2007, Article 8; Royal Government of Cambodia 1994, Chapter 5).

The incentives for the Kamchay investment are: tax-free/duty-free import of construction equipment and materials and tax holidays. The government also guarantees that there will be no price controls and no nationalisation until the end of the BOT contract period and that the company will have the option of renewing the contract at the end of the BOT period should the government be unable to operate the dam.[14]

All energy projects, including output capacity and years of operation, have to be approved by Electricité du Cambodge and the Ministry of Industry, Mines and Energy.[15] This also means that the line ministries send their letters of approval for the granting of individual permits and licenses to the project proponent through the Ministry of Industry, Mines and Energy rather than through the Cambodian Investment Board.

Having considered government coordination in hydropower development, Sam Chamreoun argues that the "institutional structure and arrangements in Cambodia are highly compartmentalized and they lack mechanisms for coordination and feedback among key agencies dealing with numerous water resources management activities. In addition, there is no clear definition of the regulatory and development functions" (Sam no date, p. 14).

When the applicant submits the project application, it needs to indicate that the firm has secured financial support.[16] The procedure is defined in the 2007 Concession Law, which states that the concessionaire must finance the infrastructure project (Royal Government of Cambodia 2007, Article 16). According to Article 20 of the Concession Law, the project owner is also responsible for the design and construction of the infrastructure project (Royal Government of Cambodia 2007). By its own admission, the Cambodian government lacks the skills for the permanent and thorough monitoring of the quality of construction in such areas as civil works and electrical engineering. It therefore commissions an independent expert to perform this task. The expert is active during the construction period.[17]

[13] Interview P10092010a.

[14] Interview P08092010.

[15] Interview P10092010a.

[16] Interview P08092010.

[17] Interview P10092010a.

BOT projects require two agreements: an Implementation Agreement (i.e. the concession agreement) and a Power Purchase Agreement.[18] For the expropriation involved in the concession project and the conclusion of the Implementation Agreement and Power Purchase Agreement, the project company has to carry out a preliminary study, which includes "a small paragraph"[19] on environmental effects, followed by a pre-feasibility study, including an Initial EIA, and then a feasibility study. All three studies must be approved by the General Department of Energy in the Ministry of Industry, Mines and Energy. The Initial EIA requires separate approval by the Department for Environmental Impact Assessment Review (EIA Department) in the Ministry of Environment. After the feasibility study has been submitted, the General Department organises a workshop attended by the ministries responsible (most importantly, the Ministries of Land Management, Water Resources and Meteorology, and Environment, and the Departments of Forestry and Land of the Ministry of Agriculture). Also represented are local authorities (district and commune) and communities. The project company is also present and is required to present its findings on the effects the dam will have on hydrology, ecosystems, fisheries, navigation, geology, the land-holding situation, etc. If the presentation is not approved, the company must revise the study, another workshop then being convened. Once the report is accepted, the Ministry of Industry, Mines and Energy issues a letter of approval.[20]

The next step is for the Ministry of Industry, Mines and Energy, the Ministry of Environment, Electricité du Cambodge, the Ministry of Finance and the Electricity Authority of Cambodia to join with the project company in negotiations on the Implementation Agreement and the Power Purchase Agreement. For this, the project company must submit a financial proposal and a technical proposal, including a full EIA report, if so required by the EIA Department of the Ministry of Environment.[21]

The technical proposal must specify the type of dam to be built (such as roller compacted concrete, rock fill or land fill), clarify such procedures as reservoir clearance before construction, ensure that the equipment installed complies with such international standards as those set by the International Organisation for Standardisation (ISO) and implement procedures to ensure dam safety.[22]

After the Cambodian Investment Board has approved the investment, it initiates a monitoring mechanism to ensure that the project is being implemented in accordance with the permits, including the EIA, and the investment incentives are being properly used[23]:

---

[18] Interview P10092010a.

[19] Interview P10092010a.

[20] Interviews P10092010a and P16092010.

[21] Interview P10092010a.

[22] Interview P10092010a.

[23] Interview P08092010.

- If the company breaches any permits, such as those attached to the environment permit, the monitoring department and an inter-ministerial task force write a report to the government and may impose a penalty and revoke the investment licence.[24]
- There is also a post-audit mechanism which scrutinises how a project has used the investment incentive. The Cambodian Investment Board approves everything associated with the construction work, namely the use of materials and natural resources. The post-audit mechanism is designed to monitor their use and the company's adherence to the master plan.[25]

However, when the project becomes operational, monitoring ceases.[26]

9.5 Environmental Impact Assessment Procedure

The legal basis for EIA in Cambodia comprises the 1996 Law on Environmental Protection and Natural Resources Management, the 1999 Sub-decree on Environmental Impact Assessment Process and the 2009 *Prakas* on General Guidelines for Conducting Initial and Full Environmental Impact Assessments.[27]

The first step in creating a legal basis for EIA was taken in 1995 when the EIA Department was established in the Ministry of Environment with support from the Asian Development Bank (ADB) as part of a capacity-building Technical Assistance programme.[28] In 1996 the National Assembly then passed the Law on Environmental Protection and Natural Resources Management. This was followed by the government's approval of the Sub-decree on the Environmental Impact Assessment Process, which entered into force in 1999. The Annex to the Sub-decree lists public and private projects that are subject to Initial and Full EIA.[29]

The Annex stipulates that EIA is required for all hydropower stations producing more than 1 MW (Royal Government of Cambodia 1999). An interviewee in the Ministry of Environment explained that hydropower stations of less than 50 MW must undergo Initial EIA, those larger than 50 MW Full EIA. Some large projects have particularly serious problems in that they entail resettlement of the local population. It is these projects in particular that need a Full EIA.[30] However, it is not clear whether these categories are rigorously applied.

---

24 Interview P08092010.

25 Interview P08092010.

26 Interview P10092010a.

27 Sub-decrees are instruments for implementing laws. They are approved by the Council of Ministers but, unlike laws, do not require parliamentary approval. *Prakas* are regulations adopted by a ministry and are binding only on the ministry in which they are adopted. For a brief overview of the hierarchy of Cambodia's legal norms see, for example, McCallum (2008, pp. 81–82).

28 Interview P16092010.

29 Interview P16092010.

30 Interview P16092010. See also Royal Government of Cambodia (1996).

As hydropower stations require EIA, the project developer (here: Sinohydro) must conduct an Initial EIA. If the Ministry of Environment determines that the project will have a serious environmental impact, a Full EIA is required (Sam no date).

The 2009 *Prakas* lays down substantive requirements for the Initial and Full EIA, including a description of the legal framework, the project, the environmental resources (i.e. the physical, biological and socio-economic conditions), public participation, social and environmental impacts and mitigation measures, and the environmental management plan (environmental protection measures, required training, monitoring schedule and required budget). The Full EIA must also include an economic analysis and an analysis of environmental value (Ngy 2010).

Once the EIA documents have been approved, the Ministry of Environment sends the letters of approval to the project owner and to the Ministry of Industry, Mines and Energy, which gives final project approval.[31]

Article 3 of the Sub-decree states that, during the construction period, the company must comply with the Environmental Management Plan.[32] Both the Initial and Full EIA studies must be accompanied by such a plan. It must include a monitoring schedule and details of an endowment fund financed by the project company for reforestation and construction of infrastructure (roads, etc.), of skills training for people affected by resettlement (for fishing and breeding such animals as pigs, for example) and of a community credit fund.[33]

The Ministry of Environment monitors the Environmental Management Plan—"according to project activity".[34] If problems arise for local people or the environment, the Ministry of Environment will monitor the situation—environmental pollution, for instance—at various intervals, depending on the gravity of the problem. The Environmental Management Plan defines the frequency of monitoring: once or twice during the construction phase, at least every three months during the operation phase and during the closure phase.[35]

As for the standards applied, the EIA Sub-decree refers to those laid down by the Ministry of Environment on solid waste, water pollution, air pollution and noise.[36] An interviewee in the Ministry of Environment claimed that the EIA for Kamchay and all other EIAs in Cambodia comply with ADB standards.[37] This would seem obvious, since Cambodia's EIA institutions were set up with the ADB's assistance. However, whether SAWAC Consultants, the company hired by Sinohydro to conduct the EIA, and other domestic EIA companies refer to ADB standards in their EIA reports could not be ascertained. According to another

[31] Interview P20092010a.

[32] Interview P07092010.

[33] Interview P16092010.

[34] Interview P20092010a.

[35] Interview P20092010a.

[36] Interview P20092010a.

[37] Interview in the Ministry of Environment, 16 September 2010.

official in the Ministry of Environment, the application of international standards may depend on the involvement of the ADB and World Bank, which would automatically apply their own standards.[38]

9.5.1 How Effective is the EIA Regime?

While it looks good on paper, a number of significant problems are associated with the EIA process. They are due to two factors: first, the low status of the Ministry of Environment in the governmental hierarchy, resulting in poor enforcement of environmental legislation; second, a lack of implementing legislation for key laws.

Technically, the Ministry of Industry, Mines and Energy is required to approve a hydropower project only when all government agencies have issued their technical letters of approval. However, an interviewee in a government ministry said that, while all hydropower projects have so far undergone an EIA process, the Ministry of Industry, Mines and Energy does not necessarily wait for the Ministry of Environment to approve the (Initial) EIA of small projects, such as (Chinese) garment factories.[39] In addition, because the project proponent is required to pay for the EIA,[40] the report is often lacking in thoroughness.

Only five Full EIAs had been required before the interviews were conducted (September 2010). The first of these was for the Vietnamese-built Lower Sesan 2 Dam. Further Full EIAs (e.g. for the Chinese-built Kamchay and Kirirom III Dams) were being carried out at the time. A government interviewee claimed that the Stung Treng and Sambor Dams would probably need a Full EIA, because they were on the Mekong mainstream[41] and the rules of the Mekong River Commission therefore applied.[42] One reason for the low number of Full EIAs so far is that they take more time and require a larger investment budget. An interviewee in the Ministry of Environment claimed that if the Ministry required a Full EIA for all projects, it would be difficult for investors to come to Cambodia. The government clearly does not want the EIA report to be an obstacle to investment.[43]

Problems encountered during EIA reflect the general difficulty of enforcing Cambodia's environmental laws. It appears that the government often places the emphasis on speed rather than environmental thoroughness. Referring to the progress made by Huadian, the Chinese company that is building the US$ 412 million Lower Stung Russei Chrum Dam in Koh Kong province with finance from the China Exim Bank, Ith Prang, Secretary of State in the Ministry of Industry,

[38] Interview in the Ministry of Environment, 20 September 2010.

[39] Interview P07092010.

[40] Interview P10092010a.

[41] Interviews P20092010a and P07092010.

[42] Comment by an OECD embassy official, 9 August 2011.

[43] Interview at the Ministry of Environment, 7 September 2010 and 20 September 2010.

Mines and Energy, said: "We are trying to push the company to speed up and finish its construction so that power can be generated to respond to the power shortage in the country" (Dyer and Chun 2010).

This is also partly the reason why the Ministry of Environment cannot exploit its full sanctioning potential. Articles 28 and 29 of the EIA Sub-decree allow the Ministry of Environment to halt construction activities and penalise the project owner for violations of the Environmental Management Plan. However, Article 28 makes it clear that, to halt activities, the Ministry must cooperate with other competent agencies. An interviewee in the Ministry of Environment said that no firm had yet been penalised for failing to comply with EIA procedural requirements or the requirements set out in the Environmental Management Reports.[44]

The Environment Ministry's low status is also the reason why the 1999 EIA Sub-decree has yet to be revised, even though the EIA Department has identified the need for reform (see Table 9.1). Revision is one of the main goals for the EIA Department's Strategic Plan 2008–2013.[45]

The need for reform goes together with a need to raise the profile of the EIA process in the investment approval procedure. According to the Ministry of Agriculture, for example, the Fisheries Administration is brought into the process at a very late stage. During the EIA process, the Fisheries Administration works with the Ministry of Environment and is asked to comment on the impact of the dam on fisheries.[46]

However, when the EIA report is sent to the Fisheries Administration, the decision for a dam has already been taken in principle by the Ministry of Industry, Mines and Energy. The Fisheries Administration would therefore like to see investment proposals for hydropower dams not reaching the EIA stage before it has had a chance to comment on them. By the time it is asked to comment on the EIA report, the decision-making process is already too far advanced and it can only recommend mitigation measures.[47]

What is more, the section of the EIA report that concerns fisheries is very weak. The Fisheries Administration has therefore been developing a Fisheries Impact Assessment (it has completed a draft). EIAs often assess only the local impact, but as fish migrate between their various habitats, the impact may be more widespread. This must be considered in the EIA format. But, so the argument goes, it is not possible to discuss this with the Ministry of Environment, because it abides by the letter of the 1999 EIA guidelines. Nor does the Fisheries Administration have a mandate to work with the project developers' engineers on the establishment of management plans.[48]

---

[44] Interview P07092010.

[45] Interview P07092010.

[46] Interview P20092010b.

[47] Interview P20092010b.

[48] Interview P20092010b.

Table 9.1 Strengths and weaknesses of Cambodia's 1999 EIA regime (Ngy 2010)

Strength	Weakness
EIA is legally required for all planned and ongoing development projects	Penalties are complex and difficult to enforce
The EIA process is defined in the government sub-decree (1999)	The sub-decree on the EIA process has many shortcomings
There is a general guideline on the drafting of EIA reports	The general guideline has many deficiencies, and it has only just (September 2009) been declared officially applicable
	The are no sectoral guidelines
Officials have undertaken to perform their duties	Not enough staff
	Staff capacity is limited
	No incentive for staff
The Ministry of Environment and the Ministry of Economy and Finance have issued a joint declaration on the review and monitoring of fees	The joint declaration has many deficiencies, and money collected can be used directly for work associated with EIA reviews and monitoring
Project owners are required to contribute some money to the environmental endowment fund	While some project owners do contribute as required by their contracts, there is no legal framework for fining them if they do not
Capital and provincial environment departments are delegated some EIA-related duties and responsibilities by MoE declaration	Provincial officials do not have the skills to carry out EIA-related work
	The MoE declaration has some gaps, resulting in ambiguity as to duties and responsibilities
There is an EIA training curriculum	The training curriculum is not good enough
	There is no EIA training toolkit
	There is no budget for organising training courses
There is some cooperation from line-ministries	The cooperation is very limited and usually occurs only in the context of EIA reviews

The second problem with EIA occurs when other legislation, particularly the 2007 Concession Law, the 2007 Water Resources Management Law and the 2010 Expropriation Law, comes into play during the process of dam approval and construction. None of these laws is currently accompanied by an implementing sub-decree specifying the administrative processes. For the Water Resources Management Law, for instance, this means that there is no formal procedure for the issue of water use permits (McCallum 2008, p. 62). For the same reason, considerable confusion exists over the extent and form of public participation in project planning (Sam no date, p. 21). Consequently, the entire approval and construction and post-construction monitoring process is of an informal and personalistic nature. McCallum contends that a:

> challenge posed by the Cambodia legislative environment is the degree to which different laws coordinate with each other. The main issue here is that laws and ministries often over-lap in areas of jurisdiction, a situation that leads to contradictory interpretations of how things, such as natural resources, should be managed and used (McCallum 2008, p. 82).

This results in a "failure of laws to 'talk to each other'" (McCallum 2008, p. 82) and so in unclear legal processes.

One of the main stepping stones leading to a clearer legal environment is the 2007 Concession Law. The Concession Law was drafted by international experts drawn from such international organisations as the World Bank and the World Trade Organization (WTO).[49] It was passed by the National Assembly in 2007. A World Bank Technical Assistance programme for the Sub-decree on implementation of the Concession Law was finalised in April 2009.[50]

However, in September 2010 the Draft Sub-decree was still being reviewed by an inter-ministerial committee on which the Cambodian Investment Board was represented. There was confusion over why the Sub-decree was still being reviewed at the time. One interviewee working in an international institution said that the delay might be a ploy to give the government time to amend the provisions of the Sub-decree.[51]

McCallum raises this point when he elaborates on Cambodia's legal system and the hierarchy of legal norms:

> Laws developed in Cambodia represent legal statements over the overall intent of the legislation, which provides guidance for the subsequent stages of implementation and enforcement. What is required, further on, are technical clarifications on how this will occur. This system has created a hierarchy of laws and regulations, with each deriving its validity and authority from the one above it; while those below offer further technical details on how those above them should be interpreted and implemented (McCallum 2008, p. 81).

From a practical point of view, laws can be passed more quickly under this system since they do not have to contain any technical details. However, it also allows the executive to delay the enforcement of laws significantly and to apply a wide interpretation (McCallum 2008, p. 82).

In 2010, the National Assembly passed an Expropriation Law, but it has yet to enter into force because the sub-decrees are still being prepared. However, expropriation must be considered during the EIA phase.[52] The next section will detail expropriation and resettlement requirements.

---

[49] Interview P14092010.

[50] Interview P10092010b.

[51] Interview P14092010.

[52] Interview P23092010.

9.6 Expropriation and Resettlement

To determine whether people qualify for compensation when facing expropriation, it is first necessary to ascertain whether they own or possess land. Cambodia's 2001 Land Law distinguishes between owners and possessors of land. The distinction was made because the cadastral system was destroyed during the Khmer Rouge period and land frequently changed hands during the successive stages of the Cambodian wars (Royal Government of Cambodia 2001; Chan et al. 2001; NGO Forum on Cambodia 2009).

Legal ownership is officially corroborated by a land title. Article 44 of Cambodia's constitution states that the state may expropriate land only in the public interest and only following "just and fair compensation." Article 5 of the 2001 Land Law specifies this:

> No person may be deprived of his ownership, unless it is in the public interest. An ownership deprivation shall be carried out in accordance with the forms and procedures provided by law and regulations and after the payment of fair and just compensation (Royal Government of Cambodia 2001).

In contrast, while possessors do not have a land title, they have a legal right to live on and use the land. Under the 2001 Land Law, individuals may legally possess land if they satisfy a number of conditions:

- They must have moved on to the land before the Land Law entered into force in August 2001. If they moved on to the land after that date, they must prove that they purchased it from someone who legally possessed it before August 2001 (Article 30).
- In addition, the 2001 Land Law sets five conditions: possession must be "unambiguous, non-violent, publicly known, continuous and in good faith" (Article 38).

Importantly, the state cannot expropriate the land of legal possessors, even in the public interest. Before expropriation can take place, legal possession must be turned into legal ownership, i.e. the state authorities must issue a land title. Possession is transformed into ownership after a period of possession of five years. Where people legally possessed land for less than five years prior to the entry into force of the 2001 Land Law, the period of possession can be extended until the five years have been completed (Articles 30 and 31).

Proving legal possession of land is often difficult, and land grabbing by politically well-connected companies is frequent. Indeed, despite being enshrined in the Land Law:

> possession rights are not consistently recognized by local or national authorities. [...] In many cases where untitled residents have found themselves lying in the path of development projects they have been labelled squatters, irrespective of whether or not they have rights as legal possessors under the Land Law (Grimsditch 2012, p. 34).

State public, state private and communal land cannot be owned or possessed by individuals. Individuals living on such land can be evicted without compensation (Royal Government of Cambodia 2001, Articles 16, 18, 19, 20–28, 43, 248).

As for resettlement, no clear legal framework exists. The following procedures are currently in place: in the case of publicly financed projects (such as World Bank, ADB and bilateral aid projects), the donor develops a resettlement plan in cooperation with the Inter-Ministerial Resettlement Committee under the Ministry of Economy and Finance. The donor also monitors compliance with the plan. In the case of private projects, the developer is responsible for planning and implementation, but the Inter-Ministerial Resettlement Committee has to review and approve the resettlement plan. It is the project owner's responsibility to conduct socio-economic surveys to identify the people affected by the project and to pay them compensation.[53]

Projects in the public interest are governed by the 2010 Expropriation Law, the Ministry of Economy and Finance being the responsible institution. However, there is still no sub-decree specifying the procedure. At the request of the Ministry of Economy and Finance the ADB approved Technical Assistance 4490 (*Enhancing the Resettlement Legal Framework and Institutional Capacity*) in 2004 and, together with the Ministry of Economy and Finance, published a draft Sub-decree on Land and Property Acquisition and Addressing the Socio-Economic Impacts by the State's Development Projects in May 2007 (ADB 2004; Centre on Housing Rights and Eviction 2008, p. 3). However, the draft Sub-decree was never adopted. At present, procedures similar to those applicable to private projects therefore seem to apply to public-interest projects (personal communication from an independent consultant, 7 July 2012).

People facing expropriation have the following compensation options: first, cash equivalent to the market or replacement price of their land, house and crops[54]; second, compensation in kind for the land and the house on it and for damages; third, compensation paid by the investor to people for any items partly affected.[55] However, according to an independent consultant, people whose property has been expropriated are usually offered different terms, but none the true market value. Indeed, the developer's interest in compensation seems to determine what terms are offered. The government does not play much part in the proceedings, tending to do no more than act as the middleman between the company and the people affected (personal communication, 30 April 2012).

While the 2010 Expropriation Law specifies the types of compensation, it does not say anything about the amount, except that it should be "fair and just". A ministerial official argued that the total compensation per household is US$

[53] ADB 2007, p. 55; interview at the Ministry of Environment, 20 September 2010; interview P20092010a; interview P23092010; personal communication from an independent consultant, 7 July 2012.

[54] Interview at the Ministry of Industry, Mines and Energy; interview P20092010a; and Royal Government of Cambodia 2010, Articles 22–24.

[55] Interview P20092010a and Royal Government of Cambodia 2010, Articles 22–24.

20,000–30,000 for economic trees, land, houses and transport. Compensation for economic trees is based on their age and economic output. A mango tree, for example, yields US$ 100–200 a year when the mangos are sold at the market. Production is then projected to last 5–10 years, and the investor supports the household for 1–5 years.[56]

Where projects do not involve resettlement, communities may still be affected. However, according to one interviewee, even though dams, such as Kirirom III and Kamchay, affect downstream communities, there is no need to compensate them, because compensation is linked to the direct impact of the reservoir. The project owner must nonetheless mitigate downstream effects.[57] According to an independent consultant, however, there is no clear legal framework to show how this should be done (personal communication, 30 April 2012).

9.7 Planning and Financing for the Kamchay Dam

When the Cambodian and Chinese governments agreed on the construction of the Kamchay Dam in 2006, it was the single largest foreign direct investment project in Cambodia. The Kamchay Dam is situated in the Bokor National Park on the Kamchay River in the Elephant Mountain Range in the Province of Kampot. Kamchay had been planned since the 1950s, first by the Soviet Union, then by Japan in 1992 and by Canada in 1995. The Canadian International Development Agency (CIDA) had funded the feasibility study by Experco International (Ministry of Industry, Mines and Energy 2003, p. 5; Malmquist and Sigfridsson 2002, p. 7). Following international protests, including Probe International's 1996 Mekong Campaign, CIDA withdrew its support for the project (Labelle 1997). SAWAC Consultants also blamed high costs for the withdrawal of all three countries from the project (Cheang 2007).

Following this lack of support, the Cambodian government, represented by the Ministry of Industry, Mines and Energy, held another international bidding contest for the Kamchay Dam between June 2004 and January 2005. Around seventeen companies from Cambodia, Korea, Japan and China submitted offers. Sinohydro won the contract,[58] beating China Guodian among others (Middleton 2008, p. 56).

On 27 April 2005, the Ministry of Industry, Mines and Energy and Sinohydro signed a contract (Sam 2007, p. 1). On 4 July 2005, during the Second Greater Mekong Subregion Summit in Kunming, China's Prime Minister, Hu Jintao, and Cambodia's Prime Minister, Hun Sen, signed an agreement approving Sinohydro's plan for the construction of the Kamchay Dam. On 23 February 2006, the Ministry

[56] Interview P20092010a.

[57] Interview P20092010a.

[58] Middleton (2008, p. 19); Sam (2007, p. 1); interview at the Ministry of Industry, Mines and Energy, 10 September 2010.

of Industry, Mines and Energy and the Ministry of Economy and Finance signed the BOT agreement with Sinohydro, and Sinohydro and Electricité du Cambodge signed the Power Purchase Agreement (Sam 2007, p. 1; Cheang 2007; Sinohydro 2008, p. 5). Both documents were signed following a letter of authorisation from Prime Minister Hun Sen (Middleton 2008, p. 56). On 26 July 2006, the National Assembly voted to guarantee Sinohydro financial compensation should the dam underperform (Middleton 2008, p. 56). As the Initial EIA was not finalised or approved until October 2006 (Grimsditch 2012, p. 37), the EIA procedure played no part in the investment approval process. Construction of the dam began in September 2007 (Xinhua 2011).

It was claimed at the time of interview that much of the energy generated by Kamchay will be used for the Special Economic Zones in Sihanoukville.[59] One of these Special Economic Zones is being developed by the Chinese company Jiangsu Taihu in a joint venture with the Cambodia International Investment Development Group (ASEAN Affairs 2008), which is owned by Lao Meng Khin, a Senator in the ruling Cambodian People's Party. While the energy from Kamchay is unlikely to benefit Lao Meng Khin's Special Economic Zone on account of the construction of a large coal-fired power station producing 405 MW in the Zone (personal communication from an independent consultant, 21 June 2012; Council for the Development of Cambodia no date), Lao Meng Khin's interests appear to be served by the Kamchay investment in other ways: according to company registration documents, he and his wife are members of the four-person board of directors of Sinohydro Cambodia, an affiliate of the Sinohydro Corporation. One of the directors of Sinohydro Cambodia, Chen Decai, is also the sole director of Sinohydro Kamchay, the company in charge of building the Kamchay Dam. Both companies are registered under the same office address in Phnom Penh. At the opening ceremony of the Kamchay Dam, Prime Minister Hun Sen awarded Lao Meng Khin a medal for his contributions to the country (Caminfoweb 2012). Chinese investment in the Kamchay Dam is thus closely linked to leading politicians in the governing party with access to the Prime Minister. The project is therefore intertwined with the economic interests of Cambodia's political elite.

The Power Purchase Agreement approved by the National Assembly for Kamchay included a guarantee from the government to offset losses and overproduction during the dry and rainy seasons. According to an interviewee, this provision conforms to international standards for Power Purchase Agreements. The World Bank, the Asian Development Bank and UNIDO have advised the Cambodian government on private participation in infrastructure projects.[60] Table 9.2 summarises the main features of the Kamchay Dam project.

The Ministry of Industry, Mines and Energy is the main government department involved in the negotiations on the Implementation Agreement and the Power Purchase Agreement. All other departments (the Ministries of Water

[59] Interview P08092010.

[60] Interview P08092010.

Table 9.2 Main features of the Kamchay Dam project (Sinohydro 2008, p. 3)

Purpose	Wall height	Installed capacity	Average annual output	Total reservoir area	Licence period
Energy production	112 m	194 MW	508 GWh	20.06 km^2	44 years, including 4 years of construction

Resources, of Land Management and of Environment) send their licences to the company through the Ministry of Industry, Mines and Energy, and the company also submits its applications for licences through the Ministry of Industry, Mines and Energy.[61]

In this process, the Council for the Development of Cambodia approves such tax incentives for the investment as tax holidays and grants the investment licence.[62] The Ministry of Water Resources and Meteorology approved the water use permit (Middleton 2008, p. 20) and the Ministry of Environment the environment licence for the Initial EIA.

According to an interviewee in the Ministry of Industry, Mines and Energy, the government has attached no conditions to Kamchay except to require payment of compensation and the replacement of any lost economic trees by Sinohydro, which will also restore any areas of forest affected by the project to their original state. There are no fees associated with the project. According to the interviewee, any fees levied by the government would raise electricity tariffs, and the country's development would be impeded. An annual licence fee payable to the government would also pose problems for economic development. The interviewee went on to say that the Ministry of Environment, for instance, is not therefore allowed to charge for environment licences. If fees were charged, the investment costs would rise.[63]

9.7.1 *Financing for the Kamchay Dam*

Kamchay is funded as part of a Chinese US$ 600 million aid and investment package announced in April 2006 (Sam 2007, p. 1). Of the US$ 600 million, US$ 280 million makes up Sinohydro's investment in Kamchay (Box 9.1). In an interview conducted in Phnom Penh in September 2010, a representative of the Ministry of Industry, Mines and Energy confirmed that the investment was backed by a loan from the China Exim Bank.[64]

---

[61] Interview P10092010a.

[62] Interview P10092010a.

[63] Interview at the Ministry of Industry, Mines and Energy, 10 September 2010.

[64] Interview on 10 September 2010.

Box 9.1 China's US$ 600 million aid-investment package for Cambodia
Pledged by China during Wen Jiabao's visit to Cambodia in April 2006 (Associated Press 2006b), the package consists of the following:

- 11 bilateral agreement, including a low-interest loan, grant aid and 30 fire trucks (Associated Press 2006a)
- Sinohydro's investment of US$ 280 million for the construction of the Kamchay Dam (Associated Press 2006a; Perlez 2006; Ryou 2009)
- a fibre-optic network connecting Cambodia's telecommunications network to those of Thailand and Vietnam (Perlez 2006)
- a loan of almost US$ 200 million for infrastructure to promote economic development in remote areas (Associated Press 2006b)
- two roads totalling 147 miles in the east and northwest (Associated Press 2006b)
- two bridges near Phnom Penh (Prek Kdam Bridge over the Tonle Sap River and Prek Tamak Bridge over the Mekong River). Both bridges were financed through fully tied concessional loans at a fixed interest rate of 2 % for each loan, grace periods of 7 years and amortisation periods of 13 years (Associated Press 2006b; Perlez 2006; Cambodia ODA Database 2013; Ryou 2009).

Later in 2006 and in addition to the US$ 600 million package, China pledged US$ 195.7 million in preferential loans for roads and bridges, and an agreement on an interest-free loan of US$ 12.5 million was signed (Ryou 2009).

The government guarantees the investment costs in two cases: the purchase of electricity (if Electricité du Cambodge is unable to buy the electricity, the government steps in), and force majeure (the government buys out the company, the buy-out being calculated on the basis of the remaining years of the concession). Both cases are detailed in the Implementation Agreement and the Power Purchase Agreement.[65]

According to the Ministry of Industry, Mines and Energy, the Power Purchase Agreement specifies an internal rate of return of 6 %. The Ministry said that these rates are Exim Bank standards. To achieve such an internal rate of return, the loan from Exim Bank bears an interest rate of 6.2 or 6.5 %. The equity is 70–30 %, meaning that Exim Bank holds 70 % and Sinohydro 30 % of the investment. The Ministry said this conformed to international investment practice.[66]

According to a project design document drawn up to register Kamchay under the Kyoto Protocol Clean Development Mechanism, Sinohydro gives an internal

[65] Interview P10092010a.

[66] Interview on 10 September 2010.

rate of return of 7.48 %, taking US$ 0.08/kWh as the highest margin for electricity tariffs (Sinohydro 2008, p. 12). The project design document claims that this does not make the investment financially viable unless the internal rate of return reaches 9.5 %. Consequently, only registration under the Clean Development Mechanism will make the project financially viable (Sinohydro 2008, p. 10, 12).

This indicates that Chinese investment is often based on marginal profits, meaning that other competitors are outbid with low prices and subsidised government loans.[67] Given that the Kamchay investment was publicly announced at the 2005 summit meeting of the Greater Mekong Subregion, it can be argued that the Chinese government either actively supports or at least tolerates this.

9.7.2 Environmental and Social Effects of the Kamchay Dam

The environmental effects largely consist of a reduction in downstream water supply, water pollution and the impact on the Bokor National Park. Sinohydro claims that it will provide alternative habitats for animals affected by flooding and will replant 2,000 ha of forest near the Bokor National Park to compensate for the 2 % of the national park area that will be submerged. Sinohydro will further provide US$ 400,000 a year for a period of 30 years to implement the environmental management plan (Sinohydro 2008, p. 25; Middleton 2008, p. 61). It is not possible at this stage in the operation of Kamchay to assess whether Sinohydro is honouring its commitments.

When choosing a consulting firm to carry out the EIA, the project owner must confer with the EIA Department.[68] A member of a NGO expressed the opinion that Chinese companies seem to prefer local consultants, because international consultants may be too expensive. Furthermore, local consultants know the area and do not need extra time to acquaint themselves with local conditions.[69]

According to the Ministry of Environment, the EIA Department approved the Initial EIA conducted by SAWAC Consultants in October 2006 without any revisions and sent the letter of approval to the Ministry of Industry, Mines and Energy and to Sinohydro. Although the EIA Department required Sinohydro to carry out a Full EIA, the approval of the Initial EIA was sufficient for construction to begin.[70]

The Full EIA, also prepared by SAWAC Consultants, was necessary because the forest area that would be flooded was not addressed in the Initial EIA. To assess the forest area that would be submerged, an inter-ministerial evaluation of

---

[67] Comments made during a meeting with representatives of the German construction industry at the German Development Institute in Bonn, 1 October 2010.

[68] Interview at the Ministry of Environment on 16 September 2010.

[69] Interview P23092010.

[70] Interviews on 16 September 2010 and 20 September 2010.

the forest density was carried out. This evaluation was led by the Ministry of Economy and Finance and included the Ministry of Agriculture and the provincial government of Kampot (but not the commune and district governments). The reason for this was that the reservoir area had to be deforested before flooding in order to avoid water quality problems in the future reservoir. This needed to be addressed in the Full EIA.[71] In addition, the Full EIA must provide for an Endowment Fund for reforestation and a Social Fund for compensating the owners of plantations of the durian tree.[72] In March 2011, "a consultation workshop was held to discuss the draft Full EIA", but the only local representatives attending were government officials (Grimsditch 2012, p. 38). In September 2011, an inter-ministerial meeting was held to discuss a revised draft Full EIA. This was attended by some local village representatives, the NGO Forum and the human rights NGO Adhoc. However, invitations were issued at short notice, leaving civil society and community representatives no time to review the EIA in advance of the meeting (Grimsditch 2012, p. 38). The Ministry of Environment did not approve the Full EIA until 7 October 2011, two months before Kamchay became fully operational (International Rivers 2012).

Various problems should be mentioned at this juncture: first, given that the Full EIA was not approved until October 2011, four years after construction had commenced, its significance must be questioned. In the absence of a Full EIA, the full Environmental Management Plan also remained unknown to the public, yet affected people throughout the construction period (Grimsditch 2012, p. 38). Given Cambodia's semi-authoritarian political system, the involvement of elite interests in Chinese investment, the little importance attached to EIA in the investment approval process and the EIA Department's lack of capacity for monitoring construction, it will be virtually impossible for the environmental bureaucracy to impose significant penalties, if necessary, in accordance with Articles 28 and 29 of the EIA Sub-decree. Once it has been agreed in principle at the highest level and has prime ministerial approval (as announced at the Greater Mekong Subregion summit), a project will not be stopped by ministries that rank low in the government hierarchy.

Similarly, the implementation of the Environmental Management Plan will depend on Sinohydro's corporate social responsibility policy, since there will be almost no post-construction monitoring. The direct link between elite economic interests and the Kamchay investment is likely to give Sinohydro ample space to operate in a situation where crucial laws have yet to be followed by implementing instruments. However, Box 9.2 shows what pressure an informed and organised populace can have on plans to build dams in Cambodia.

[71] Interviews on 16 September 2010 and 20 September 2010.

[72] Interviews at the Ministry of Environment, 16 September 2010.

Box 9.2 Popular pressure and the Environmental Impact Assessment process

Reportedly the only example so far of popular pressure forcing the government to review the EIA process occurred during planning for the Vietnam-invested Lower Sesan 2 Dam, which appears to have caused government planners and the Vietnamese investor some serious headaches. Officials interviewed at the Ministry of Industry, Mines and Energy and the Ministry of Environment in September 2010 constantly reverted to the subject of Lower Sesan 2. A member of an NGO in Phnom Penh also stated that the construction of the dam had not yet started, because the Vietnamese company had had to revise the EIA several times after locals in the affected areas became aware of the effects that dams can have. They are already affected by the Yali Dam on the Vietnamese side of the border. Local communities therefore raised their concerns vociferously, and the EIA process took one year. Other communities, such as those in the Kamchay area, have no prior experience of the negative impact of dams.[73]

9.7.3 Public Participation

According to a Ministry of Environment official, public participation is required by Ministry of Environment guidelines, which are forwarded to the project developer and local people. Small projects are the subject of provincial-level consultations. For large projects, these are joined by national consultations led by an inter-ministerial committee and including the media and NGOs, particularly the NGO Forum on Cambodia, because it is the most influential NGO organisation and regularly cooperates with the Ministry of Environment. Sometimes international NGOs are invited, too. Flora Fauna International and Conservation International, for example, are observing a hydropower project in the Cardamom Mountains.[74] WildAid/Wildlife Alliance was invited to consultations on the EIA of the Lower Russei Chrum Dam.[75]

However, as already pointed out above, public participation in the EIA process is very limited. During the scoping phase, there is no public participation, only what an interviewee referred to as "closed scoping": the project owner hires a consultant and submits the Initial EIA report without local consultations. However, the reviewing team discusses the EIA report with local people during the review of the Initial EIA.[76]

---

[73] Interviews P23092010 and P16092010.

[74] This is the Stung Atay hydropower project.

[75] Interview on 16 September 2010.

[76] Interview P07092010.

Information-sharing between the government and communities and between the company and communities is limited. Sensitive information in particular is not shared, and public meetings for EIAs often resemble information meetings at which the community is presented with a positive picture of new or improved access to electricity, the provision of cheaper electricity or enhanced means of employment generation, while such negative impacts as a decline in the amount of fish caught, losses of income, uneven flows and threats of flooding are not mentioned.[77]

In the case of Kamchay, some government officials were worried about water pollution, while villagers seemed mainly concerned about dam safety, the possibility of the dam collapsing and landslides (Sinohydro 2008, pp. 31–39). The last point was confirmed during interviews with workers in the project area.[78] There was also some concern about losses of income in some communities, particularly Mak Prang village and among food vendors at Teouk Chhou (Sinohydro 2008, p. 37, 39). It is unclear, however, how far villagers were informed of the full extent of income losses due to the destruction of the bamboo forests. Information seems to have been selective, with the emphasis placed on benefits from electricity provision and employment generation as a result of a potential stimulus to business activity due to better access to cheaper electricity: the project design document expressly states that public consultations were held with the aim of presenting to the communities the improvements the dam would entail for the lives of local communities (Sinohydro 2008, p. 30).

What benefits local communities will eventually derive from the Kamchay Dam is unclear. Villagers in the Kamchay area mentioned rumours they had heard about their villages being connected to the electricity grid free of charge and about the provision of cheaper electricity. However, they conceded that no one from the provincial or central government had come to confirm these rumours.[79] Indeed, as mentioned above, the electricity from Kamchay seems chiefly to benefit the Special Economic Zones in Sihanoukville.

A local councillor explained that, although Sinohydro planned the project for his commune area, it had not interacted with local authorities, only with provincial and central branches of the Ministry of Industry, Mines and Energy and the Ministry of Environment. Nor, he said, had there been any consultations with local authorities during the planning of the dam; only provincial and central government authorities had been involved,[80] although Grimsditch (2012, p. 28) refers to one district-level meeting attended by a Sinohydro representative and by commune and village chiefs. During the meeting, villagers were told that they would lose access to the bamboo forest, but that they would be able to apply for jobs at the construction site. However, apart from this meeting, the discussion of planning and

[77] Interview P23092010.

[78] Interviews K18092010c.

[79] Interviews K18092010c.

[80] Interview K18092010a.

design was almost entirely confined to national government agencies and Sinohydro.[81] In addition, when Sinohydro encounters problems, it usually contacts the provincial government.[82]

According to Sam Chanthy of the NGO Forum on Cambodia:

> [the] initial Environmental Impact Assessment prepared by Sinohydro prior to construction lacked any significant public participation, as required by the 1996 Law on Environmental Protection and Natural Resources Management. According to the law, [the project] is not legal at all (Vong and Strangio 2008).

Sinohydro, on the other hand, claims that public consultations on Kamchay took place during a period of 14 days from 3 to 16 September 2006 and were conducted by SAWAC Consultants. The meeting was attended by:

representatives of the Bokor National Park, Kampot Environment Department, Kampot Department of Industry, Mines and Energy, Kampot Water Supply, Kampot Department of Agriculture Forestry and Fisheries, Kampot Department of Water Resource and Meteorology, Kampot Tourism Department and the district office of the Forestry Administration of Kampot Province, the Deputy Governor of Kampong Bay District in Kampot Province, the Deputy Governor of Kampot District in Kampot Province, the First Assistant to the head of Makprang commune, representatives of Kampong Kraeng Health Centre, villagers from Math Pear, Andoung Chi Meun, Makprang, Kampong Krong, Kampong Kraeng, Prey Tnoat and Snom Prampir, food sellers at Teouk Chhou resort, and the NGOs Coastal Zone Management and Cambodia Global Action (Sinohydro 2008, pp. 30–31).

There is further evidence that Sinohydro made contact with Cambodian NGOs during the planning and construction of Kamchay. One NGO member mentioned that Sinohydro had contacted her NGO in 2008 to ask for its cooperation. However, Sinohydro had not followed up on this, and several requests addressed by the NGO to Sinohydro were left unanswered. In 2010, Sinohydro again contacted the NGO, but the interviewee felt that it had done so because it had been applying for carbon credits for the Kamchay Dam under the Kyoto Protocol and was therefore trying to improve its image.[83]

9.7.4 Livelihoods and Compensation

Even though there is no need for resettlement at Kamchay, some people lost land because of flooding, road construction and the erection of poles for transmission lines. However, according to Grimsditch (2012, pp. 21–22), none of the people who lost land was compensated, because Kamchay is located inside a national park and therefore on state land. However, where construction necessitated the clearing

[81] Interview K18092010a.

[82] Interview K18092010a.

[83] Interview P23092010.

of the plantations of durian trees for the second power house and associated transmission lines, the project developer compensated the plantation owners.[84]

Low water flows have also affected the Teouk Chhou recreation area. Downstream water supply is also affected,[85] and there are reports of the Kamchay River being polluted by construction materials and faeces from construction site toilets, making the water unusable for washing and consumption.[86] Polluted water has also affected the health of durian trees.[87] Grimsditch (2012, pp. 20–21) reports that water quality improved again markedly after peak construction.

Fisheries do not appear to be a major loss. An interviewee in the Ministry of Agriculture said that, since the Kamchay River does not have very productive fisheries, the situation is not a national issue, as the planned Sambor and Stung Treng Dams on the Mekong River will be.[88] Nevertheless, the Initial EIA for Kamchay included a section on reservoir fishing, which detailed the species of fish that could live in the reservoir and those that could not. The Initial EIA also outlined options for fish migration.[89] Overall, local people's livelihoods are affected in four ways:

First, income generation from tourism in the Teouk Chhou recreation area is already much reduced since construction began. This affects the small restaurant owners and food vendors who once made a living from tourists visiting Teouk Chhou.

Second, income generation from durian plantations will be made difficult if the compensation for lost trees is inadequate.[90] Given that it takes seven to eight years for these trees to bear fruit, their loss can be particularly detrimental in the absence of adequate compensation.[91]

Sinohydro, however, does pay compensation for lost durian trees. According to a local councillor, the first and second of the three dams that make up the Kamchay cascade dam were constructed on public land. The third and lowest dam is still on public land, but some people used to grow durian trees there without government permission. However, Sinohydro is paying some compensation to these people: US$ 300 for a large durian tree and US$ 150 for a small one. The different levels of payment are based on the economic yield of the trees.[92] This is in addition to compensation to legitimate durian farmers, whom Sinohydro has offered US$ 600 per tree (Middleton 2008, p. 64).[93] Crucially, compensation was paid before the

84 Compensation was presumably paid for the loss of the economic trees, not for the loss of land.

85 Interview P16092010.

86 Interviews P11092010 and K18092010a.

87 Interview K18092010a.

88 Interview P20092010b.

89 Interview P10092010a.

90 Interview K18092010b.

91 Interview K18092010a.

92 Interview K18092010a.

93 Grimsditch (2012, p. 22) reports slightly different numbers.

trees were destroyed (Grimsditch 2012, p. 22). The company is also paying subsistence farmers US$ 3 per square metre of land.[94] These payments go to the provincial branch of the Ministry of Industry, Mines and Energy, from where they must be distributed to the people affected.[95]

Yet evidence found in collaboration between the Heinrich Böll Foundation in Cambodia and the Chinese Academy of Social Sciences in Beijing showed that, while Sinohydro transferred the compensation to the government for all the eligible people, the government did not hand it on to them until local protests forced it to release the funds.[96]

Third, the crops, land and houses of people living close to the quarry from which building material was obtained for the dam were damaged by the blasting of rocks. Following complaints by villagers to the Cambodian NGO Licadho, Sinohydro entered into negotiations with the affected villagers and agreed to rent (and, in one case, purchase) the land for the duration of construction (Grimsditch 2012, p. 23).

Fourth, income generation in surrounding villages depends on bamboo basket-making. All the bamboo in the reservoir areas was cut to prevent it from being submerged in the water. Villagers now have to travel around 100 km to the bamboo forests along National Road No 4. This entails three to four days of work, including travel, considerable expense and sleeping outdoors in the mountains. It is therefore expensive, time-consuming and dangerous. To replace their lost income, some villagers took jobs on the Kamchay construction site. Villagers received no compensation for the lost bamboo, and employment on the dam site, being temporary, did not restore their livelihoods in the long term.

According to the Ministry of Industry, Mines and Energy, 1,200 workers were employed on the Kamchay site during the construction of the dam, 1,000 of them Cambodians. As they performed simple tasks, they did not receive any training, but learnt on the job. Only engineers of Electricité du Cambodge have been trained to monitor and operate the dam after the end of the concession period. Electricité du Cambodge staff are trained in China. The training is paid for by Sinohydro.[97]

This procedure appears to be consistent with the 2007 Concession Law, which states that the project owner "shall provide training to Cambodian nationals [...] prior to the expiry of the Concession Period" to enable them to operate the facility after the concession terminates (Royal Government of Cambodia 2007, Article 33). However, there is no training designed to equip villagers with the new skills they need to replace their former income from weaving bamboo baskets. The Ministry of Industry, Mines and Energy insists that the company must teach resettlers new skills. It must, for example, train fishermen to become farmers. This provision must be part of the EIA study. Thus the Chinese investor in the Lower Russei Chrum Dam was required to build new roads, hospitals and other

---

[94] Interview K18092010a.

[95] Interview K18092010a.

[96] Personal communication, Phnom Penh, 9 September 2010.

[97] Interview on 10 September 2010.

infrastructure in the resettlement areas. Resettlement and compensation cost US$ 50 million, leading to complaints from the project company about the high investment costs resulting from the expense of compensation.[98] Indeed, Chinese companies have repeatedly complained to the Ministry regarding the high investment cost of hydropower projects and asked for a rise in the electricity tariffs, but the Ministry has brushed off these requests.[99]

It appears that, as resettlement is not part of the Kamchay project, Sinohydro, as project owner, is not required to teach Cambodians employed on the construction site the skills they need to earn a living.

Workers received the wages of casual labourers. Sinohydro paid between 900 and 1,800 riel per hour for two days of work. For a workshop worker with special skills, Sinohydro's subcontractor, also a Chinese company, paid US$ 4–7 or 8 according to skills and responsibility. However, workers who stayed with Sinohydro longer were paid more. This was not true of the subcontractor.[100] Unlike the subcontractor, Sinohydro also paid overtime and double-time for Saturday work and implemented robust safety measures in accordance with Article 32 of the Concession Law.[101]

According to a local councillor, the wages paid by Sinohydro are not enough to support a family. At the time he was interviewed, however, there had been no complaints from workers and no discussions with Sinohydro about increasing the wages. Nor was there any provision for putting Cambodian workers on contracts, any union representation in the project area, any group or formal structure among the employees or, therefore, any negotiations about wage increases or the formation of a union.[102] This does not come as a surprise, since the situation of unions in Cambodia is very difficult, as Box 9.3 shows.

Box 9.3 The political situation of Cambodia's trade unions

Unions in Cambodia are in a weak position and illustrate civil society's virtual inability to influence the elite's political decision-making process under Cambodia's semi-authoritarian, neo-patrimonial system (Pak et al. 2007). During widespread strikes in the garment sector in September 2010, Prime Minister Hun Sen authorised legal action against strike leaders at a time when negotiations between the government and trade unions were making little headway. In March 2011, trade unions announced they were opposed to a draft Trade Union Law which they said "would allow the government to block protests, jail leaders, disband unions and prevent new bodies from forming" (Reuters 2011).

[98] Interview on 10 September 2010.

[99] Interview at the Ministry of Industry, Mines and Energy, 10 September 2010.

[100] Interviews K18092010c.

[101] Interviews K18092010c and Royal Government of Cambodia (2007, Article 32).

[102] Interview K18092010a.

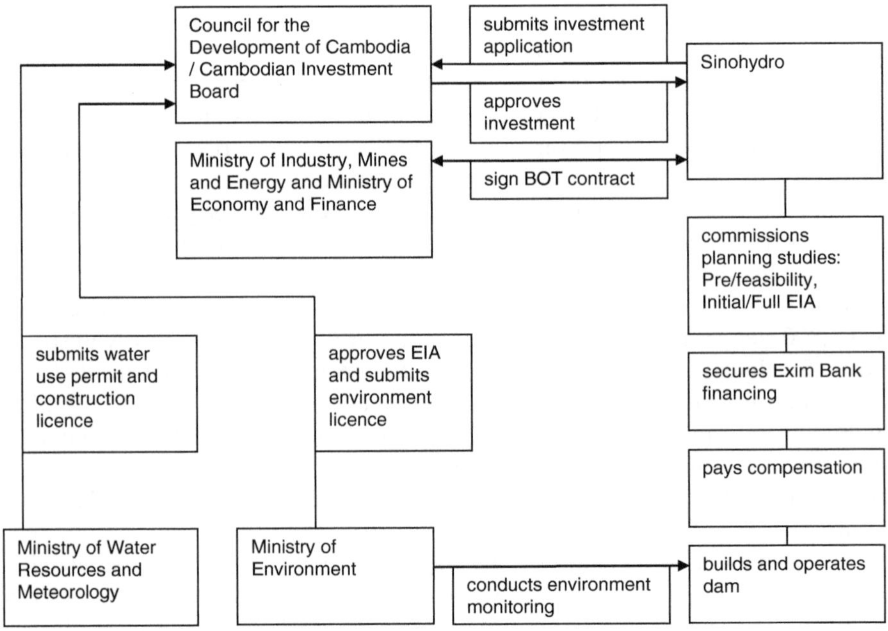

Fig. 9.1 Lines of responsibility of Sinohydro and the Cambodian government during the planning and construction of the Kamchay Dam (author's own compilation)

Many of Cambodia's trade union members are outspoken critics of the government. Prime Minister Hun Sen's Cambodian People's Party has an absolute majority in the National Assembly and habitually passes laws without any parliamentary discussion, thus ignoring requests by the opposition parties for a debate. The planned Trade Union Law is therefore viewed by civil society organisations and outside observers as an attempt by the government to continue stifling the opposition and strengthen governmental authoritarianism. In the context of the garment sector strikes, but also in such other areas as land disputes, law courts have regularly given rulings that support the government's interpretation of the law (Reuters 2011; Cheang 2011; Lewis and Mom 2011).

The unions are also characterised by fragmentation, rivalry and corruption and are in some cases co-opted by the government,[103] which compounds the problematic situation of unions in Cambodia.

Figure 9.1 shows the lines of responsibility between the Chinese and Cambodian agencies in the construction of the Kamchay Dam.

[103] Comment by an OECD embassy official, 9 August 2011.

9.8 Conclusion

The legal and political situation in Cambodia is made difficult by a number of factors: a neo-patrimonial, authoritarian polity in combination with elite economic interests produces a legally problematic situation in which personal interests and perceptions outweigh abstract administrative processes. This explains the failure of environmental and social regulation in Cambodia.

The Cambodian government applies only a minimum of safeguards to hydropower BOT investments. When the focus is narrowed to the economic and technical context, two reasons can be identified: first, the government has devised a liberal investment regime which is aimed at attracting Foreign Direct Investment and relegates EIA to a formality. Second, the government lacks the technical and financial capacities needed to monitor EIA implementation during construction.

By the time the EIA process for obtaining an environment licence begins, the project has already been approved in principle by the Cambodian Investment Board and has received political backing from the Prime Minister. The EIA Department of the Ministry of Environment can then do nothing more than advise on mitigation measures rather than using EIA as a tool to influence decision-making. In this decision-making framework, however, the strength of the mitigation measures may be questioned, since they require strict oversight by the government and especially the technical department of the Ministry of Environment. They must also be accompanied by the thorough administration of such penalties as the cancellation of a project or the levying of credible fines.

However, it would seem that both the government's emphasis on a quick expansion of hydropower capacity and the direct involvement of elite interests, here in the person of Senator Lao Meng Khin, are preventing this. Once all the licences are granted and the project enters the construction phase, the government acts largely as the consumer of the electricity through Electricité du Cambodge. Environmental monitoring does exist, but it does not seem to carry much weight in this political setting.

As regards the division of responsibilities between the Cambodian government and Sinohydro, the government is responsible for approving all licences and the investment project. Licences are granted free of charge. For its part, Sinohydro is responsible for carrying out EIA, paying compensation, implementing the Environmental Management Plan and constructing and operating the Kamchay Dam. The government does not exercise strict control over any of these four areas of company responsibility.

On the contrary, compensation payments made by Sinohydro were delayed by local government bodies until protests by the people affected forced the government to distribute the money.

Sinohydro appears to be abiding by Cambodian compensation and EIA regulations. However, the fact that Sinohydro started to construct Kamchay with only the letter of approval for the Initial EIA shows a lack of responsibility for and sensitivity to social and environmental problems associated with large hydropower plants.

At Cambodian government level, both the Ministry of Industry, Mines and Energy and the Ministry of Environment are aware of the social and environmental impact of large dams. The Ministry of Industry, Mines and Energy, though conscious of the impacts, was very much in favour of large hydropower dams as the way to develop Cambodia's energy capacities.

Consequently, the highest level of government has so far shown no interest in institutionalising a strong mechanism for public participation in EIA. Awareness-raising is often left to NGOs, but they do not have access to the decision-making process.

Cambodia's government is set to go ahead with the expansion of hydroelectric capacity. If all dams currently on the drawing-board are built, it will be necessary to find mechanisms for peaceful conflict prevention and arbitration that reconcile the government's economic development policies with the livelihood requirements of local communities. The events surrounding the Vietnamese-invested Lower Sesan 2 Dam clearly show the need for such a mechanism. At the same time, those events and the protests against the withholding of compensation for the damage done by the Kamchay Dam show that well-informed and sufficiently well-organised local communities are, to some extent at least, capable of holding the government to account.

Acknowledgments The author would like to express his gratitude to the following people and institutions who made the research for this study possible: Hady Riad, Counsellor at the German Embassy in Phnom Penh; In Veasna at the German Embassy in Phnom Penh; the Ministry of Industry, Mines and Energy; the Ministry of Environment; the Cambodian Investment Board; the Ministry of Agriculture; Katrin Seidel, then Head of the Heinrich Böll Foundation in Phnom Penh; Jean-Yves Dekeister; and Petra Schill, then MRC-GTZ Cooperation Programme Coordinator at the GTZ/GIZ Office in Vientiane. I am also grateful to Mark Grimsditch for his detailed comments on a previous draft of this article.

References

ADB (Asian Development Bank) (2004) Enhancing the resettlement legal framework and institutional capacity: Cambodia. Technical assistance 4490. http://www.adb.org/projects/37535-012/main. Accessed 17 July 2013

ADB (Asian Development Bank) (2007) Compensation and valuation in resettlement: Cambodia, People's Republic of China, and India. ADB, Manila

ASEAN Affairs (2008) SEZ foundation laid in Sihanoukville, 26 Feb 2008. http://www.aseanaffairs.com/page/cambodia_china_infrastruicture_development_sez_foundation_laid_in_sihanoukville. Accessed 17 July 2013

Associated Press (2006a) Chinese premier to pledge more aid during visit to Cambodia, 7 April 2006

Associated Press (2006b) Cambodia gets development loan from China, 24 Oct 2006

Cambodia ODA Database (2013) Council for the Development of Cambodia. http://cdc.khmer.biz/. Accessed 17 July 2013

Cambodian Rehabilitation and Development Board of the Council for the Development of Cambodia (2011) The Cambodia development effectiveness report 2011. Royal Government of Cambodia, Phnom Penh

Caminfoweb (2012) Zhong zi cheng jian Ganzai shuidianzhan jungong fadian. Hong Sen zongli Pan Guangxue dashi deng chuxi qingdian yishi [Chinese-invested and Chinese-built Kamchay hydropower station completed and producing electricity. Prime Minister Hun Sen and Ambassador Pan Guangxue attended the festive ceremony], 2 Feb 2012. http://www.caminfoweb.com/ReadNews.asp?NewsId=1709. Accessed 17 July 2013

Centre on Housing Rights and Evictions (2008) Annex: land and housing issues. Additional information and supplementary questions. COHRE

Chan S, Tep S, Acharya S (2001) Land tenure in Cambodia. A data update. Working paper 19. Cambodia Development Resource Institute, Phnom Penh

Cheang Sohka (2007) Kampot poised to enter world of hydroelectric power. In: Phnom Penh Post, 12–25 Jan 2007. http://ki-media.blogspot.com/2007/01/locals-concerned-about-building.html. Accessed 17 July 2013

Cheang S (2011) Unions unite on draft law. In: Phnom Penh Post, 8 July 2011. http://www.phnompenhpost.com/index.php/2011070850278/National-news/unions-unite-on-draft-law.html. Accessed 17 July 2013

China Exim Bank (2007) Issuance notice of the guidelines for environmental and social impact assessments of the China export and import bank's loan projects. China Exim Bank, Beijing

Council for the Development of Cambodia (no date) List of special economic zones. http://www.cambodiainvestment.gov.kh/list-of-sez.html. Accessed 17 July 2013

Council for the Development of Cambodia (2010) The price of power. In: Invest in Cambodia Magazine

Council for the Development of Cambodia (2012) Cambodia investment guidebook 2012. Royal Government of Cambodia, Phnom Penh

Dyer E, Chun S (2010) Huadian secures financing for hydropower plant in Koh Kong. In: The Phnom Penh Post, 19 July 2010

Grimsditch M (2012) China's investments in hydropower in the Mekong Region: the Kamchay Hydropower Dam, Kampot, Cambodia. Bank Information Center, Washington DC

Hensengerth O (2011) Interaction of Chinese institutions with host governments in dam construction: the Bui Dam in Ghana. Discussion paper 03/2011. German Development Institute, Bonn

International Rivers (2012) Comments to CF Carbon Fund II Ltd regarding the Kamchay hydropower project (Cambodia), 20 Jan 2012. http://www.internationalrivers.org/resources/comments-to-cf-carbon-fund-ii-ltd-regarding-the-kamchay-hydropower-project-cambodia-3059. Accessed 17 July 2013

Kubny J, Voss H (2010) The impact of Chinese outward investment: evidence from Cambodia and Vietnam. Discussion paper 16/2010. German Development Institute, Bonn

Labelle H (1997) CIDA responds to probe's 1996 Mekong Campaign. In: Probe International, 29 Jan 1997. http://journal.probeinternational.org/1997/01/29/cida-responds-probes-1996-mekong-campaign-1/. Accessed 17 July 2013

Lewis B, Mom K (2011) Critics take aim at union law. In: Phnom Penh Post, 17 Jan 2011. http://www.phnompenhpost.com/national/critics-take-aim-union-law. Accessed 17 July 2013

Malmquist P, Sigfridsson M (2002) The Kamchay hydropower project: hydropower development in Cambodia and Eastern Asia. Bachelor thesis, Malmö University

McCallum W (2008) Before the dam: a study of environmental impacts and community rights associated with the construction and operation of the approved Kirirom III Hydropower Scheme, Sre Ambel District, Southwest Cambodia. American Friends Service Committee, Rivers Coalition in Cambodia

Middleton C (2008) Cambodia's hydropower development and China's involvement. International Rivers and Rivers Coalition in Cambodia, Berkeley and Phnom Penh

Ministry of Industry, Mines and Energy (2003) Cambodia National Mekong Committee National Sector Review 2003: Hydropower. Ministry of Industry, Mines and Energy, Phnom Penh

NGO Forum on Cambodia (2009) Statistical analysis on land dispute occurring in Cambodia 2009. NGO Forum on Cambodia, Phnom Penh

Ngy L (2010) Current status of environmental impact assessment (EIA) in Cambodia. In: Presentation held at the regional workshop on EIA in Asia: good practices and capacity building needs, Manila, 9–10 June 2010

Ouch C (2010) Assessing the impacts of China's official development assistance on poverty reduction in Cambodia: the case of transportation projects. In: Presentation at the 2010 development research forum, Phnom Penh, 9–10 Sept 2010

Pak K, Vuthy H, Netra E et al (2007) Accountability and neo-patrimonialism in Cambodia: a critical literature review. Working paper 34. Cambodia Development Resource Institute, Phnom Penh

Perlez J (2006) China competes with west in aid to its neighbors. In: The New York Times, 18 Sept 2006, Section A, Column 2, Foreign Desk:1

Reuters (2011) Cambodian unions threaten strikes over new laws, 15 March 2011. http://uk.reuters.com/article/2011/03/15/uk-cambodia-unions-idUKTRE72E27520110315. Accessed 17 July 2013

Ryder G (2009) Powering 21st century Cambodia with decentralized generation: a primer for rethinking Cambodia's electricity future. NGO Forum on Cambodia and Probe International, Phnom Penh, Toronto

Ryou H (2009) A matter of give and take: China's infrastructural investment in South East Asia's poorest countries will pay back. In: Business Forum China, Feb 2009. http://businessforum-china.com/article_detail.html?articleid=270. Accessed 17 July 2013

Royal Government of Cambodia (1994) Law on investment of the Kingdom of Cambodia. Royal Government of Cambodia, Phnom Penh

Royal Government of Cambodia (1996) Law on environmental protection and natural resources management. Royal Government of Cambodia, Phnom Penh

Royal Government of Cambodia (1999) Sub-decree on environmental impact assessment process. Royal Government of Cambodia, Phnom Penh

Royal Government of Cambodia (2001) Land law. Unofficial translation prepared by the Ministry of Land Management, Urban Planning and Construction with the support of the World Bank, ADB TA 3755-CAM, GTZ and Finnmap

Royal Government of Cambodia (2007) Law on concessions. Royal Government of Cambodia, Phnom Penh

Royal Government of Cambodia (2010) Law on expropriation. Royal Government of Cambodia, Phnom Penh

Sam C (no date) Scoping study of existing frameworks related to the world commission on dams strategic framework: Cambodia. The article was later published in Oliver RAR, Moore P, Lazarus K (eds.) (2008), Mekong region water resources decision-making, national policy and legal frameworks vis-à-vis world commission on dams strategic priorities. IUCN, Bangkok, Gland

Sam C (2007) The Kamchay Hydroelectric Dam. NGO Forum on Cambodia, Phnom Penh

Sinohydro (2008) Clean development mechanism project design document form for Kamchay hydroelectric BOT project, 12 July 2008

Vong S, Strangio S (2008) High hopes for hydropower. In: Phnom Penh Post, 29 May 2008. http://www.phnompenhpost.com/special-reports/high-hopes-hydropower. Accessed 17 Jul 2013

Xinhua (2011) Cambodia largest hydroelectric dam begins operation, 7 Dec 2011. http://www.chinadaily.com.cn/xinhua/2011-12-07/content_4607116.html. Accessed 17 July 2013

Chapter 10
German Development Cooperation and the World Commission on Dams

Kirsten Nyman, Klaus Horstmann and Christiane Rudolph

Abstract More than 10 years after the German government welcomed the recommendations of the World Commission on Dams (WCD) and made them the yardstick for all dam-related activities in its development cooperation, this article reviews how those recommendations were integrated into German development cooperation and what mechanisms were used to promote their implementation. The first part of the article considers the background to the work of and results achieved by the WCD, the policy response and four interrelated mechanisms for promoting the implementation of the WCD's recommendations: (1) support for dialogues on implementation at national level; (2) integration into bilateral development cooperation; (3) advocacy of implementation in multilateral development cooperation, and (4) the promotion of international standard-setting and sector dialogues. The second part of the article takes a practical view and discusses challenges to the implementation of the WCD's recommendations at project level on the basis of experience in German development cooperation. The aim of the article is to paint a comprehensive picture showing how and why German development cooperation has supported the work of the WCD and helped to implement its recommendations.

K. Nyman (✉)
Gesellschaft fuer Internationale Zusammenarbeit (GIZ), Bonn, Germany
e-mail: kirsten.nyman@giz.de

K. Horstmann
KfW Development Bank, Frankfurt, Germany
e-mail: klaus.horstmann@kfw.de

C. Rudolph
German Investment and Development Company (DEG), Cologne, Germany
e-mail: Christiane.Rudolph@deginvest.de

W. Scheumann and O. Hensengerth (eds.), *Evolution of Dam Policies*,
DOI: 10.1007/978-3-642-23403-3_10,

10.1 Introduction

The establishment of the World Commission on Dams (WCD) was the culmination of a fierce debate on large dams during the 1990s. Protest groups and environmental activists put pressure on the World Bank and other organisations to stop supporting the construction of large dams financially. They criticised the adverse environmental and social impacts associated with such dams. However, the debates and conflicts over large dams are far more complex than they may appear at first glance and cover fundamental questions about the process of development in societies. Opponents and proponents of dams are motivated by various factors, ranging from national development objectives, business interests and political influence to the protection of the environment and the rights of indigenous people (Fink and Cramer 2008). The debate surrounding large dams thus concerns political interests and decision-making rather than scientific or technical considerations.

In 1996, the World Bank evaluated 50 dam projects it had financed and found that only a quarter of them met its own ecological, social and economic standards. This evaluation (World Bank and OED 1996) prompted the World Bank and the International Union for Conservation of Nature (IUCN) to organise and host a workshop in the Swiss town of Gland in April 1997, which gave rise to the idea of establishing the World Commission on Dams. The new organisation became operational in May 1998. Its mandate was

- to review the development effectiveness of dams and to assess alternatives for water resources and energy development, and
- to develop internationally accepted criteria, guidelines and, wherever possible, standards for decision-making in the planning, design, construction, monitoring, operation and decommissioning of dams.

Twelve commissioners, who included both advocates and critics of dams, spent 2.5 years conducting hearings and consultations, carrying out case studies and analysing more than 100 dam projects. The commission was chaired by the former South African Minister of Water Affairs and Forestry, Professor Kadar Asmal. It was advised by a WCD forum representing 68 interest groups.

A WCD report entitled *Dams and Development. A new framework for decision-making* was published in November 2000. It contains a comprehensive analysis of the social, economic, technical and ecological problems and impacts associated with large dams, followed by recommendations which are summarised in the form of five core values, seven strategic priorities, and 26 guidelines (Box 10.1). Particular importance is attached to environmental protection, the rights of people affected by dam projects and their participation in decision-making processes.

The five core values are equity, efficiency, participatory decision-making, sustainability and accountability. They form the general basis for decision-making. On the basis of these values the WCD report presents a concept for acknowledging the rights of those affected by a dam project, assessing the risks they involuntarily face and incorporating these factors in dam project planning and implementation. The seven strategic priorities form the cornerstones of the planning process; guidelines support their implementation. The overall objective of the planning procedure is to analyse options carefully, to gain public acceptance and to ensure fair negotiation and the equitable distribution of the benefits of a dam project.

Box 10.1: World Commission on Dams: strategic priorities and guidelines

Strategic Priority 1: Gaining public acceptance

Stakeholder analysis//Negotiated decision-making process//Free, prior and informed consent

Strategic Priority 2: Comprehensive assessment of options

Strategic impact assessment of environmental, social, health and cultural heritage issues//Project-level impact assessment of environmental, social, health and cultural heritage issues//Multi-criteria analysis//Life-cycle assessment//Greenhouse gas emissions//Distributional analysis of projects//Valuation of social and environmental impacts//Improving economic risk assessment

Strategic Priority 3: Addressing existing dams

Ensuring operating rules reflect social and environmental concerns//Improving reservoir operations

Strategic Priority 4: Sustaining rivers and livelihoods

Baseline ecosystem surveys//Environmental flow assessment//Maintaining productive fisheries

Strategic Priority 5: Recognising entitlements and sharing benefits

Baseline social conditions//Impoverishment risk analysis//Implementation of the mitigation, resettlement and development action plan//Project benefit-sharing mechanisms

Strategic Priority 6: Ensuring compliance

Compliance plans//Independent review panels for social and environmental matters//Performance bonds//Trust funds//Integrity pacts

Strategic Priority 7: Sharing rivers for peace, development and security

Procedures for shared rivers

Source WCD 2000, 213-257

10.2 Policy Responses

10.2.1 Support During the Preparation of the WCD's Recommendations

Through the Federal Ministry for Economic Cooperation and Development (BMZ), the German government welcomed the creation of the WCD and supported its work in different ways from the outset. On the one hand, a substantial financial contribution of around EUR 1 million was provided to back the work of the Secretariat. The Dams and Development Project (DDP) subsequently implemented by the United Nations Environment Programme (UNEP) with the aim of disseminating the WCD's recommendations and promoting dialogue on best practices and initiatives for their implementation was also supported financially. On the other hand, the German development cooperation community shared its dam portfolio experience with the WCD Secretariat by commissioning a study (Hartje and Horstmann 1999) of the following aspects:

- the role of dams in the spread of water-borne diseases and the ability to mitigate these health effects
- the impact of dams on river ecology and experience of environmental appraisal and mitigation methods
- the impact and socio-economic effects of the construction period on the environment
- the social impact on affected persons and their hosts and the ability to turn displacement into a development process
- appraisal of alternatives to dams for electricity generation and
- alternatives to dams for the provision of drinking water and water for agriculture.

The study included a desk appraisal of 24 dam projects, twelve of which were launched before, twelve after 1990. This desk study revealed the need for greater attention to be paid to water-borne diseases in project planning and implementation. Decision-making with respect to river ecology had improved, particularly after the formal environmental appraisal process (environmental impact assessment, EIA) was introduced in 1988, and the reports to be submitted to the BMZ were changed to include an environmental annex. Only in recent projects had the environmental and socio-economic impacts of construction sites and camps been dealt with comprehensively. In earlier dam projects the scale of the displacement process and the resulting losses suffered by the affected population had been underestimated. It was realised that the resettlement process should be regarded as an integral part of the development process that the dam is meant to support. As an alternative to hydroelectricity, an increase in the energy efficiency of existing and expanding electricity systems had also been identified as part of the German development cooperation programme. Direct funding of irrigation projects changed to smaller projects and more upgrading of existing perimeters und included more technical advice on water conservation; on institutional reforms related to the management of irrigation schemes by water users associations, and on designing and implementing tariff systems used to finance operation and maintenance of irrigation systems.

The study concluded that the success of a dam project cannot be measured solely in such technical terms as megawatts generated or area irrigated, but should also consider the "winners" and "losers".

10.2.2 Response After Publication

The German Ministry for Economic Cooperation and Development is one actor in the international arena to have endorsed the WCD recommendations as the guide for all dam-related activities (BMZ 2007). The Minister at that time, Heidemarie Wieczorek-Zeul, said at the Dialogue Forum on the WCD Report: "Hydropower will still be needed in the future. However, dam projects will have to satisfy strict environmental impact and social acceptability criteria and will have to provide development benefits for the people who are immediately affected" (GIZ 2001, p. 1; GIZ and KfW 2004, p. 12). In the same context, the Ministry welcomed the findings of the WCD report in January 2001: "This way, confrontation between opponents and proponents of large dams can be overcome in the best interest of developing countries" (GIZ 2001, p. 1; GIZ and KfW 2004, p. 13).

However, not all the responses to the WCD's recommendations were positive. Such other stakeholders as industry and international development banks were less supportive. One reason for this more negative response was the lack of a final discussion in the report of the positive, beneficial aspects of dams. Some stakeholders rejected the recommendations out of fear that they would put an end to all dam-building activities. Others felt that the recommendations were too abstract and would therefore be difficult to put into practice. Some recommendations, such as that calling for strategic priority to be given to the free, prior and informed consent of indigenous people, might not be practicable without further guidance.

10.2.3 The WCD in the Policy Framework of German Development Cooperation

German development cooperation aligns its objectives primarily with the overall concept of sustainable development and the Millennium Development Goals (MDGs). If the MDGs are to be achieved, access to clean energy and substantial electricity generation will be needed. This is particularly true of MDG 1 (eradication of poverty and hunger) and MDG 7 (ensuring environmental sustainability). Seen in that light, hydropower provides opportunities for contributing to sustainable development. Hydropower, whether on a large or a small scale, is without doubt a renewable energy source that produces fewer emissions than coal-based electricity generation, for example. In addition, multi-purpose dams can give access to clean drinking water and secure irrigation for food production. However, it is important to balance the positive and negative effects of hydropower and multi-purpose dams in an economically, socially and ecologically friendly manner. German development cooperation is committed to contributing to these processes and to respecting the various social, ecological and economic requirements.

It is against that background that German development cooperation formally accepted the WCD's recommendations as a framework for decision-making on dams. It endorsed not only the WCD's strategic priorities, as many other actors did, but also its core values.

In its Sector Strategy "Sustainable Energy for Development", the BMZ states that "dams are an important source of renewable energy, as well as providing protection against flooding, irrigation and drinking water. In the light of population growth, economic development, climate change and, in part, untapped potential of hydropower, dams can in many cases be a sensible option. Nevertheless, numerous dams have in the past caused devastating social and ecological consequences and have also proved to be unprofitable. Superior alternatives are, for example, restoring existing dams or building run-of-river power plants with adequate compensation flows and bypasses. Large-scale storage dam projects must meet the WCD's criteria as regards planning, construction and operation. The criteria should also be applied to smaller dams" (BMZ 2007, p. 23).

Thus German development cooperation clearly aligns its norms with the three dimensions of sustainable development: social, environmental and economic. It supports only dam projects in which these three dimensions are promoted and balanced.

10.3 Implementation Channels

The implementation of the WCD's recommendations is promoted at four main levels:

- national dialogues for implementation
- integration into bilateral cooperation
- support through multilateral cooperation
- promotion of international standard-setting and sector dialogues.

10.3.1 Promoting National Dialogues on Implementation

For the global dam debate to have an impact at local level, dialogue must continue within the national context. This task was assumed by the Dams and Development Project. Established in November 2001, the DDP operated under the auspices of the United Nations Environment Programme. Its core activities included the dissemination of information on the WCD and its follow-up processes and the initiation of national dialogues on the implementation of the WCD's recommendations. A major forum of stakeholder representatives served as an advisory board to the DDP. Among them were the governments of China, India and Turkey, who had at first sharply criticised the WCD recommendations, but also government agencies from developing countries, such as the Volta River Authority. Multi-stakeholder dialogues have been supported in various countries, including Ghana, Uganda, Togo, Nepal, Vietnam and Thailand.

The BMZ supported the DDP financially and represented the interests of the donor countries on the 14-member DDP Steering Committee. It contributed to the development of planning and management tools, such as the Compendium (GIZ and KfW 2004, p. 14). The BMZ also facilitated the sharing of experience, the dissemination of information and networking among stakeholders.

Following the publication of the WCD report, the BMZ set up a project known as "Support for the implementation of the WCD's recommendations?", which was renamed "Policy Advice on Sustainable Hydropower" in 2008. This project, which was implemented by the German Society for International Cooperation (GIZ), included pilot studies on the development of procedures and concepts and technical support for governments in the application of the WCD's strategic priorities. It also backed a number of national and regional multi-stakeholder dialogues as a follow-up to the UNEP Dams and Development Project. The overarching objective of these national dam dialogues was to improve decision-making on dams and to address country-specific, action-oriented policy guidelines. Dialogues facilitated by GIZ include those with the Southern African Development Community (2003–2005), Cameroon (2003–2004), Nepal (2003–2005), Uganda (2006–2007), Togo (2007–2008) and Ghana (2006–2011) (Box 10.2). These dialogue processes have achieved a number of successes, for example, by bringing together stakeholders who would not have met otherwise, raising awareness of special issues and needs of stakeholders (particularly resettled communities), mediating in conflicts between dam-affected communities, operators and governments and developing policies on and procedures for the implementation of dam projects. However, their success ultimately depends on whether the governments concerned take account of the outcome of dialogues in their policy-making procedures and possibly in project implementation.

Box 10.2: The Ghana dams dialogue

Goal: To build capacities and provide tools for improved decision-making on dam-related issues, leading to the equitable, transparent, participatory and sustainable development of dams in Ghana.

The dialogue process has been successful in that it has resulted in

- the creation of a 60-member Forum and a 15-member National Coordinating Committee to act as a sounding board and advisory group for the implementation process,
- the identification of experts and analysts in a wider stakeholder community,

- one National Consultative Meeting and three Forum Meetings, which brought together a wider network of interested parties, served as platforms for all voices to be heard and were attended by more than 250 participants from all stakeholder groups,
- the initiation of five independent, in-depth studies of dam-related issues in Ghana,
- the resolution of two conflicts between affected communities, governmental authorities and operators and
- the strengthening of the negotiating positions of communities directly affected by the newly constructed Bui Dam on the Black Volta river.

10.3.2 Dialogue in Germany on the Integration of the WCD's Recommendations into Bilateral Cooperation

To foster the integration of the WCD's recommendations into German development cooperation and their implementation by German-based actors involved in dam-related activities, the BMZ convened a dialogue forum 2 months after the publication of the WCD report and a number of panel discussions for an exchange of opinions on the contents of the report. Some 35 representatives of the private sector, federal ministries, non-governmental organisations (NGOs), the scientific community and public development cooperation attended these events.

A German WCD working group representing ten of these bodies adopted a joint position paper on the WCD's recommendations indicating where there was a consensus and where agreement had not been reached on the contents of the WCD report. The latter refers, for instance, to the powers of stakeholder forums, the rights of indigenous people, the kind of alternatives to be considered and the distribution of responsibilities for existing dams. The process of negotiation with affected communities, for example, was controversial, as Box 10.3 shows.

Box 10.3: Controversies within the German WCD working group

Agreement

Efforts should be made to involve those affected in the dialog process concerning the project as a whole and to provide for adoption of affirmative counterbalancing and compensatory measures by way of negotiations with those affected.

Disagreement

There is disagreement about the extent of the affected people's negotiation mandate.

Position A: Following consultations with the stakeholders, the competent national authority should make the decision. The stakeholders must have the right to appeal.
Position B: The solution must always result from negotiations between equal partners.
Recommendation
The German stakeholders involved in the conception of the project should seek to achieve that a decision concerning affirmative counterbalancing and compensatory measures be arrived at preferably by mutual consent between the government and the affectees (recipients), but at least according to appropriate rules."

10.3.3 Integration into Bilateral Development Cooperation

The WCD report specifically addresses the role of bi- and multilateral development assistance and requires institutions to review "internal processes and operational policies in relation to the Commission's recommendations to determine changes needed in the selection of projects for lending portfolios; the appraisal process; and implementation, monitoring, and evaluation" (WCD 2000, p. 316). As mentioned above, the BMZ formally adopted the WCD recommendations as the framework for dam decision-making and subsequently integrated the need for compliance with those recommendations into its Water Sector Strategy, which states that "the construction of large dams is a controversial issue worldwide. Due to the often negative impacts on the eco-balance and the communities affected by resettlement, the issues of sustainability and human rights compatibility must be given particularly careful consideration in the context of dam construction. Dams can provide solutions for energy production, irrigation, drinking water supply and flood protection. However, dam construction projects must always be contingent on a comprehensive evaluation of alternative options and must satisfy sustainability criteria. Often, the rehabilitation of existing dams is a better option than new construction. Large dam projects must satisfy the World Commission on Dams' criteria for planning, implementation, construction and operation; these criteria should also be applied accordingly to smaller dam projects." (BMZ 2006, p. 27). As this policy is binding on all German development cooperation agencies, acceptance of the WCD recommendations is formally integrated into bilateral development cooperation. The implementing agencies most involved in dam-related activities are the GIZ, whose assistance consists primarily of advisory services, process facilitation and capacity development, Germany's development bank, Kreditanstalt fuer Wiederaufbau (KfW), which provides financial assistance, and Germany's Investment and Development Company (DEG), which is active in the field of private investment.

Numerous actors were involved in the dialogue on how best to implement the WCD recommendations. The governments of countries with plans to build dams bear a special responsibility, of course, because they are the ones who have to investigate and clarify their own countries' specific requirements when it comes to implementing the recommendations. It is their task to consider the available options for putting the recommendations into practice—possibly through the amendment of legislation. However, although development cooperation accounts for less than 15 % of all investment expenditure on large dams in developing countries, its importance should not be underestimated.

Basic national decisions on such matters as meeting demand and power generation alternatives, though made before the planning of any concrete project begins, call for suitable sector policies, on which the WCD also has recommendations to make. The development cooperation community may advise and assist partner countries to develop such policies.

10.3.3.1 Integration of the WCD's Recommendations into Technical Cooperation

The GIZ bases its work with partner countries on the WCD framework. Dams sometimes play an important role in a number of water, energy and environmental projects. The GIZ then cooperates not only with government institutions, but also with civil society organisations, the private sector, the scientific community and international organisations. In recent years, the GIZ has enlarged its transboundary water management portfolio. Advice is currently being given to river basin organisations in Africa, Central Asia and the Mekong region. In all these river basins, dam construction is prominent and is affecting the distribution of water rights both within and among countries.

The GIZ's advisory services are provided when it comes to laying the necessary foundations in the shape of appropriate water, environment and energy policies, legislation and capacity-building for institutions and key stakeholders. In many cases, the process of bringing about such changes takes many years and is affected by numerous external factors. Drawing the attention of the general public and decision-makers to the risks and opportunities associated with dam projects is a suitable way of helping to achieve effective reforms. Even though the environmentally and socially sound implementation of dam projects can stimulate such movement, only structural and strategic change in the environmental, water and energy sector can lay durable foundations for the long-term implementation of the WCD principles. That, in turn, requires the creation of conditions conducive to development. A uniform sector policy and effective sector strategies need to be developed, a suitable legal framework must be established, capable institutions developed, and transparent licensing procedures introduced.

10.3.3.2 Integration of the WCD's Recommendations into Financial Cooperation

The KfW and the DEG act in accordance with the KfW's environmental and social principles when a general policy is being translated at operational level into sustainability guidelines that meet international standards. In processing any dam project, the KfW makes it a rule to verify its compatibility with the WCD's recommendations. On the basis of these checks it is then decided whether or not additional activities are required to ensure full compliance with the WCD recommendations.

The KfW and DEG employ international experts to investigate whether international standards are being applied and to monitor their application closely during the construction and operation phases. Like the World Bank, the KfW also backs up major projects with a panel of experts, including environmental and social specialists.

The more banks apply good-practice standards in their financing, the greater the pressure on companies to take the sustainability path. Financial institutions can help to ensure compliance with WCD criteria and principles. They have various effective means to this end: (i) they can require environmental and social reviews and the identification of any deficiencies and of means of overcoming them, (ii) they can insist on sound environmental and social management, (iii) they can include the cost of mitigating or compensatory measures in the project budget and ensure that the necessary action is taken by providing for it in the project contract, where it becomes legally binding, and (iv) they can carry out external monitoring to supervise developments on the ground. Any breach of contract, including non-compliance with environmental and social protection requirements, may result in a demand for premature termination and repayment of the loan. This is in no way desirable, but it may be a more effective incentive to ensure compliance than any fine imposed by a national supervisory authority.

Since 1988, all KfW projects have had to undergo in-depth environmental and social impact assessments based on the bank's sustainability guideline and a BMZ directive. Even earlier, attention had been paid to ecological and social factors, though perhaps not to the same extent as today. Until well into the 1970s, however, all resettlement matters were the sole responsibility of the partner countries. Gradually, over the following two decades, it became internationally accepted practice—for the OECD Development Assistance Committee, for instance—to include a resettlement plan in dam project documentation and to set quality standards for its implementation. That approach can now be expanded on the basis of the WCD recommendations. It is now considered, for example, how those affected can benefit from such projects in addition to merely safeguarding their living standards and incomes. Other aspects have also become easier to address, one example being the need for a minimum environmental flow of water to sustain ecosystems and livelihoods downstream of dams.

Whenever the KfW participates in the financing of a large dam project, the national authorities responsible for planning and operating the dam are supported in numerous ways. Additional funds have been provided in a number of cases specifically for assistance in environmental and social matters. They have been used for additional studies in these areas, for advice on the establishment or upgrading of environmental and social units within implementing organisations, for training activities or for additional social investments (schools, health stations etc.) in the vicinity of the construction site. Such financing is usually provided in the form of non-repayable grants. The DEG, which finances private projects, requires sustainable environmental and social management contractually and promotes it by providing expertise and support. Both institutions not only take an interest in planning and construction, but also seek to ensure lasting and sustainable operation.

10.3.3.3 Integrating the WCD's Recommendations into Export Credit Guarantees

Another area of support is insurance and export credit guarantees for projects. Euler-Hermes Deutschland is mandated by the German government and has provided export credit guarantees for German companies investing in developing countries since 1949. An inter-ministerial committee, on which the Ministry of Economics and Technology, the lead body, the Ministry of Finance, the Foreign Office and the BMZ are represented, is responsible for the budget, basic issues of cover policy and the granting of guarantees. In general, export guarantees must comply with the OECD's Common Approaches on Environment and Officially Supported Export Credits. However, the BMZ advocates the WCD's recommendations whenever a dam project is considered for an export guarantee (see the case study of Atzl on Turkey's Ilisu Dam in this volume).

10.3.4 Promoting and Supporting the WCD's Recommendations in Multilateral Institutions

The German government is also a shareholder in (regional) development banks, such as the Asian Development Bank, and in the World Bank. Germany is represented by the BMZ on the boards of directors of these banks, often with one seat and vote. In this role the BMZ participates in the development of new strategies, portfolio development, policy-making and standard-setting processes and formally votes on individual projects. As the commitment to the implementation of the WCD recommendations also applies to multilateral cooperation, the BMZ is an advocate in multilateral financial institutions of the implementation of those recommendations and of the progressive development and implementation of environmental and social policies, safeguards and standards of the institution concerned. The policy is based on the BMZ's water and energy sector strategies (outlined above), which not only apply to bilateral cooperation but also form the basis on which positions are developed in multilateral contexts.

Another area in which the BMZ has championed the WCD's recommendations is the European Union (EU): the EU Linking Directive requires the application of international criteria and guidelines, including those of the WCD, when emission reduction certificates arising from hydropower projects exceeding 20 MW under the Kyoto mechanisms are to be used in the European emission trading system (EU 2004). However, the EU member countries have complied with this requirement in different ways. This has prompted the BMZ and the German Ministry for Environment, Nature Conservation and Nuclear Safety (BMU) to hold workshops and to support the establishment of recommendations and the harmonisation of guidelines aimed at the development of a common understanding on how to apply and implement the WCD recommendations. This activity culminated in the adoption of "Guidelines on a common understanding of Article 11b (6) of Directive 2003/87/EC as amended by Directive 2004/101/EC", which have been applied by all EU Member States since April 2009.

10.3.5 Promotion of International Standard-Setting and Sector Dialogues

In 2008 the International Hydropower Association (IHA)[1] asked the BMZ to contribute to the forum set up to revise the Hydropower Sustainability Assessment Protocol (HSAP). In the aftermath of the WCD recommendations the IHA had developed its Sustainability Guidelines (2004) and the Sustainability Assessment Protocol (2006). The original HSAP was a self-assessment tool used by industry to rate the performance of a project in economic, social and environmental terms. The original IHA Protocol, however, was developed entirely by industry, with little stakeholder interaction. It therefore reflected a strong commitment by industry, since it was considered a practical assessment tool. The Hydropower Sustainability Assessment Forum (HSAF) was an initiative that brought together industry, governments of developed and developing countries, commercial and development banks as well as environmental and social NGOs. The objective was "to develop a broadly endorsed sustainability assessment tool to measure and guide performance in the hydropower sector" for a period of 2 years ending in early 2010 (IHA 2010, p. 16).

If performance in the hydropower sector is to improve, all the key players—not just the group of like-minded supporters of the WCD report—will need to comply with sustainability criteria. Hence the BMZ's decision to support the HSAF approach with financial and advisory contributions. This approach broadens the debate on the sustainability of dam projects and provides guidance on how to measure and implement sustainability criteria. The GIZ sector project "Policy Advice for Sustainable Hydropower" was assigned on the BMZ's behalf to participate in the HSAF process as an observer. Building on industry's strong support,

---

[1] IHA is a non-profit organisation, working with a network of members and partners to advance sustainable hydropower. Membership is open to individuals and organisations who support its mission, for details, see http://www.hydropower.org. Accessed 22 July 2013.

the GIZ increased its efforts to foster broad and balanced stakeholder participation and to make it easier for NGOs, people directly affected and the governments of developing countries to be heard. For this reason, two rounds of consultations were supported financially. In Ghana, the GIZ joined with partners from the Ghana Dams Dialogue in organising consultations with communities affected by the Akosombo, Kpong and Bui Dams (see Hensengerth in this volume). This event was one of the very few occasions on which directly affected communities were consulted during the HSAF process.

The recommended final draft Hydropower Sustainability Assessment Protocol (November 2010) was submitted to Forum member organisations for adoption and support. The response was a letter of support or acknowledgement from the representatives of the banks, the environmental NGOs, Transparency International, which was one of two representatives of social NGOs, and the BMZ. The BMZ valued the HSAP as a new approach to broadening the debate on the sustainability of dam projects and providing guidance on measuring and implementing sustainability criteria, but explained that the WCD's recommendations would remain the benchmark for German development cooperation policy in the dam sector. This concluded the first phase, protocol development, which was followed by the second phase, protocol implementation.

During the transitional phase between the launch of the Protocol in June 2011 and the creation of the Hydropower Sustainability Assessment Council, with a governing committee to be elected by the time the next IHA Congress was held in May 2013, governance issues and questions relating to the terms and conditions for the application of the Protocol were addressed. The terms and conditions state: "Use of the Protocol is to be governed to promote its widespread use, protect its integrity, ensure appropriate qualification of trainers and assessors and provide quality control" (IHA 2011, p. 2). This also sets the framework for the publication of assessment results, with a distinction made between the "official" use of the Protocol, followed by the publication of the findings, and its "unofficial" use, meaning project assessments for internal use by someone who is not accredited. Unofficial assessments can also be published and used, for example, for dialogues with stakeholders, but this requires publication with the relevant disclaimer.

To ensure broad stakeholder input and confidence in the Protocol's contents and application, a permanent multi-stakeholder governance structure is being established: the Hydropower Sustainability Assessment Council will consist of chambers to be populated by stakeholder groups similar to those attending the Forum, representing industry, financial institutions, NGOs and governments. The Council, with its elected governing committee, is the body to tackle governance issues for future Protocol implementation and its further development.

The idea is that, with industry endorsing the Sustainability Assessment Protocol, such assessments will be carried out with active stakeholder involvement more systematically and widely and possibly even as part of regular business planning and operations. And the BMZ hopes that the organisations responsible for each project will be encouraged to increase inclusive multi-stakeholder engagement for the sake of improved performance, transparency and joint learning.

10.4 Experience of Implementing the WCD's Standards

Implementing the WCD's recommendations at project level is a daunting task, there being no blueprint available. On the other hand, the recommendations permit a wide range of approaches, not all of which are unanimously regarded as consistent with the recommendations. As the recommendations are based on specific legal and practical requirements related to hydropower project preparation and operation, such factors as the absence of planning laws (e.g. land registry, development freeze) in partner countries and assessment criteria (e.g. decisions in favour of energy production versus irrigation) are crucial.

The following discusses problems that will arise if one tries to apply the WCD's seven strategic priorities at project level.

10.4.1 Strategic Priority: Gaining Public Acceptance

In general, participation is a normal part of the planning process, since it is an integral component of environmental and social impact assessments and is widely required in dam/hydropower projects. Organising public hearings entails the risk of being accused of failing to comply with the spirit of the WCD recommendations, whether due to the participants' definition, timing, place, the language in which documents are written or the presence of officials, to name but a few factors. Public information and consultation is often limited to a public hearing announced in the local newspaper and inadequately tailored to stakeholder needs, especially with regard to the people affected. In some countries pre-meetings are held to inform these people about projects and their rights, but they are positive exceptions. Problems are associated with participation in societies where such groups as women are excluded from public life or where such strangers as foreign consultants are traditionally distrusted in the project area. Special problems arise when indigenous people are involved: this starts with their legal authentication and the WCD requirement that free, prior and informed consent be obtained.

10.4.2 Strategic Priority: Comprehensive Options Assessment

The problems begin with the definition of what electricity generation options the country in question really has, and whether demand-side management is an option to be favoured for both water and electricity. If the national policy is to develop national hydropower potential, the range of options is limited. Any analysis is often limited to a comparison with a hypothetical thermal generation option. Furthermore, if decisions date back a very long time, they can no longer be questioned.

10.4.3 Strategic Priority: Sustaining Rivers and Livelihoods

Ideally, national rivers are divided into those which should not be modified because of their ecological nature and those that can be modified through the construction of multi-purpose water storage infrastructure, for example. However, hydropower potential is often the main criterion in river classification. Conservationists often face the problem of being unable to argue about the ecological quality of rivers because of their remote location and lack of relevant data. When financial institutions come into play, such basic decisions have, as a rule, already been made at national level. They can, and should, however, refrain from providing funds if rivers and livelihoods are not sustained in an acceptable manner.

Projects are often located in areas where data on the ecological situation are unavailable or incomplete. They are remote from the capital and remote from university studies, and it is simply too expensive, too time-consuming or technically impossible to go there to conduct scientific inventories. During the feasibility phase of a project there is normally no time for multi-year assessments. Assessments have thus to be based on limited investigations and additional secondary data obtained from local fishermen, for instance. The same applies to social data: dam sites are often remote and the inhabitants are poor, rely on subsistence farming and—depending on the region—belong to ethnical minorities. Many such regions are economically and politically disadvantaged.

There are always competing interests in land and water, hydropower generation, irrigation, water supply and shipping being three examples. Irrigation reduces hydropower generation; storage for water supply reduces irrigation water, and so on and so forth. There are no universally agreed assessment tools for resolving such conflicts.

Another issue concerns cumulative impacts and assessments of the best location for a dam from both an economic and an environmental and social point of view. Rivers with hydropower potential attract various developers, resulting in different projects being at different stages, from early planning to construction and operation. The authorities responsible are often unaware of what constitutes an appropriate approach to sustainable river basin management. On some rivers attempts at cumulative impact assessment and joint management of environmental and social impacts are being made, with various stakeholders involved. Such efforts are still rare, however.

10.4.4 Strategic Priority: Addressing Existing Dams

There is no doubt about the need to mitigate or compensate for negative impacts of dams in the past which have yet to be adequately addressed. Such issues can be raised if the partner in a new project was also responsible for a previous project where environmental and/or social problems remain unsolved, as in the case of a public utility. It is more difficult to deal with such issues when two parties are involved. The problems become insurmountable when one party is a private company normally unwilling (and virtually unable) to solve the problems of what are very likely to be public utilities.

10.4.5 Strategic Priority: Recognising Entitlements and Sharing Benefits

Respecting rights, to land and water, for example, be they formal or informal, is a major concern. Taking stock of such factors forms an integral part of every environmental and social impact assessment study. Informal rights may pose problems, since they are often not legally recognised and are of a traditional nature, an example being usufruct rights to leaves or fruits, and more difficult to compensate for. The problems start with the identification of the right-holder, who may not be resident in the project area, but arrives when the fruit is ripe.

Compensation is a problem almost everywhere. There are always people who feel, or really have been, mistreated, and there is no universal gauge for assessing whether compensation has been appropriate and based on fair rules. Land-for-land compensation in agricultural settings is a must, but the situation is more difficult where replacement land is not available or is of a lesser quality. An approach often adopted is compensation at market values. It must be realised, however, that, as a way of avoiding taxes, the purchase value entered in the land registry is sometimes less than the amount actually paid. Does the improvement of the living conditions of the people affected go beyond mere compensation and require a development approach that exceeds the responsibility of the project? The only possibility is to adopt "good" compensation practices, but people will often be left dissatisfied.

Expropriation and compensation is a difficult area even if sophisticated regulations exist. It becomes even more complicated where existing systems are weak or deal with rights in an unsatisfactory manner. This is often found to be the case when it is considered who is eligible for compensation. Normally, this is limited to title-holders. But many other people may be affected by a project and should be compensated for losses, examples being tenants, holders of traditional rights and seasonal workers. A project will often be affected by various rights to goods. A number of rights may be associated with a tree, for example: the tree itself, the leaves, the fruit and the land where the tree stands. This makes compensation complicated, and project-specific regulations need to be developed by the competent authorities with the participation of the people affected.

In remote areas in particular, but in cities too, there may be no official land register. Land may be owned by a clan and distributed among the clan members by the chief. It may also be owned by a community, or it may not be cultivated. It is always difficult to identify the individual owner. If there is no individual owner, compensation must be so designed as to benefit the whole group using the land.

Where projects of public interest are planned, German planning law provides for the possibility of prohibiting any changes in the area; this is known as a development freeze. When land is identified for such a project, no investment may be made in existing structures, such as the construction of an extension, with a view to increasing the value of the property and, therefore, the compensation due. Many countries do not have such legal provisions, or they may not be enforceable. On the other hand, if such a development freeze remains in place for too long, it may lead to the impoverishment of the owners, since they are prevented from making the investment needed if they are to maintain their livelihood.

Benefit sharing as the WCD sees it goes beyond compensation and requires affectees to share the benefits of the project. Giving them priority when workers are recruited at the construction stage and—as far as possible—after the dam goes into operation is among the benefit-sharing mechanisms most commonly used. However, the WCD and other sources document a range of such mechanisms as preferential electricity and water tariffs for affectees, a percentage of the revenue from the electricity sold set aside for developmental activities in the project region, contracts for erosion control in the watershed area and the allocation of fishing rights to those resettled beside the newly created reservoir (Égré et al. 2002; Égré 2007).

There is, of course, always a debate on what constitutes compensation for lost assets and on approaches going beyond compensation (Cernea 2003; Cernea and Mathur 2008; Hensengerth et al. 2012). Some countries have benefit-sharing mechanisms in the form of taxes on the use of natural resources or special environmental or communal funds to which companies must contribute a fixed percentage of their profits. In other projects, companies voluntarily commit themselves to contributing a certain percentage of their benefits to environmental and social programmes.

10.4.6 Strategic Priority: Ensuring Compliance

Various public and private players are involved in designing, licensing, building and operating hydropower plants. Their interests, capacities and knowledge vary significantly. To ensure compliance with regulations, guidelines, and project-specific arrangements proper feasibility studies, environmental and social impact assessments and environmental and social management and resettlement plans are needed. The content and quality of these studies vary from one country and one project to another, from a very rough approximation to WCD standards to full compliance. Once water and land right issues are settled and a project has environmental and social clearance, requirements laid down in permits may still not be satisfied.

Many environmental and social impacts occur during the construction phase, when the contractors are mainly responsible for compliance with environmental and social requirements. They often lack the necessary information, knowledge and capacity, nor do they have a long-term interest in sustainable management. A lesson that has been learnt is that the company which will be operating the hydropower project must form its own experienced environmental and social management team as early as the construction phase and that concrete environmental and social management plans, including the required budgets, need to be in place. It should be possible to adapt management plans and budgets to any unforeseen events occurring during lengthy construction periods. The authorities responsible also need to improve their supervision. In some countries on-site supervision by environmental and social officials (mostly paid by the company) are required by law, but in many others local authorities are failing to perform the critical task of supervision.

Proper environmental and social management is also required during the operation of dams. Sustainable project management entails more than conventional pollution abatement and must take account of such matters as sedimentation, erosion, watershed management, downstream hydrology, biodiversity aspects and community health, safety and security.

10.4.7 Strategic Priority: Sharing Rivers for Peace, Development and Security

Rivers crossing national borders can be the starting point for conflict or cooperation. Negotiations with neighbours in good faith can be made—and are made—mandatory in project preparation. This is in line with the requirement in European law that environmental impact assessments be undertaken in a transboundary context (Convention on Environmental Impact Assessment in a Transboundary Context in force since 1991). The results of negotiations in this context depend on the regional political situation; however, mere notification of river-related development plans is not acceptable. Dam projects on shared rivers should be used to initiate a process of on-going cooperation through consultations among the riparian states.

10.5 Résumé

Many of the problems referred to above are caused by incomplete framework conditions and the absence of assessment standards. Like other international standards adopted in recent years, the WCD recommendations are no blueprint. They need to be tailored to the specific situation. A commitment to sustainable hydropower requires expertise, capacities and financial resources to be provided by various stakeholders.

Many challenges must be overcome to make the WCD work. Patience will be needed. The WCD has celebrated its 10th anniversary. Designing, building and operating a hydropower plant is a lengthy process, possibly continuing for decades. Potential projects are haggled over by various project developers before a decision is taken. Most projects currently being implemented were launched more than 10 years ago. Before the WCD recommendations are considered in such cases, it must first be decided whether the project in general can be sustainably managed. If so, environmental and social action plans help to close gaps identified when existing assessments are compared to the WCD recommendations. However, it is very difficult or impossible to address some other matters retrospectively, examples being ecological and social baseline data, alternatives, location assessments and consultations with the affected population.

For one thing, the WCD addresses national governments that are planning their electricity generation sector. The earlier development cooperation is integrated into this planning process, the greater its influence. This may begin with the development of capacities for sector policy elaboration, a stage at which the idea of sustainable hydropower can be included. By the time a specific project is financed, many basic decisions have already been made.

The WCD has opened the eyes of public and private stakeholders and triggered wide-ranging revisions and improvements of guidelines and processes at international, national and sector level. Today governments, industry, consultancies, civil society, the research community and, not least, financial institutions have growing capacities and expertise of the type needed to address the issues raised by the WCD.

Ten years on, there is greater awareness of the need to reduce greenhouse gas emissions and to combat accelerated climate change through the increased use of renewable energy sources, which include as yet untapped hydropower potential. At the same time, such "new" players as China and Brazil are increasing their investments in hydropower projects in developing countries. This makes it necessary to move forward and engage further in the harmonisation of rules and frameworks that seek to improve sustainability in the hydropower sector throughout the world rather than in just a few development projects.

References

BMZ (Bundesministerium für wirtschaftliche Zusammenarbeit und Entwicklung) (2006) Water Sector Strategy. Strategies 152. http://www.bmz.de/en/publications/topics/environment/konzept152.pdf. Accessed 22 July 2013

BMZ (Bundesministerium für wirtschaftliche Zusammenarbeit und Entwicklung) (2007) Sustainable Energy for Development. Sector Strategy Paper. Strategies 154. http://www.bmz.de/en/publications/type_of_publication/strategies/konzept154.pdf. Accessed 22 July 2013

Cernea MM (2003) for a new economics of resettlement: a sociological critique of the compensation principle. Int Soc Sci J, no 175. UNESCO, Blackwell, Paris

Cernea MM, Mathur HM (2008) Can compensation prevent impoverishment? Reforming resettlement through investment and benefit-sharing. Oxford University Press, Oxford

Égré D (2007) UNEP Dams and Development Project: compendium on relevant practices, 2nd stage. Revised final report, benefit-sharing issue. UNEP

Égré D, Roquet V, Durocher C (2002) Benefit sharing from dam projects. Phase 1: desk study, final report. World Bank, Washington, DC

EU Linking Directive 2004/101/EC of the European Parliament and of the Council of 27 October 2004 amending Directive 2003/87/EC establishing a scheme for greenhouse gas emission allowance trading within the Community, in respect of the Kyoto Protocol's project mechanisms

Fink M, Cramer A (2008) Towards implementation of the World Commission on Dams Recommendations. Experiences and reflections after 5 years. In: Scheumann W et al (eds) Water politics and development cooperation. Local power plays and global water governance. Springer, Berlin, pp 33–53

GIZ (Deutsche Gesellschaft für Internationale Zusammenarbeit) (2001) Dialogforum zum Bericht der World Commission on Dams. http://www2.gtz.de/dokumente/bib/02-5051.pdf. Accessed 22 July 2013

GIZ (Deutsche Gesellschaft für Internationale Zusammenarbeit), KfW (Kreditanstalt für Wiederaufbau) (2004) Implementation of the WCD recommendations within German Development Cooperation Experiences of GTZ and KfW. https://www.kfw-entwicklungsbank.de/Download-Center/PDF-Dokumente-Sektoren-Berichte/2004_06_WCD-Staud%C3%A4mme_E.pdf. Accessed 22 July 2013

Hartje V, Horstmann K (1999) Dams and development co-operation. Experiences and perspectives in German development co-operation with selected topics. Berlin, Dortmund

Hensengerth O, Dombrowsky I, Scheumann W (2012) Benefit-Sharing in Dam Projects on Shared Rivers. Discussion Paper 5. German Development Institute, Bonn

IHA (International Hydropower Association) (2011) Hydropower Sustainability Assessment Protocol: Terms and Conditions for Use. http://www.hydrosustainability.org/IHAHydro4Life/media/PDFs/AssessmentReports/Hydropower-Sustainability-Assessment-Protocol_T-C_FINAL.pdf. Accessed 22 July 2013

IHA (International Hydropower Association) (2010) Activity Report. http://www.hydropower.org/downloads/ActivityReports/IHA_Activity_Report_2010.pdf. Accessed 22 July 2013

WCD (World Commission on Dams) (2000) Dams and development. A new framework for decision making. The Report of the World Commission on Dams. Earthscan Publ. Ltd, London

World Bank, OED (World Bank Operations Evaluation Department) (1996) The World Bank's Experience With Large Dams. A Preliminary Review of Impacts, no 28695, 1. September

Chapter 11
Have International Sustainability Norms Reached the Emerging Economies? Evidence from Dams in Brazil, China, India and Turkey

Waltina Scheumann, Oliver Hensengerth and Nirmalya Choudhury

The chapters in this book are an in-depth examination of the processes of norm diffusion and contestation. They examine four emerging economies and their positioning in international norm production processes, the role of non-governmental organisations (NGOs) and the norms Chinese companies have to comply with in their operations abroad. As explained in the introduction, the origin of this book stems from a large research project that examined how global norms reach decision-makers in the emerging economies of Brazil, China, India and Turkey. While we have expanded this original focus considerably following the conclusion of the project, this chapter returns to the origins of this research and compares the findings of the four country case studies by using our original hypotheses.

While taking a historical perspective, the four country cases used the World Commission on Dams (WCD) report "Dam and Development: A New Framework for Decision-Making" (WCD 2000) as an important point of reference in the international debate on sustainability standards for dams. However, the methodological rigor in these country case studies required that the WCD does not become the sole point of reference. This is because it was an extreme event, and any analysis based on an extreme event could result in a faulty causal attribution (Geddes 2003). Although one objective was to probe if, over the years, the WCD's strategic priorities and guidelines had any discursive, normative, regulatory and material effects (Dingwerth 2007), the frame of analysis included a much longer time period.

W. Scheumann (✉)
Department of Environmental Policy and Natural Resources Management, German Development Institute—Deutsches Institut für Entwicklungspolitik (DIE), Bonn, Germany
e-mail: Waltina.Scheumann@die-gdi.de

O. Hensengerth
Department of Social Sciences and Languages, Northumbria University, Newcastle, UK
e-mail: Oliver.Hensengerth@northumbria.ac.uk

N. Choudhury
Council on Energy, Environment and Water, New Delhi, India
e-mail: Nirmalya.Choudhury@ceew.in

W. Scheumann and O. Hensengerth (eds.), *Evolution of Dam Policies*,
DOI: 10.1007/978-3-642-23403-3_11,

The WCD also faced the problem that the governments of the four countries rejected its recommendations (India, China and Turkey), or at best responded with reticence (Brazil). Yet, already before the WCD published its report, all four countries had experienced considerable domestic reform efforts. The country studies investigated this apparent contradiction in order to position the role of the WCD in global norm production more precisely and trace the evolution of domestic sustainability standards in the four countries.

Da Costa found that the reticent response of the Brazilian government towards the WCD was largely because of the first strategic priority—gaining public acceptance—which was perceived as a threat to the decision-making power of the Brazilian government. Of course, the WCD norms did resonate discursively and normatively among the non-governmental organisations that were critical of dam building in the country. Basing her assessment on a longitudinal view of the macro-political and economic development in the country, da Costa argued that rather than diffusion of WCD norms, progressive norms have emerged incrementally since the late 1980s due to the gradual democratisation in the country and the drafting of a new constitution. By the time the WCD norms came into being, Brazil had already implemented many of those norms as a part of the dam decision-making process.

Hensengerth argues that although the Chinese government rejected the WCD report, there has been steady progress in terms of social and environmental standards governing dam building in China since the early 1980s. Relevant reform measures therefore precede the WCD. As a consequence, it is difficult to attribute the post-WCD changes in Chinese law to the existence of the WCD. At best, there could be an indirect attribution, whereby the WCD norms might have had an impact on Chinese norms via their influence on the standards and safeguards of the multilateral financial agencies such as the World Bank, the International Finance Corporation (IFC) and the Asian Development Bank (ADB), through which international norms were then diffused into China. However, such an indirect causal attribution remains to be explored.

The Turkish State Hydraulic Works (DSI), which is home of the national chapter of the International Commission on Large Dams, outright rejected the WCD process as well as its report. The DSI argued that the WCD standards would hamper hydropower development and, as a consequence, economic and social development. In this respect, Turkey voiced the same concerns as many other developing countries regarding the WCD. However, the DSI and the Turkish government have always referred to the World Bank standards to which they would adhere—in legislation and in practice. NGOs and civil society organisations—in unison with their international partners—also almost exclusively referred to the Bank's safeguards. This has, in principle, not changed with the liberalisation of the energy sector; points of reference for private and transnational actors involved in Turkey's dam business are the Equator Principles and the Organisation for Economic Cooperation and Development's (OECD) Common Approaches, which in turn are based on the World Bank's safeguards.

In India the WCD continues to be a red herring within policy circles. Sections of civil society focus their endeavours on using the leverage provided by multiple domestic legislations—mostly dating back to the pre-WCD era—to strengthen their implementation and compliance as well as reform them. Some groups—for instance, community-based organisations—refer to the WCD norms more explicitly. However, Choudhury argues that such normative references cannot be construed as a normative impact of the WCD. They are a mere continuation of norms that civil society in general has discursively and normatively been referring to since the late 1980s. Choudhury further argues that, regarding the India case, the WCD norms were not diffused in India. Rather, the norms that were circulating in the domestic policy domain became an international anchor when the WCD was established. The discursive dominance of Indian civil society—catalysed by the presence of a prominent civil society leader as one of the Commissioners—and the lackadaisical approach of the government of India during the WCD process, ensured that the domestic norms became anchored at a global forum.

Given this ambivalent role of the WCD in influencing policy in the four case study countries, we need to gain a better understanding of the complexity of the triggers, pathways and drivers for norm diffusion in Brazil, China, India and Turkey. To do so, we now examine our research hypotheses in more detail.

Hypothesis_1 International Embeddedness

A country's embeddedness in global politics and its efforts to be, or become, a global player have repercussions for its decision of whether or not to refer to (dam-related) international norms and standards, not only rhetorically but in practice. As understood here, embeddedness acquires different forms, including a country's participation in international regimes, its ratification of international treaties, the socialisation of its elites in international epistemic communities and its membership in international governmental organisations (see e.g. Johnston 2007).

The assumption was that, with increasing international embeddedness in an issue area, governments would refer to the international norms and principles valid in that issue area and could therefore be held accountable for complying with international norms. In this respect, international embeddedness would be one of the possible pathways through which international norms would "actually reach a domestic area" and "get on the domestic agenda", as Checkel put it (Checkel 1997, p. 476; pp. 478–479).

Hypothesis_2 Social Mobilisation

In the case of social mobilisation, the hypothesis was that countries with a thriving civil society are susceptible to international social and environmental standards in policy and in practice through various combinations of information, symbolic politics, and leverage and accountability politics (see e.g. Tarrow 2011). The access that NGOs and civil society have to the political and juridical systems and to decision-making processes in various policy fields may differ with the nature of the political regime.

In addition, it is necessary to understand the motivation of public involvement, which can be distinguished by three rationales: democratic, instrumental and management. Public involvement, meanwhile, can be decomposed into public information, public consultation and public participation to reflect the different degrees to which the public can be involved in decision-making processes (Rowe and Frewer 2005).

Hypothesis_3 Reliance on Foreign Financial Services and Technology
If the implementation of national dam programmes or projects relies on international financial services and expertise, international financial institutions and corporations can act as norm carriers and standard setters. Based on the premise that these international actors attach conditions to their financial services, we put forward the hypothesis that they would be able to act as drivers in the diffusion of more ambitious social and environmental norms in various target countries by applying technical and financial leverage.

11.1 International Embeddedness

Brazil's response to the World Commission on Dams report and its follow-up does not indicate international embeddedness as a determinant factor in the institutionalisation of international norms. A historical trend analysis undertaken by da Costa shows that the gradual diffusion and adoption of progressive social and environmental norms governing the dam arena is more an outcome of an internal policy learning process, aided by the evolving political democratisation that the country has witnessed since the drafting of the Federal Constitution in 1988. Rather than compliance requirements towards international treaties, agreements and conventions, it was the reaction to the former authoritarian regime—with its undemocratic and centralised decision-making processes—that triggered norm changes. Apart from this establishment of a set of environmental and social norms, what is also very similar to international standards is a risk-minimising strategy, as the environmental licensing process in Brazil is perceived as uncertain, unpredictable and risky. In addition, such standards are implemented to provide a secure environment for private investors, who in recent years have emerged as important players in the hydropower business.

Hensengerth's chapter on China depicts the paradoxes that exist in contemporary China when it comes to its relationship with the international political and economic environment. The emergence of China as a global economic power now defines how China views the international arena *vis-à-vis* itself. Although the Chinese government criticised the WCD, it became involved in another global norm formation process—the International Hydropower Association's Hydropower Sustainability Assessment Forum (IHA-HSAF). Hensengerth argues that China's involvement in a global forum is an instance of "direct Chinese participation in norm-building". This is motivated by the realisation that, as China's national interests have begun to extend beyond its borders, China should not only be seen as a responsible international player, but it should also actively participate in global norm-setting processes to safeguard its interests.

This is to say that although the Chinese government was represented in the IHA-HSAF process, this representation also aimed at preventing the new IHA protocol from incorporating norms that would undermine China's authoritarian system of governance. Also, by participating in the IHA-HSAF process, the Chinese government's representatives were able to convey their understanding of international norms. As a consequence, different understandings of international norms were present at the HSAF.

The Chinese case therefore shows that international embeddedness should not automatically be viewed as a straightforward process of norm diffusion. Governments also participate in international norm-setting forums to safeguard their national interests. As such, international embeddedness can also be a rational process in which governments consciously engage to exert influence on norm-setting processes. The hypothesis links here to the social mobilization hypothesis: the Chinese government has greatly improved environmental impact assessment and resettlement legislation, particularly with regard to public involvement. Although this is primarily driven by domestic learning processes, policy reform has more recently also become responsive to increased levels of domestic discontent. During the process of reform, the Chinese government has borrowed from ADB and World Bank policies and has used the Aarhus Convention and US legislation as a model. However, much of this is driven by a rationale of maintaining and enhancing regime legitimacy.

In addition, as Wang (2000) has shown, where government bureaucrats are influenced by international norms and try to implement them—such as in the case of the Chinese Ministry for Environmental Protection—they may be relatively powerless to do so and therefore might not have much sway over norm-setting processes. It is therefore important to distinguish between different forms of international embeddedness, motivations for engaging with the international community and the importance of domestic governance settings as intervening variables that determine the domestic outcomes of international processes.

In India, environmental international norms became part of the domestic debate in the early 1970s. By the mid-1980s they had already reached a *prescriptive status* (Risse and Sikkink 1999) through the enactment of the Environmental Protection Act 1986 and the establishment of the Ministry of Environment and Forest. This norm empowerment can be seen as an outcome of India's commitment to the United Nations Conference on the Human Environment in Stockholm 1972, followed by a domestic policy learning process through the setting up of a high-powered committee to mainstream environmental issues in the policy discourse. Subsequently, several environmental laws were drafted that normatively referred to the United Nations Conference on Environment and Development. The resettlement norms mostly evolved through an organic process of internal policy learning. Over the years the government of India has constituted several committees on resettlement. Starting in the 1990s, different ministries drafted multiple policies, but they did not see the light of day till 2004, when the country adopted its first resettlement policy.

Regional integration (European Union) and integration in global energy markets have inspired the Turkish government to push for developing the country's renewable energy potential. Its long-standing efforts to exploit the country's hydropower potential as well as the prospect of participating in global carbon markets and accessing the Clean Technology Fund and loans from the International Bank for Reconstruction and Development—which are available to help transform fast-growing developing countries into low-carbon economies—have triggered change.

Turkey not only has provided incentive packages for private renewable energy generators—among them hydropower—but also removed environmental and social barriers to accelerate investments: since December 2010, an amendment to the Renewable Energy Law allows for the construction of dams in and close to protected areas, thereby undermining Turkey's key biodiversity areas. In this respect, being a signatory to the UN Convention on Biodiversity and the Ramsar Convention on Wetlands of International Importance had no profound impact. Furthermore, in 2010 the cabinet supports hydropower projects at the expense of property rights and democratic procedures which had been strengthened in 2001 as a reaction to lawsuits brought to the European Court of Human Rights because of the government's expropriation practices. Whether this is also a response to the EU's policy of slowing down the negotiation process with Turkey remains speculation.

Scheumann et al. show that aspirations to become an EU member have brought partial alignment with the EU environmental acquis, namely the EU Environmental Impact Assessment (EIA) Directive, but not with substantial regulations concerning the protection of the environment and water resources management, as prescribed by the EU Water Framework Directive. The Turkey case is similar to that of China in that state bureaucracies, which have tended to take European/international norms seriously, are the less powerful parts of the bureaucracy. If compared with the water and energy bureaucracies, the environmental bureaucracy is disadvantaged by status, capacity and the weak status that EIAs have in decision-making. But it also seems that the regional water bureaucracy is more responsive to local demands and cooperates with local authorities who are responsible for protecting nature reserves and who refer to international frameworks (for example the Ramsar Convention).

11.2 Social Mobilisation

Social mobilisation has been one of the key drivers of diffusion of international environmental norms in Brazil under an overall macro-political environment that has been geared towards increasing democratisation since the late 1980s. The growth in environmental awareness as well as in social mobilisation on environmental issues have strengthened each other. Da Costa portrays a unique mechanism through which social mobilisation in Brazil has been able to bring progressive changes. Rather than operating outside the ambit of the government, social mobilisation in Brazil has taken place within the larger political domain of democratisation. This has resulted in the same individuals moving across various governmental and non-governmental organisations and assuming different roles at various points in time: sometimes working "within and with" the government administration, sometimes "opposed" to particular organisations. This mobility of both governmental and non-governmental actors across the governmental and non-governmental spheres also breaks down the otherwise organisational silos and increases organisational porosity, through which new norms can be diffused. The increased importance of the Ministério Público within the cabinet has also acted as an enabling factor in allowing civil society movements to be more influential.

Brazil is the only country among the four case countries in this book that allows public involvement, not only with individual project decision-making but also during the preparation of sector plans. At the same time, da Costa's study highlights the paradox that often is associated with involving the public. Those in power tend to promote public involvement, but there is an instrumental rationale behind the idea of public involvement: namely, to ensure that the public, which otherwise has little trust in the official decision-making process, is not "susceptible to ideologies and arguments". At the same time, the Brazilian government is reluctant to relinquish any decision-making power to the public. Da Costa argues that one of the reasons behind the Brazilian government's reticence towards the WCD report was the concern that norms such as Gaining Public Acceptance or Free Prior and Informed Consent would shift the decision-making power from the constitutionally mandated authorities to the people affected by the project. This dissonance was then resolved by partially opening up the decision-making process—more at the level of public information and public consultation—while retaining the power to take final decisions.

Although the role of social mobilisation in a political setting such as China is expected to remain limited, Hensengerth portrays a fragmented polity that allows social actors to influence decisions under certain circumstances. NGOs and policy entrepreneurs who engage in pragmatic activism as well as spontaneously organised protests play a role in the implementation of environmental and resettlement laws and their associated norms. The dispute over the Nu River project is a case in point. The case studies on hydropower projects in China also portray the fissures that exist at various levels of governance and the conflicts between civil society, local governments and hydropower companies concerning environmental protection laws and regulations laying down provisions for public involvement.

However, the reason why the Chinese government has increased options for public participation—such as through NGO registration regulations, the seminal 2003 EIA law and subsequent legislation on public participation and public access to information—is largely motivated by an instrumental rationale. Beijing is not averse to promoting participation at lower levels of governance and enhancing transparency at that level, as participatory processes are seen as conduits through which local governments can be forced to implement national laws in the absence of a central enforcement mechanism. In addition, better implementation of national laws will enhance regime legitimacy. Therefore, public participation in

China—from the perspective of the central government—is not motivated by the democratic rationale, but it forms a mechanism through which the government can maintain and enhance its legitimacy.

Social mobilisation is, by far, the most prominent driver for international norm diffusion concerning dam projects in India. Choudhury argues that at the policy level—during the drafting of the various resettlement and rehabilitation policies in India during the 1990s and during the drafting of the EIA Notification of 2006—and at project level, civil society organisations form coalitions through which they undertake various combinations of informational, symbolic and leverage politics (Keck and Sikkink 1999). At the policy level, civil society organisations since the late 1980s have formed national coalitions and deliberated widely among various stakeholders on the resettlement and rehabilitation of the populations displaced by the dam and then engaged with the relevant ministries. The case studies on the Lower Subansiri project, the Dibang Multipurpose project and the Allain Duhangan project showed that civil society played the most important role in enforcing international social and environmental standards within each of these projects. Through informational, leverage and symbolic politics—and by approaching the judiciary and filing lawsuits—they would force the application of social and environmental standards within a project.

However, Choudhury argues that the formal space for public involvement is limited. The public hearing mechanism within the environmental clearance process—an outcome of the 1997 amendment of the EIA Notification of 1994—is the only formal space where public involvement in dam projects can take place. Given that this is the only space where a project is presented to the public, there are often sharply contrasting ideologies that transform a public hearing into one of the most contentious parts of the decision-making process.

Choudhury further argues that in India, public involvement is restricted to consultation at best, and information at worst. The EIA Notification of 1994 contained some progressive provisions guiding the public hearing, but their implementation was always a challenge. In order to make the environmental clearance process faster, the EIA Notification of 2006 strengthened some procedural components, such as video recordings of the public hearing process—a process also followed in Turkey, as Scheumann et al. argue in their paper—but weakened the public hearing mechanism, thereby preventing it from becoming a forum for deliberative discourse. While formally the space for public involvement in India is consequently limited, the decision-making process is long and involves multiple actors at various stages, each with specific responsibilities, thus resembling a polycentric decision-making process. Such a process allows the non-governmental stakeholders to approach the various decision-makers. Evidence from various hydropower cases also shows that the non-governmental actors have often tried to approach and lobby the government-appointed committees and have not shied away from approaching the judiciary as a last resort.

Environmental awareness has been low in Turkey until recently, and mass protests against large dams—for instance the Atatürk Dam (start of operation in 1992)—were unknown. This has changed. What started as local protests confined to individual dam projects has meanwhile turned into a country-wide movement that concerns inadequate expropriation and resettlement practices, impacts on the environment, cultural heritage, people's rights and their ways of life. Representative of this trend is the Turkish Water Assembly, founded in January 2010, which unites local people and their representatives, NGOs, lawyers and scientists who all advocate a "new water policy". The Union of Chambers of Turkish Engineers and Architects, with more than 300,000 members, condemns "the privatization of the country's rivers". Elected mayors of large metropolitan cities and provincial capitals complain that they have no voice in deciding on the purpose of dams, their locations and whether they want them, but that they still have to bear costs which occur when people migrate from the reservoir areas to the district towns, without getting financial support for infrastructure. All these statements show that reforms of dam-related decision-processes are not confined to the single-issue area of dam-decision-making but affect rules that touch upon systemic issues entrenched in the political system, and the range of manoeuvre the private sector enjoys.

NGOs and civil society groups have referred to international frameworks and allied with transnational and European partners who were able to exert pressure on international consortia via their home countries/companies (see Atzl in this volume), but they have not found influential allies within the government. Instead, the attitude of Recep Tayyip Erdogan's AKP government has hardened. His key allegations are: that the protests are externally controlled; that they are linked to Kurdish separatism and nationalism; that they impair the nation and its hydro-based modernisation.

In Turkey, the EIA Directive is the only statutory mechanism for public involvement that includes, in essence, public information and public hearings. The EIA process is not meant to change overall decisions on dams but to integrate environmental—and to some extent social—concerns. The Turkey case study argues that, in practice, even access to information is low and the conditions attached to environmental clearance and reports on monitoring compliance are rarely available in the public domain. Meanwhile, civil society groups have approached the judiciary as a last resort: 83 lawsuits have been filed against dams and hydropower plants. The overwhelming majority of them have resulted in the cancellation of dam projects in the first instance, on the grounds that the submitted EIA reports were inadequate, misleading or that the projects were to be built in protected areas.

11.3 Leverage Through International and Domestic Technology and Development Partners

Da Costa's historical approach of understanding the diffusion pathways in Brazil offers two different periods when it comes to the leveraging of international technology and finance partners in diffusing environmental and social norms. As in many other developing countries, international environmental standards were incorporated into domestic practices because Brazilian companies had to comply with international standards to fulfil conditions in order to access international finance and technology from multilateral development agencies. Creation of an environmental unit within Eletrobras (a major Brazilian electric utilities company) was a clear example of how international technology and financial partners acted as important drivers in diffusing international standards. The Brazilian study argues that in the hydropower plant (UHE) Itaparica, the influence was direct. Over the years, as the Brazilian economy became more stable and as the domestic financial sector—and private capital in particular—became more important, the direct effects of international finance decreased. At the same time, new pathways emerged, as in the case of UHE Santo Antônio, where private capital was a dominant mode of financing. Importantly, however, the Equator Principles became relevant for UHE Santo Antônio, as the foreign banks funding the project were signatories to them.

Hensengerth argues that the evolution of resettlement and environmental norms in China has been largely an outcome of a domestic policy learning process, assisted by the increased vitality of social actors. At the same time, multilateral aid agencies, particularly the Asian Development Bank, have influenced resettlement legislation through technical assistance programmes at the request of the Chinese government to assist in the improvement of the domestic legal environment in this issue area. Furthermore, the World Bank's monitoring and evaluation rules governing the resettlement process had a material effect on domestic policy in 2006, when the 1991 Rules of Land Compensation and People Resettlement in Medium and Large Hydraulic and Hydroelectric Projects were updated and the concepts of monitoring and evaluation—which until then were applicable to internationally funded projects only—became universal within the country. Whereas the ADB had a direct influence on Chinese law, the World Bank norms were incorporated into Chinese law by using them as models that could help to improve domestic law.

For India, Choudhury observes that the role of international development partners, mostly the World Bank, as drivers of international norm diffusion at the policy level is profound in individual projects that the Bank funds. Although the various resettlement and rehabilitation policies drafted in the 1990s refer to the World Bank requirements, Choudhury argues that the influence of international actors does not reach beyond the project level. This was clearly observed in the case of the Allain Duhangan project, where the IFC of the World Bank Group acted as financier. As the IFC had an environmental and social safeguard policy that was superior to that of the government of India, the project participants witnessed a decision-making process that was unprecedented in India. However, Choudhury argues that the presence of international actors with international standards does not imply that the same standards would be complied with in practice. Yet, such standards would provide leverage to civil society organisations, which could then ensure that the same standards are complied with in a project. The case of Allain Duhangan clearly reflected the fact that the presence of international actors (in this case the IFC) results in a superior planning process.

But the planning may not translate into better execution, as was reflected by the repeated failures of the project to comply with environmental clearance conditions.

With the liberalisation of the electricity sector, successive Turkish governments have invited the private sector to finance (and operate) hydropower dams and opened the sector to foreign direct investments. In cases where international companies and private banks are involved, they insisted on EIAs and resettlement action plans. International financiers would often find that EIA reports were of low quality, and the export credit agencies would often ask for higher environmental standards, improved baseline studies and the strict implementation of mitigation measures.

In the Yusufeli and Ilisu dam projects, which, due to their large sizes, will have tremendous social and environmental impacts, export credit agencies ensured that EIAs were carried out despite the projects being exempted from legal EIA requirements. In both projects, international consortia carried international norms into negotiations and guaranteed adherence to them as long as they negotiated the terms of financing. The moment they withdrew from financing because environmental standards and resettlement plans did not adhere to international standards, they lost the possibility of exerting influence. A Rapid Resettlement Action Plan was developed for the Ilisu Dam but did not become legally binding, meaning that people affected have no legal entitlement. While individual dam projects may improve in environmental and social terms when international financiers are involved, general repercussions on sector policies are not observed. Since the withdrawal of the expert credit agencies, both projects are now being carried out using domestic financial sources.

11.4 Conclusion

The hypotheses have been discussed separately, but they are to some extent interlinked. The following section discusses some of these linkages and their limitations, as observed in the country case studies.

Social mobilisation (hypothesis 2) can be increased if a country is dependent on foreign (Western) technical expertise (hypothesis 3). India and Turkey, for instance, have seen an increase in civil society action where international organisations and Western donors have been involved in dam financing. The presence of international actors in project development or finance provides additional leverage to civil society. Civil society action in such cases is often geared towards ensuring that the international norms are actually complied with *in practice*. In China, however, international actors had no such effect, as social control exerted by the government is not relaxed while foreign financiers or technical advisors are present. Regime type and policy style seems to be the intervening variable that explains different outcomes in different countries.

Similarly, international embeddedness (hypothesis 1) may lead to increases in the scope of civil society activities (hypothesis 2), as governments are exposed to international normative frameworks. Among the countries studied here, Turkey is perhaps an extreme case, given the influence of the EU on Turkish domestic laws. In addition, given the democratic nature of the Turkish polity, Turkish social movements are freely able to strike alliances with social movements across the EU and can make use of the European Court of Human Rights. However, Turkey's relative embeddedness in the EU's normative frameworks has had no systemic effects, as can be observed with the authoritarian style of the Erdogan government. Again, regime type and policy style seem to be the decisive factors.

Hypothesis 1 needs to be qualified further. The cases of Turkey and China showed that the form of embeddedness and the motivation for becoming embedded are important factors when evaluating whether international norms have effects on domestic policies. It is therefore important to understand why and to what degree foreign norms were incorporated, or localised, into domestic legislation. In the fields of environment and resettlement, the bureaucracies in question may consciously subscribe to international norms, but in the four countries examined in this book, elites in China and Turkey promoting these norms are in the minority. This has to do with the fact that international standards for EIA and resettlement are geared towards democratisation of the political process. The Chinese government tries to avoid this. The political-administrative system in Turkey where dam-related water resources and energy planning is highly centralized and organised along sectoral lines with strong central bureaucracies, does not (want to) open to civil society and/or locally elected representatives.

If elites' beliefs do not persuade them to adopt international norms, then norm adoption can occur as a rational process. For example, governments may include foreign standards in domestic laws to increase their international legitimacy. They may also become a member of an international organisation or sign up to international treaties to safeguard national interests by participating in the formation of international norms, as has happened in the case of China joining the IHA-HSAF.

In addition, governments may find it useful to adopt international norms in order to increase domestic legitimacy. In the fields of EIA and resettlement, China belongs in this category. Although China is not interested in opening the political process to democratisation, the government is using legal reforms to enhance its legitimacy, thus causing a disconnect between the original intention of the norms and their localisation in China.

Furthermore, internationally competing norm sets allow governments to pick and choose the international norm they find convenient and better serves national interests. The Turkish government has chosen to subscribe to the low-carbon norm to drive its energy agenda, but it ignores environmental protection and democratic rights.

Given the interlinkages between the hypotheses, it is important to note that processes of norm diffusion are multidimensional. Different actors follow different pathways, and detailed knowledge of the domestic settings is important to understand why norms are adopted, and if and how they are implemented.

As far as the norms of the WCD are concerned, they not only affect single-issue areas such as the environment and resettlement, but they profoundly change the way in which decisions are made in political systems. WCD norms therefore operate on a macro level. As a consequence, a direct implementation of WCD norms would likely alter the core norms of country constitutions, especially in more authoritarian settings. It should come as no surprise, therefore, that the WCD report was met with such criticism, given the lack of government representation. At the same time, the country studies show an intense and highly complex process of local–global interactions, in which norms are constantly adapted to new contexts and translated to accommodate local perspectives.

References

Checkel JT (1997) International norms and domestic politics: bridging the rationalist-constructivist divide. Eur J Int Relat 3(4):473–495

Dingwerth K (2007) Designing institutions for water governance: lessons from the world commission on dams. In: EAWAG seminar on water policy and governance, Dübendorf, Switzerland

Geddes B (2003) Paradigms and sand castles theory building and research design in comparative politics. The University of Michigan Press, Ann Arbor

Johnston AI (2007) Social states: China in international institutions, 1980–2000. Princeton University Press, Princeton

Keck ME, Sikkink K (1999) Transnational advocacy networks in international and regional politics. Blackwell Publishers, UNESCO, Oxford

Risse T, Sikkink K (1999) The socialisation of international human rights norms into domestic practices: introduction. In: Risse T, Ropp SC, Sikkink K (eds) The power of human rights. International norms and domestic change. Cambridge University Press, Cambridge, pp 1–38

Rowe G, Frewer L (2005) A typology of public engagement mechanisms. Sci Technol Human Values 30(2):251–290

Tarrow SG (2011) Power in movement: social movements and contentious politics, 3rd edn. Cambridge University Press, New York

Wang H (2000) Multilateralism in Chinese foreign policy: the limits of socialisation. Asian Surv 40(3):475–491

WCD (World Commission on Dams) (2000) Dams and development. A new framework for decision-making. Earthscan Publications Limited, London

Index

1

[1] Note: Page numbers followed by "f" and "t" indicate figures and tables respectively.

W. Scheumann and O. Hensengerth (eds.), *Evolution of Dam Policies*, DOI: 10.1007/978-3-642-23403-3,

D

E

F

S

T

MIX
Papier aus verantwortungsvollen Quellen
Paper from responsible sources
FSC® C105338

If you have any concerns about our products,
you can contact us on
ProductSafety@springernature.com

In case Publisher is established outside the EU,
the EU authorized representative is:
Springer Nature Customer Service Center GmbH
Europaplatz 3, 69115 Heidelberg, Germany

Printed by Libri Plureos GmbH
in Hamburg, Germany